Seventh Edition

POPULATION GEOGRAPHY

Problems, Concepts, and Prospects

Gary L. Peters
California State University
Chico

Robert P. Larkin
University of Colorado
Colorado Springs

KENDALL/HUNT PUBLISHING COMPANY
4050 Westmark Drive   Dubuque, Iowa 52002

Chairman and Chief Executive Officer   Mark C. Falb
Vice President, Director of National Book Program   Alfred C. Grisanti
Editorial Development Supervisor   Georgia Botsford
Developmental Editor   Liz Recker
Vice President, Production Supervisor   Ruth A. Burlage
Prepress Service Manager   Kathy Hanson
Prepress Editor   Michele Steger
Permissions Editor   Renae Heacock
Design Manager   Jodi Splinter
Designer   Suzanne Millius
Senior Vice President, College Division   Thomas Gantz
Vice President and National Feids Manager   Brian Johnson
Managing Editor, College Field   John Coniglio
Associate Editor, College Field   Janice Samuells

Maps by John Harner

To Carol, Jason, Erika, Joyce, Michael, and Christine

CONTENTS

PREFACE

The scientific study of the human population is an interdisciplinary one involving demographers, geographers, sociologists, anthropologists, psychologists, political scientists, economists, biologists, physicians, and even philosophers. The dramatic growth of the human population in recent decades compels us to pay attention to the causes and consequences of population growth.

Our purpose is to provide students with an introduction to population geography, a task that requires drawing upon materials from many disciplines and integrating them into a readable text. We begin with population growth in an effort to generate an interest in the study of population. Following that, we look at demographic data, which is so essential to helping us understand population processes. Population distribution and composition are considered, followed by discussions of theories of population growth and change. After that, the focus turns to the basic demographic processes—mortality, fertility, and migration. The final two chapters examine relationships among population, environment, and food supply.

ACKNOWLEDGMENTS

It would be impossible to acknowledge individually each of the people from whom we have learned about population, though citations give credit to many of them. Without the continued research and writing efforts of many people, this book could not have been written. We would also like to thank all of those instructors who continue to use this text; we encourage you to send us your comments.

Though this book has evolved considerably since the first edition appeared in 1979, it continues to reflect our own graduate learning experiences. We would still like to thank Professor Paul Simkins for stimulating our initial interest in population geography during our student years at Penn State. Many of the ideas in this book can probably still be traced back to Paul, and we hope that he approves of the current edition. Of course, we alone remain responsible for whatever errors and shortcomings this book may have.

ABOUT THE AUTHORS

Gary L. Peters is currently a professor of geography at California State University, Chico. He received his B.A. in geography at Chico State University and his M.S. and Ph.D. in geography at The Pennsylvania State University. His interest in population studies began when he was a graduate student at Penn State. He is especially interested in migration, regional dimensions of population change, and the changing demography of the United States.

He has been teaching geography since 1971. He is co-author of *California*, which is published by Kendall/Hunt Publishing Company. He has also authored several other books including *Wines and Vines of California* and *American Winescapes*.

Robert P. Larkin is a professor of geography and Environmental Studies at the University of Colorado, Colorado Springs. He received his B.A. in Social Studies Education from the State University of New York, Cortland College, his M.A. in geography from the University of Colorado at Boulder, and his Ph.D. in geography from the Pennsylvania State University.

He has been teaching at the college and university level for nearly thirty years. His interests are in population geography, gerontology, and the teaching of geography. He has published seven books and a variety of articles on geographic topics.

Introduction

In 1999 the earth's population reached 6.0 billion. That number cannot help but further strain this planet's ability to provide support for its many people in the new millennium. It is difficult to argue with Sharon Camp's remark (1993, 126) that "The 1990s are clearly the decade of decision for one of the most basic environmental problems: the exponential growth in human populations." Have we done enough?

We're adding nearly 220,000 people to the planet every day (including weekends and holidays!); geographers need to be involved in communicating the causes and consequences of that growth to as wide an audience as possible. As writer Erla Zwingle (1998, 38) commented, "Population: We know it's a problem. In fact most of the problems facing individuals, families, communities, and nations are affected by the quantity of humans sharing the planet." A geographic perspective is especially valuable in clarifying and interpreting the considerable variations that exist from place to place with respect to birth and death rates, as well as major patterns of population movements. Most of the world's additional people are added in the developing countries, whereas most of the developed countries are growing slowly or not at all; some are even declining. Such demographic disparities are likely to set in motion other processes, especially international migration, which could easily become the new century's most nagging demographic quandary.

Furthermore, as demographer Geoffrey McNicoll (1999, p. 435) noted, ". . . the demographic relativities of the future will differ from today's in the certain diminution of the West—and, more generally, the North—giving still greater cause for a new international order."

It is hardly surprising to find that more and more attention is being focused on population issues. From scholarly journals to the popular press, a burgeoning literature on population dynamics and population problems challenges us. Business has discovered the importance of demographics as well, as is apparent in publications such as *American Demographics*. Our view of population, the geographic perspective, requires a brief look at the intellectual development of population studies as a geographic concern.

POPULATION GEOGRAPHY

Because of the interdisciplinary nature of population studies, it is not always easy to distinguish **population geography** from other disciplinary contributions.

■■■

Because of the interdisciplinary nature of population studies, it is not always easy to distinguish **population geography** from other disciplinary contributions. However, it seems that we should briefly consider ways in which some important population geographers have defined their realm of study.

Population geography is a relatively young subdiscipline of geography, though geographers have long been concerned with population characteristics as a part of broader regional studies. Though the roots of the subdiscipline developed somewhat earlier, most would agree that Glenn Trewartha (1953), a noted climatologist as well as population geographer, provided the first definitive statement on population geography. Therein he noted that population geography had been, and continued to be, a neglected aspect of the discipline. Furthermore, he argued that population was a pivotal element in geography, and that its continued neglect would seriously affect the development of geography.

Some geographers would still agree with Trewartha (1953, 87) that "... the geographer's goal in any or all analyses of population is an understanding of the regional differences in the earth's covering of people. Just as area differentiation is the theme of geography in general, so it is of population geography in particular." Other geographers, as we shall see, would find different approaches to the study of population geography more appealing.

The response to Trewartha's plea for increased attention to population geography was neither rapid nor overwhelming, at least not at first. In the year following the publication of Trewartha's presidential address, geographer Preston James (1954, 107) wrote that "... the fundamental problem of population geography is the search for a systematic method of outlining enumeration areas that are meaningful in terms of the question being asked." On the same page he went on to note that "... in the field of population geography the need is to find a way to generalize the distribution of people without obscuring the relationships of man to the other phenomena with which he is areally associated." Additionally, James argued that demographic studies could benefit from the geographic perspective, especially from the regional concept and regional methodology. He also provided a useful review of population studies by American geographers, suggested numerous new research directions, and even offered a discussion of training requirements for population geographers.

After an embryonic decade or so, two books on population geography appeared in 1966, one by American geographer Wilbur Zelinsky (1966) and the other by French geographer Jacqueline Beaujeu-Garnier (1966). Zelinsky (1966, 2) commented that a ". . . rich harvest of facts and ideas has not yet been reaped by the population geographer because this is still the period of germination in the history of his discipline." Furthermore, he offered the following definition (Zelinsky, 1966, 5):

> Population geography can be defined accurately as the science that deals with the ways in which the geographic character of places is formed by, and in turn reacts upon, a set of population phenomena that vary within it through both space and time as they follow their own behavioral laws, interacting one with another and with numerous nondemographic phenomena.

Perceptively noting the burden that this definition placed on the reader, he offered a shorter version (Zelinsky, 1966, 5) in which he stated that ". . . the population geographer studies the spatial aspects of population in the context of the aggregate nature of places." He contended that population geography was concerned with three "distinct and ascending levels of discourse": (1) simple description of the location of population numbers and characteristics, (2) explanation of the spatial configurations of these numbers and characteristics, and (3) the geographic analysis of population phenomena. As did Trewartha and James, Zelinsky emphasized areal differences in populations.

Beaujeu-Garnier's concept of population geography was perhaps best expressed in the following statement (1966, 3):

> If the demographer measures and analyzes the demographic facts, if the historian traces their evolution, if the sociologist seeks their causes and repercussions by the observation of human society, it is the business of the geographer to describe the facts in their present environmental context, studying also their causes, their original characteristics, and possible consequences.

In her view the geographical study of population had three foci:

1. the distribution of people over the globe,

2. the evolution of human societies, and

3. the degrees of success that those societies have achieved.

In his major work on the topic Trewartha (1969, 1-2) stated that ". . . the geography of population (or population geography) is concerned chiefly with one aspect of population study—its spatial distribution and arrangements." As he had in 1953, Trewartha again emphasized the importance of population geography within the broader discipline of geography, noting that ". . . population serves as the point of reference from which all other geographic elements are observed, and from which they all, singly and collectively, derive significance and meaning."

In the following year Demko, Rose, and Schnell (1970, 4) offered yet another definition, conditioned by changes in methodology that were affecting most fields of geography by that time (spatial analysis, logical positivism, quantitative methods):

> Population geography is, therefore, that branch of the discipline which treats the spatial variations in demographic and nondemographic qualities of human populations, and the economic and social consequences stemming from the interaction associated with a particular set of conditions existing in a given areal unit.

Their emphasis was somewhat broader than those of previous definitions, leaning more toward process than simple areal differentiation, more toward hypothesis-testing than simple description.

Soon afterward, geographer John Clarke (1972, 2) wrote that population geography was "... concerned with demonstrating how spatial variations in the distribution, composition, migrations, and growth of populations are related to spatial variations in the nature of places." A few years later Courgeau (1976, 261) added that:

> Demography and population geography are concerned with the same topic: the study of human populations. Basically, they are quantitative disciplines that mainly use statistical data, but they also employ qualitative approaches. The main difference between the two sciences is the fact that the demographer places his emphasis on time whereas the geographer places his emphasis on space.

At least to some degree this distinction remains acceptable.

In the following decade Clarke (1984) noted the success that population geography had experienced as a subdiscipline, especially since the 1970s, and pointed out that since Trewartha's seminal paper in 1953 nearly ten percent of all published geographical papers had dealt with some aspect of population. He also noted the diversity of studies under the population geography umbrella, along with a look at population and political changes, as well as data and advances in quantitative methods. However, he ended up noting that, despite many of the methodological advances of the 1970s and early 1980s, "The search for general methods and laws has often obscured the complexities of reality." (Clarke, 1984, 9)

DEMOGRAPHY

Definitions of **demography** also vary, both over time and among practitioners.

■ ▪ ■

Definitions of **demography** also vary, both over time and among practitioners. According to demographer Donald Bogue (1969, 1), "... demography is the empirical, statistical, and mathematical study of populations." In his view "formal" demography focused on: (1) changes in population size, (2) the composition of the population, and (3) the distribution of population in space.

Demographer William Peterson (1975) defined demography as ". . . the systematic analysis of population phenomena. . . ." However, he differentiated between formal demography and population studies, arguing that formal demography was concerned with the gathering, collating, statistical analysis, and technical presentation of data, whereas population studies were concerned with population characteristics and trends in their social setting. Along this latter line, Thomlinson (1976, 4) stated that ". . . population study involves the number and variety of people in an area and the changes in this number and variety."

Along with formal demography and population studies, a third dimension has emerged in recent years and deserves mention, namely "housing" demography. Describing **housing demography,** Dowell Myers (1992, 6) commented that, "This theory stresses the detailed interconnections between populations and their housing stocks." Both urban and population geographers have found this concept useful and have incorporated it into their work (Gober, 1992).

TRENDS IN POPULATION GEOGRAPHY

Since 1980 several new texts have appeared in both demography and population geography. Among them are Myers (1992), Murdock and Ellis (1991), Overbeek (1982), Weeks (1999), and Weller and Bouvier (1981) in demography, along with Hornby and Jones (1980), Plane and Rogerson (1994), Schnell and Monmonier (1983), and Woods (1982) in population geography. Though they are collections of readings rather than texts, Coleman and Salt (1992), Lindahl-Kiessling and Landberg (1994), Turner, Hyden, and Kates (1993), Menard and Moen (1987), Noin and Woods (1993), Pacione (1986), and Woods and Rees (1986) provide considerable material of interest to both demographers and population geographers. Books with a regional population focus include Newman (1995) and Clark and Noin (1998). The array of new texts in recent years affirms the growing interest in population issues in the social sciences, issues that will become more prominent as we enter the new millennium.

Obviously geographers and demographers share many common and overlapping interests in their respective studies of population. However, population geographers tend to place more emphasis on the spatial patterns of population characteristics and processes, whereas demographers, though often interested in population distributions, seldom make spatial patterns their central interest.

Focus on the spatial dimensions of demography and demographic techniques have been central in much of the recent literature of population geography. Findlay (1991, 64) noted that, "The 1980s may well be recorded by historians of population geography as the decade in which the subdiscipline became strongly demographic and moved in the direction of being redefined as spatial demography." This trend, exemplified by such works as Woods and Reese (1986) and Congdon and Batey (1989), was especially pronounced in the United Kingdom. Findlay and Graham (1991) also looked at the nature of recent studies in population geography, first noting the emphasis on the spatial perspective, then criticizing population geographers for their neglect of postmodernist debates that took place in geography in the 1980s. "That these issues were not taken up and discussed in population geography in the 1980s," Findlay and Graham

Describing **housing demography,** Dowell Myers (1992, 6) commented that, "This theory stresses the detailed interconnections between populations and their housing stocks."

(1991, 156) argued, "meant that population geographers were failing to participate in the mainstream methodological debates within geography, but clung instead to the surer, but dated terrain of the 1960s and 1970s." Perhaps even more clearly, Findlay and Graham (1991, 158) commented that "A narrow definition of population geography as spatial demography is quite simply inadequate to face the challenge of what others expect geographers to contribute to the understanding of population."

Clearly, in population geography, as in most other pursuits in academia, we need to be more involved in solving key problems in society. With respect to geography and the work being done by geographers, Brunn (1992, 2), from his perspective as editor of the *Annals of the Association of American Geographers,* lamented:

> I am sometimes concerned that, unlike other disciplines, we are apparently unwilling to tackle global problems on a global scale. . . we are not prepared to look at global and macroregional problems, processes, and models . . . . The time is propitious for us to get some idea of what we are doing and how we might proceed as a sound and worthy discipline into the next millennium.

Given their roots in geography, population geographers should not lose track of how the subdiscipline fits into, and interacts with, other branches of geography. Whether looking at international flows of refugees or crosstown movements within metropolitan areas, whether considering the relationships between people and global warming or the supply and demand for food in the Sahel, we should be able to find some solid geographic foundations upon which to build. In that process, people, and their relationships to places and environments at all scales, must remain central. As Stoddart (1987, 331) noted, for geographers "The task is to identify geographical problems, issues of man and environment within regions—problems not of geomorphology or history or economics or sociology, but geographical problems: and to use our skills to work to alleviate them, perhaps to solve them." Perhaps the summary by White, et al (1989, 282) provides as clear a picture as any of where we should be going:

> Geographers might disagree about the most significant topics within population geography, the most appropriate research questions, and certainly the most appropriate research methodologies. However, it is clear that despite these honest differences, population-geography research has grown rapidly since 1953, and shows no signs of slowing. Undoubtedly, population geography is pivotal to an understanding of the cultural landscape.

Findlay (1993) urged population geographers to pursue more specific tasks. For example, he suggested the need for studies of the demographic disorder that has resulted from the end of the Cold War and the search for a "new world order." Rapid changes in birth rates, for example, have occurred in Russia and other former East Bloc countries. New refugee flows have resulted as well, especially in the region that once was Yugoslavia.

In addition, geographer Alan Nash (1994, 386) beseeched population geographers to improve their scholarship (a message often given to other geographers as well), noting that "The argument advanced here is that we must concentrate our efforts on particular issues or areas in which we have some expertise—and then we must augment that work, deepen our knowledge and hone our skills." He recommended, especially, the pursuit of more focused studies of migration, fertility, and gender issues.

To the extent that geographers continue to be concerned with the earth as a home for humans, we also need to focus more attention on the global and national interactions between growing populations and earth's capacity to support them. The world's population has more than doubled since the end of World War II and will probably double again before it stabilizes. As then-Senator Albert Gore (1992, 295) pointed out:

> Human civilization is now so complex and diverse, so sprawling and massive, that it is difficult to see how we can respond in a coordinated, collective way to the global environmental crisis. But circumstances are forcing just such a response; if we cannot embrace the preservation of the earth as our new organizing principle, the very survival of our civilization will be in doubt.

Whether you agree with Gore's pronouncement or not, we feel that there are few problems facing humanity today that will be more easily solved with continued population growth. Though the rate of human population growth has been slowing for about three decades now, its continued slowing is hardly inevitable. It clearly depends, more than ever, on the choices that millions of individuals make about how many children they want to have.

In the 1990s geographers and other social scientists with an interest in population growth began to focus more than ever on the importance of women and their roles, both in life and in fertility decisions. We know, for example, that complications from pregnancies and various reproductive disorders rob women of many days of healthy life in developing countries. We also know that throughout the world millions of women still do not have access to family planning and safe methods of controlling fertility. An estimated 585,000 women in developing countries die every year as a result of pregnancy; an additional 70,000 die of complications from abortions. Violence against women is a worldwide problem; the practice of female genital mutilation continues to scar millions, both physically and emotionally; and HIV/AIDS has affected men and women almost equally in Africa. Women are also disproportionately represented among the world's illiterate population and among the world's poor.

It is apparent, then, that one clear path to slowing world population growth would be to improve the lot of females everywhere, so that they could marry and reproduce if and when they want. Gender equality may not guarantee the "survival of our civilization" (which probably means very different things to different people around the world anyway) but it would go a long way toward improving the lives of individuals, families, and communities around the world. Violence against women (including rape and sexual coercion) remains a pervasive, if not always widely recognized, abuse of human rights around the world.

For more than half a century now ordinary people on every continent have embraced the notion that every individual has a claim to basic rights. Such rights were enumerated in 1948 when the General Assembly of the United Nations adopted and proclaimed resolution 217 A (III)—The Universal Declaration of Human Rights. The purpose of the declaration was to set a common standard of achievement for all peoples and all nations with respect to recognition of the rights and freedoms of each and every citizen of the planet. The declaration came out of a period following two world wars, a period that included such horrors as the Nazi atrocities. Article 1 (the first of 30) states that "All human beings are born free and equal in dignity and rights. They are endowed with reason and conscience and should act towards one another in a spirit of brotherhood."

THE GROWING POPULATION LITERATURE

Before ending this introduction, it is useful to recognize some of the periodicals in which population articles appear, so that both instructors and students can seek out more detailed and current studies with which to enrich this basic introduction to population geography. Before looking at some of the many periodicals, however, anyone interested in population needs to know about *Population Index*, a quarterly publication that indexes population literature from a vast number of both American and foreign sources. Your search for population information should start there. *Geographical Abstracts* provides another useful index, especially for geographers, though it is more limited in the scope of demographic studies that it indexes.

A number of journals are dedicated exclusively to population articles. Among the major ones in English are *Demography, Population Studies, International Migration Review, American Demographics, Population and Development Review, Population Bulletin, Population Research and Policy Review,* and *Family Planning Perspectives,* and the *International Journal of Population Geography.*

The Population Reference Bureau, located in Washington, D.C., publishes not only *Population Bulletin,* but also a number of other studies of population, as well as numerous materials of interest especially to those who teach population. Also located in Washington, D.C., is the Worldwatch Institute, which publishes an annual *State of the World* and a bimonthly, *World Watch,* along with an assortment of occasional publications that are of considerable value to those interested in population and environmental issues. Additional articles on population topics appear frequently in the following geographic periodicals: *The Geographical Review, Annals of the Association of American Geographers, The Professional Geographer, National Geographic, The Canadian Geographer, Transactions of the Institute of British Geographers,* and *Area.*

Aside from the above demographic and geographic publications, population articles of interest also appear occasionally in *Science, Scientific American, Nature, The Sciences, Urban Affairs Quarterly, International Journal of Health Science, American Economic Review, International Journal of Comparative Sociology, Research on Aging, Journal of the American Medical Association, American Behavioral Scientist, The Gerontologist, Social Forces, Foreign Affairs, American Sociological Review, Sociology and Social Research, Annals of the American Academy of Political and Social Science, Economic Development and Cultural Change, Rural Sociology, and Social Science Quarterly,* the *New England Journal of Medicine,* and the *Journal of the American Medical Association.*

Though the periodicals mentioned above are primarily scholarly, numerous population articles appear in more popular publications as well, including *Time, Newsweek, U. S. News and World Report, The Economist, The Atlantic Monthly, Business Week, and Forbes, along with newspapers such as The Los Angeles Times, The New York Times, The Washington Post, The Christian Science Monitor,* and *The Wall Street Journal.* Almost anywhere that you turn today, you are likely to come across something concerning population, so keep your eyes open and become an informed and critical consumer of demographic information.

Population Growth and Change

I n this chapter our aim is first to place the present world population into a broader historical perspective and then to consider current and future population trends. Only after we have seen where we have been, and how we got to where we are, can we begin to ponder our demographic future. The 6 billion or so people who currently share the planet face the prospect of being joined by another 4 billion before population growth ceases, unless dramatic or catastrophic changes occur.

MEASURING POPULATION GROWTH AND CHANGE

Any understanding of the ways in which population processes operate to shape or alter the size or composition of a region's population requires a knowledge of various measures of **population growth** and change.

THE BASIC DEMOGRAPHIC EQUATION

The most fundamental characteristic of any population is its size. An area's population may be increased either by a birth within the area or by the migration into the area of a person from another area. Similarly, the population may be decreased either by the death of someone within the area or by the migration of someone from the area out to

Primary population processes, births, deaths, emigration combine to produce the **basic demographic equation**.

another area. Thus, the primary population processes are births, deaths, and migration. These **basic demographic processes** may be combined to produce the following equation:

$$FP = SP + B - D + I - O,$$

where

FP = final population, some time interval beyond SP,
SP = starting population,
B = births during the interval,
D = deaths during the interval,
I = in-migration during the interval, and
O = out-migration during the interval.

It is easy to see why demographers have sometimes referred to it as the **"basic demographic equation."**

THE RATE OF NATURAL INCREASE

For any given population the rate of natural increase (RNI) equals the crude birth rate (CBR) minus the crude death rate (CDR). Thus

$$RNI = CBR - CDR.$$

The crude birth rate is the number of births per 1,000 population in a one year period, or

$$CBR = (B/P) \times 1,000,$$

where

B = number of births in one year and
P = mid-year population.

In 2000 the crude birth rate for the world was about 22 per thousand and for the United States it was approximately 15 per thousand. The crude birth rate is influenced to some degree by the age and sex structure of a population.

The crude death rate is the number of deaths per 1,000 population in a one year period, or

$$CDR = (D/P) \times 1,000,$$

where

D = number of deaths in one year and
P = mid-year population.

In 2000 the crude death rate for the world was about 9 and for the United States it was around 9 as well. The crude death rate is considerably more affected by the age structure of the population than is the crude birth rate, so comparisons among countries need to be done with caution. This should be immediately apparent when you consider that the world rate and the rate in the United States are the same.

Because the crude birth and death rates are both expressed per 1,000 population, it is obvious that the **rate of natural increase** will also be expressed in units per 1,000 population—births minus deaths. Since the crude birth rate for the world was 22 and the crude death rate was 9, the world's rate of natural increase for 2000 was equal to 22 minus 9, or 13 per thousand. Similarly, for the United States the rates were 15 and 9, respectively, and the rate of natural increase in the United States in 2000 was 15 minus 9, or 6 per thousand. Note that the rate of natural increase is not necessarily the same as the rate of population growth because the effect of migration is not included in the former. For the world the rate of natural increase equals the rate of population growth because migration to and from the earth is currently nonexistent (the sojourns of astronauts do not yet qualify as migrations and scientific evidence suggests that no aliens from space are currently living among us!). For the United States, however, the rate of natural increase is well below the actual rate of population growth because of a sizable annual net immigration (which is the subject of considerable debate today).

THE RATE OF POPULATION GROWTH

The **rate of population growth** is a measure of the average annual rate of increase for a population. Barring migration, it is possible to convert the rate of natural increase to the natural rate of population growth by simply converting the rate per 1,000 to an annual percentage rate. For the world the rate of natural increase was 13 per thousand, which is equivalent to an annual rate of population growth of 1.3 percent. Keep in mind, however, that most of the time migration must be considered, so the rate of population growth will usually differ from the rate of natural increase. To some extent the relative effects of natural increase and migration are inversely related to the size of the area under consideration. Whereas at the world scale migration plays no part at all in population growth, at the local scale migration may even be more important than natural increase in determining the overall rate of population growth.

In the United States example the rate of natural increase amounts to an annual growth rate of 0.6 percent. Actually, the 2000 rate of population growth for the United States was closer to 1.0 percent. The difference between these two rates results from net immigration in 2000. In other words, 60 percent of the growth in the United States population in 2000 was due to natural increase and 40 percent was due to net immigration.

DOUBLING TIME

The **doubling time** of a population is the number of years that would be required for a population to double in size, assuming that the population continues to grow at a given annual rate. This growth is analogous to the growth of money in a bank savings account. In both cases the "interest" is compounded. Without going into detail, it is possible to closely approximate the doubling time for a population by dividing the annual rate of population growth into the number 70. Thus, for the world, growing at 1.3 percent annually, the time required to double the present population would be 54 years. This assumes, of course, that the 1.3 percent growth rate continues over the entire period. For the United States, growing at 1.0 percent, the doubling time would be right around 70 years.

The **doubling time** of a population is the number of years that would be required for a population to double in size, assuming that the population continues to grow at a given annual rate.

■■■

WORLD POPULATION GROWTH

This section is concerned with the growth of the human population from prehistoric times to the present. Once we get into the twentieth century the focus shifts from the total world population to regional patterns of growth—mainly to the current division of the world into developing (mainly poor) and developed (mainly rich) regions—and to the differences between these regions with respect to population growth. We generally use the terms *developing* and *developed* regions with reference to levels of economic development—occasionally developing countries may also be referred to as *underdeveloped* or Third World countries.

BRIEF OVERVIEW OF WORLD POPULATION GROWTH

In order to understand the current world population situation, as well as future prospects for the world's ever-increasing numbers, it seems worthwhile first to discern how it is that the population reached its current level—a world of 6.1 billion people growing at an average annual rate of about 1.3 percent. Those figures suggest that each year around 79 million people (though some writers intimate that the annual figure may now be closer to 100 million) are added to what many already perceive to be an overcrowded planet; an increase nearly the equivalent of two South Africas each year. Every three years almost as many people are added to the world's population as currently live in the entire United States. Most geographers and demographers believe that this rate of growth cannot continue indefinitely. As Berelson and Freedman (1974, 3) noted more than two decades ago, "The rate of growth that currently characterizes the human population as a whole is a temporary deviation from the annual growth rates that prevailed during most of man's history and must prevail again in the future." This recent period of rapid population growth is unique in demographic history, both in terms of the rate of population growth and in terms of the absolute size of the world's population.

For most of human demographic history population growth was exceedingly slow; the annual rate of increase probably did not reach 0.1 percent (a doubling time of about 700 years) until sometime in the seventeenth century, after which it began to accelerate. This acceleration was gradual at first, but it became more noticeable after 1750.

Our current knowledge of historical populations remains conjectural. Clever demographic detectives, using whatever clues they can uncover (from archaeological excavations to early church baptismal, marriage, and death records), have pieced together the story of the human population's slow but inexorable expansion in both numbers and occupied territories. Our knowledge of historical population sizes and growth rates remains speculative because censuses and other organized and systematic collections of population data were nearly nonexistent before the middle of the eighteenth century. Earlier censuses had been taken in a few places, but their data were controversial at best. Even today reliable statistics don't exist for perhaps half of the world's population.

Estimates of population numbers in prehistoric times vary considerably and are generally made on the basis of assumptions about the **carrying capacity** of the land—its capacity to sustain a given human population at a given level of technology—and the distribution of the human population. Deevey (1960) estimated that the world's population around one million years ago was about 125,000. According to his estimates, this population grew very slowly to approximately 3.34 million 25,000 years ago and to 5.32 million 10,000 years ago. By *A.D.* 1, Deevey and others estimate, the world's population was in the neighborhood of 250-300 million. At that time the average annual rate of increase was probably on the order of 0.05 percent. At that growth rate it would take about 1,400 years for a population to double, compared to a doubling time of about 54 years for today's population.

The world's population did not reach its first billion until sometime around 1820. By then the annual rate of growth had increased tenfold to roughly 0.5 percent. Though all of human history had been required to reach this first billion, only 110 years were required to add the next billion; by 1930 there were 2 billion residents on our planet. In only 45 years this 2 billion doubled to the 1975 population of 4 billion, and by 1987 the world's population had grown to 5.0 billion. Only twelve years later the next billion had been added. As a species, we have certainly demonstrated our capacity for successful reproduction, but it may well be time for us to restrain ourselves before we

The **carrying capacity** of land is its capacity to sustain a given human population at a given level of technology—and the distribution of the human population.

■ ■ ■

The study of population is based largely on the collection and analysis of demographic data. These data vary considerably in reliability, so it is necessary to proceed somewhat cautiously, to develop a healthy skepticism about population information and its interpretation. In an informative article Bouvier (1976, 8-9) suggested the following warnings that you should certainly heed:

- **Warning 1**: Do not use growth rates to indicate changes in birth rates.

- **Warning 2**: Do not use natural increase to indicate population growth, except in those areas where migration is non-existent.

- **Warning 3**: Do not confuse numerical growth or decline with rates of population growth or decline.

- **Warning 4**: Do not take population figures as gospel truth, especially if they come from areas with less than adequate data-gathering facilities.

Each of these warnings should be considered carefully; errors in demographic thinking often result from a failure to consider one or more of them.

find a way to destroy our own ecological niche (though not the world—it could get along quite well without us, as it did for most of its history). As Ornstein and Ehrlich (1989, 45) noted:

> Increasing numbers is a "goal" of all organisms. But never before has there been an "outbreak" of a single species on such a global scale. Unfortunately it is not yet clear how enduring our unprecedented triumph will be, because it has created an unprecedented paradox: our triumphs can destroy us. As people strive to increase their dominance even further, they are now changing the earth into a planet that is inhospitable to civilization.

Geographer Crispin Tickell (1993, 220) noted that "All previous civilizations have collapsed." Though he recognized variations on the general theme, he suggested that each early civilization suffered from a fatal combination of population, resource, and environmental variables that turned unfavorable at some point. Furthermore, Tickell (1993, 220) added that:

> The prime engine of the recent dizzymaking rise in the human population and change generally is the industrial revolution. We have the misfortune to be perhaps the first generation in which the magnitude of the global price to be paid is becoming manifest.

(Such ideas are explored further in Chapter 9.)

Figure 1-1 provides a graphic illustration of human population history. The slow growth that characterized so much of the early history of humankind gives way, first gradually, then much more rapidly, to increased rates of population growth. These changes in growth rates involved alterations of both birth rates and death rates, with an emphasis on the latter; alterations that were in turn linked to sweeping changes in the socioeconomic fabric of societies.

Figure 1-2 shows past population growth on a logarithmic rather than an arithmetic graph, allowing us to focus more on changes in the rates of increase. The contrast with Figure 1-1 is both striking and suggestive. Rather than a single period of population growth, three periods of relatively rapid demographic increase become apparent, each of them followed by a slowing of growth rates. Deevey (1960) argued that each of these periods of accelerated population growth was a response to a revolution in which the earth's carrying capacity was dramatically increased. Each of these revolutions in carrying capacity can be viewed as a diffusion process, radiating outward from one or more origins to gradually encompass the inhabited world. The earliest of the three revolutions was the toolmaking or cultural revolution; the second was the agricultural revolution; and the third was the scientific-industrial revolution, which continues today.

The implications of Figures 1-1 and 1-2 for future population growth are dramatically different. Figure 1-1 implies a continuing rapid increase in the size of the world's population (perhaps followed by a catastrophic crash?), whereas Figure 1-2 suggests that the world should experience a tapering off of growth rates as the world population adjusts to the current technological levels and their concomitant limitations on population expansion. Barring another major

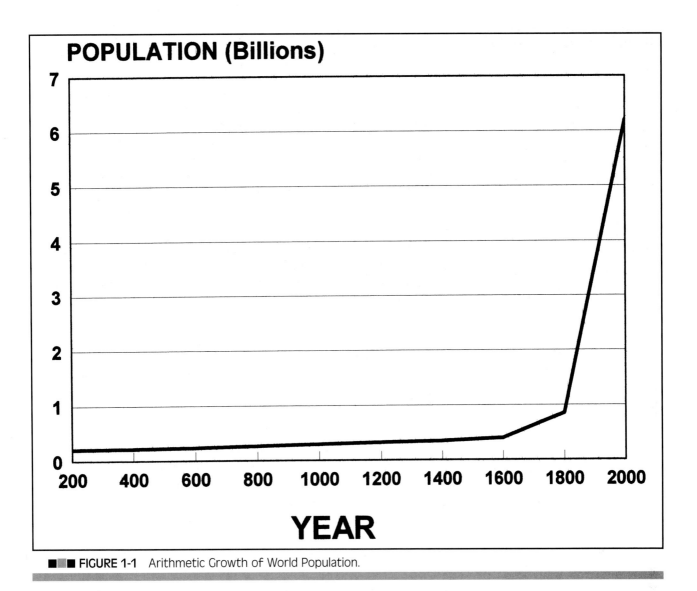

POPULATION (Billions)

YEAR

■■■ FIGURE 1-1 Arithmetic Growth of World Population.

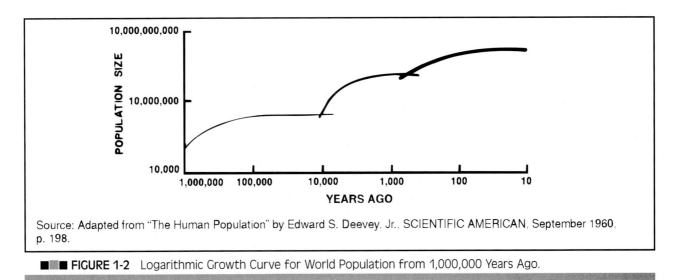

Source: Adapted from "The Human Population" by Edward S. Deevey, Jr., SCIENTIFIC AMERICAN, September 1960, p. 198.

■■■ FIGURE 1-2 Logarithmic Growth Curve for World Population from 1,000,000 Years Ago.

revolution that would again expand the earth's carrying capacity, Deevey's interpretation of population history implies a slowing of world population growth, one that we are already beginning to see. The world's population growth rate reached a peak sometime in the late 1960s at around 2.1 percent, from which it has dropped gradually to a level of about 1.3 today. Though the direction of change is encouraging, we need to keep in mind that it only represents a change in doubling time from about 33 years to 54 years. Furthermore, given the size and youthfulness of the planet's population, absolute annual population increases (as opposed to the rate of increase) will remain large for many years.

THE THREE MAJOR PERIODS OF POPULATION GROWTH

As we have already noted, Deevey (1960) proposed that population growth occurred unevenly over time, mainly in conjunction with three major revolutions in human history—the cultural, the agricultural, and the scientific-industrial revolutions. These three major revolutions serve as the basis for subdividing our discussion of population growth into three discrete periods. This is appropriate because each of the revolutions must have opened up new possibilities for population growth, mainly because each one extended the earth's carrying capacity. However, the availability of population *facts* decreases rapidly as we move back in time, hence the following discussion must be approached with a degree of caution.

THE CULTURAL REVOLUTION AND POPULATION GROWTH

The **cultural, or tool-making, revolution** occurred in prehistoric time. What knowledge exists of events in that era must be drawn primarily from the archaeological record. Seeking into the origin and evolution of the human population continues to occupy the time and energy of many researchers, mainly anthropologists. Periodically, new finds of the skeletal remains of early hominids push back the frontiers of our knowledge to yet earlier times. The geographical search for early people has established an African origin.

Three pieces of archaeological evidence have pushed the frontier of hominid development further back into the dawn of prehistory. Tim White's discovery of what he called *Ardipithecus ramidus*, a new genus, was dated back to about 4.4 million years ago. In 2001 discovery of **Ardipithecus ramidus kadabba** pushed the story of our human ancestry back to around 5.5 million years ago. Meave Leakey's Kanapoi fossils, found not far from Lake Turkana, extended *Australopithecus afarensis* back to about 4.1 million years ago.

Between **Ardipithecus ramidus** and **Australopithecus afarensis**, at an age of 4.2 to 3.9 million years ago, Leakey and Walker (1997) have added a new species, **Australopithecus anamensis,** based on further fossil evidence from sites at Kanapoi and Allia Bay.

Previously known skeletal remains uncovered in Africa suggested an age of 2.8 to 3.8 million years for *Australopithecus afarensis*, as exemplified by "Lucy" (Johanson and Edey, 1981), 1.6 to 2.2 million years for **Homo habilis,** and 1.4 to 1.7 million years ago for **Homo erectus.** Bipedalism was probably the earliest characteristic that separated hominids from their nearest relatives, the gorillas and chimpanzees. After that, increasing brain size is one of the primary features that distinguishes each one of the hominid groups from the next. However, as Ornstein and Ehrlich (1989, 37-38) have suggested, "...although the human brain appears to have enlarged rapidly (in geological time) in response to the pressures of culture, it does not seem likely that the brain crossed any real physical threshold that suddenly permitted new kinds of cultural activities."

Perhaps our arrogant view of ourselves long colored the way in which we perceived our steady evolution from ape to hominid to modern people. Like other successful animal species, the line of humans, once separated from the apes five or six million years ago, apparently went through considerable trial and error, made many false starts, and reached a few dead ends. Ian Tattersall (2000) has argued convincingly that at least four kinds of hominids lived together within a single landscape in part of what is now northern Kenya about 1.8 million years ago. Though we may never know the degree to which they interacted with each other, Tattersall argues that *Paranthropus boisei, Homo rudolfensis, Homo habilis,* and *Homo ergaster* (also known as African *Homo erectus*) occupied the same geographical area at the same time. As Tattersall (2000, 61) phrased it, human evolution has not been a simple linear pattern:

> Instead it has been the story of nature's tinkering: of repeated evolutionary experiments. Our biological history has been one of sporadic events rather than gradual accretions. Over the past five million years, new hominid species have regularly emerged, competed, coexisted, colonized new environments and succeeded—or failed. We have only the dimmest of perceptions of how this dramatic history of innovation and interaction unfolded, but it is already evident that our species, far from being the pinnacle of the hominid evolutionary tree, is simply one more of its many terminal twigs.

According to Wenke (1990), two scenarios currently set the stage for debate about the continued search for human origins. One of these scenarios suggests the migration of *Homo* ancestors, most likely *Homo erectus*, out of Africa some 1.5 million years ago, followed by gradual diversification among scattered groups but at the same time a gradual evolution toward **Homo sapiens** among all the groups because of both common genetic inheritances and similar adaptive pressures. The result of this first scenario, as Wenke (1990, 137) described it, was "that they all converged at about 30,000 years ago as one species, *Homo sapiens sapiens*." The other scenario begins in the same way, with the migration of *Homo* ancestors out of Africa some 1.5 million years ago. This scenario differs after the initial migration by suggesting considerably more divergence among groups as they spread out and adapted to different environments. Following the second scenario, according to Wenke (1990, 137), the final result is that "perhaps 140,000 years ago, *Homo sapiens* evolved in one place (probably Africa), and spread across the world, displacing most groups, driving some into extinction, and absorbing a small fraction of the others through intermarriage."

In recent years the latter scenario has received considerable attention, not from archaeological but from genetic evidence. Carr, Stoneking, and Wilson (1987) argued that the maternal lineage of all humans could be traced backed to a single African woman who was alive perhaps 200,000 years ago, a woman now referred to by many as "Mitochondrial Eve." Their data were based on studies of mitochondrial DNA (deoxyribonucleic acid), which are only maternally inherited. These DNA then form the basis for a "molecular clock," which in turn can be used to develop a branching tree. As Stringer (1990, 99) has noted, "One attempts to make a molecular clock by comparing genetic differences among various species or varieties within species and expressing their relatedness in a tree...then calibrates (or 'dates') the tree by comparing it with another group that diverged from the tree at a known date." Studies since 1987 have tended to confirm the African origin of "Mitochondrial Eve" and have given rise to considerable speculation about what that means for the Neandertals (Stringer, 1990). More recently, Barinaga (1992, 686), citing increasing questions that have been raised about the DNA methodology, noted that "...the root of the human tree has been thrown open to question once again." Current criticism is focused primarily on how the mitochondrial DNA have been analyzed and whether or not these data can identify a geographic origin for our species. Excellent summaries of the two opposing views of evolution over the past 200,000 years are Wilson and Cann (1992) and Thorne and Wolpoff (1992).

In a review of these two competing hypotheses about our origin, Tattersal (1997, p. 67) wrote that "...my strong preference is for a single and comparatively recent origin for *H. sapiens*, very likely in Africa—the continent that, from the very beginning, has been the engine of mainstream innovation in human evolution.

The emergence of anatomically modern humans, *H. Sapiens*, as the *only* surviving members of a long chain of trials and errors occurred recently—probably no more than 25,000 to 28,000 years ago. Before that, we know, for example, that 90,000 to 100,000 years ago in the region that now includes Israel modern humans lived side-by-side with Neanderthals (whose reputations have been much improved as a result of recent research).

Between 30,000 and 40,000 years ago modern humans appeared in Europe, where Neanderthals lived already. However, at least according to Tattersall (2000, 61), "Certainly the repeated pattern at archaeological sites is one of short-term replacement, and there is no convincing biological evidence of any intermixing in Europe." In a period of perhaps 10,000 years the Neanderthals disappeared.

Wong (2000), however, identifies some differences of opinion among anthropologists' interpretation of the European confrontation between Neanderthals and modern humans. For example, anthropologist Fred Smith (cited in Wong, 2000, 107) tells us that "The likelihood of gene flow between the groups is also supported by evidence that Neandertals left their mark on early modern Europeans."

We are still left with more questions than answers. It seems clearer all the time that the evolution of modern humans was a complex drama that unfolded over a period of some five million years. Trial and error, numerous wrong turns, and dead ends led to the final emergence and dominance of modern humans between 25,000 and 30,000 years ago. Tattersall argues that the final defining characteristic that gave modern humans their edge over the Neanderthals may have been the development of language and an ability to form mental symbols. As he notes (Tattersall, 2000, 62), "We do not know exactly how language might have emerged in one local population of *H. sapiens.* . . . But we do know that a creature armed with symbolic skills is a formidable competitor—and not necessarily an entirely rational one, as the rest of the living world, including *H. neanderthalensis*, has discovered to its cost."

Though these debates will undoubtedly continue, perhaps it is more important at this point for us to consider Wenke's (1990, 186) summary comment that "...we should also take note of the fact that even though we are physically very much like our ancestors of 12,000 years ago, we are enormously different culturally...in a cultural sense we are not at all the same species as the human hunter-gatherers of the late Pleistocene...to live a life for which evolution has shaped us—we should probably eat a varied diet, live closely in a small group, and walk a lot."

As fascinating as such studies of early *Homo* and earlier ancestors may be, little is known of the numbers involved, though they were undoubtedly small. As Deevey (1960, 5-6) commented, "For most of the million-year period the number of hominids, including man, was about what would be expected of any large Pleistocene mammal—scarcer than horses, say, but commoner than elephants." Not only were the numbers small, but the rate of growth in these numbers must also have been exceedingly small. The growth of *Homo sapiens sapiens* prior to the agricultural revolution remained extremely slow. Hunting, fishing, and foraging provided an existence, though undoubtedly a precarious one. Life was most likely "nasty, brutish, and short."

Population densities on the eve of the agricultural revolution were low, and populations were vulnerable to environmental changes such as climatic fluctuations. Estimates of population size are subject to wide margins of error and must be accepted with reservation; however, the order of magnitude seems reasonable. According to Deevey (1960), the world's population 10,000 years ago was 5.32 million, though others have suggested that it might have been twice that number. Even if we accept 10 million, we're talking about a small base from which we have grown in a relatively short time.

THE AGRICULTURAL REVOLUTION

An exact date for the beginning of the **agricultural revolution** is impossible to set, but it is likely that incipient cultivation and domestication developed sometime around 10,000 *B.C.* in the Near East. The rimland around the Fertile Crescent was one of the first areas to experience the agricultural changes that would slowly burgeon into a major revolution. As agricultural practices evolved and diffused, people experienced tremendous changes; they were able to turn from the wandering and tenuous life of hunter-gatherers to the more sedentary and secure life of agriculturalists. However, these changes were gradual. As Cipolla (1974, 22) noted:

> It can be stated however, with a fair degree of certainty, that the foundations of settled life in the Old World were first laid in South-West Asia between the ninth and the seventh millennium B.C. This seemingly took place where prototypes of the earliest domesticated animals and plants existed in a wild state and where the concentration on particular species as sources of food was stimulated by the ecological changes that marked the transition to Neothermal climate.

Between perhaps 7,000 *B.C.* and 5,000 *B.C.* there was some possible domestication of plants in Mesoamerica. Following 5,000 *B.C.*, a slow but consistent domestication continued there (Diamond, 1997).

The introduction of farming allowed greater population densities to exist and probably produced the first food surpluses people had ever known. In turn, a few people were then freed from the fundamental task of providing food. A multitude of inventions and innovations followed, including the development of village settlements, irrigation, metallurgy, and long-distance trade. These inventions and innovations in turn increased people's capacity to satisfy their needs from the environment and further increased the carrying capacity of the land. Trade, and the convergence of trade routes, certainly affected population distribution and the rise of early cities as well. The demographic response to agricultural and related changes was a gradual acceleration in the rate of population growth. As the agricultural revolution diffused to various parts of the earth's inhabited surface, its impact on population growth became increasingly significant. However, as Deevey's interpretation suggested, once the impact of this revolutionary increase in the earth's carrying capacity had completed its diffusion, the rate of population growth slowed again, and the population became stabilized at a new and higher plateau.

By the beginning of the Christian Era the earth's population was about 250-300 million. Though this population continued to grow, its numerical progress was slow, and the regional distribution of growth was varied. Within the overall pattern of growth, cyclical changes were typical. The rate of population growth was kept in check by food supplies (often interrupted by famines), by wars, and by epidemics of various diseases.

For the most part famines have been localized, and their impact on the death rate depends on both the severity of the famine and the links between the famine-stricken area and other locations. However, famines have occasionally been devastating. Fourteenth century China may have experienced the planet's first great famine (with deaths thought to be in excess of 4 million); they occurred in many regions over the next few centuries, including the great potato famines in Ireland (1845-1849). Since 1900 famines have been worst in China (1921-23 and 1928-29), India (1943-1944 and 1965), and Russia (1932-1934, 1941-1944, and 1947), though they have occurred also in Poland, Greece, Africa's Sahel region, Ethiopia, Bangladesh, Somalia, Nigeria, and Kampuchea. With improved transportation linkages, the effects of local crop failures diminished.

Wars have directly affected population growth rates at various times, but their impact is not always easy to assess. The simple counting of battlefield deaths alone would underestimate the demographic impact of most wars, because wars also disrupt food supplies and act as diffusion agents for numerous diseases. Of war and the latter, Zinsser (1967, 113) commented, "And typhus, with its brothers and sisters—plague, cholera, typhoid, dysentery—has decided more campaigns than Caesar, Hannibal, Napoleon, and all the inspector generals of history." The greatest losses occurred in World War I and World War II.

Epidemics and pandemics have often been devastating in their impact on regional populations. Extreme examples include the Justinian Plague of *A.D.* 541-544 and the Black Death of *A.D.* 1346-1348. The latter may have reduced the European population by 25 percent, and local death tolls reached as high as 50 percent. Recovery of the European population after this decimation was slow, and it was further hampered by the One Hundred Years War. By the sixteenth century, however, Europe had regained her lost population and was beginning a gradual acceleration in the rate of population growth, though there were still localized periods of famine, war, and disease. The rapid growth and distribution of AIDS (discussed in Chapter 4) convinced everyone who might have thought otherwise that epidemics and pandemics are still with us.

After 1492 population declined precipitously in the New World as well. Though we will never know for sure, reasonable estimates suggest that more than 50 million people lived in the Americas when Columbus first sailed westward. As journalist Lewis Lord (1997, p. 70) noted, "The 150 years after Columbus's arrival brought a toll on human life in this hemisphere comparable to all of the world's losses in World War II." Geographer William Denevan (1996) explored estimates of the Native American population in detail. Diamond (1997) suggests that 95 percent of the Native American population died as a result of diseases introduced by Europeans into the New World.

In response to the Industrial Revolution of the nineteenth century, population grew rapidly in developed countries.

THE INDUSTRIAL REVOLUTION

The **Industrial Revolution** originated in England in the latter half of the eighteenth century, though its roots may be found in earlier times.

The **Industrial Revolution** originated in England in the latter half of the eighteenth century, though its roots may be found in earlier times. From England it diffused rapidly into the countries of Western Europe and to the United States. By the beginning of the twentieth century it had reached Russia and Northern Italy. Japan was the first Asian country to experience this revolution. As countries industrialized, industry replaced agriculture as the major sector of the economy. The Industrial Revolution continues today, and it has so far only partially diffused to the developing countries. New inventions and innovations continue to pour forth almost daily.

By about 1750 in England and Wales, and soon thereafter in other countries as they began to industrialize, population growth accelerated. Prior to this time, crude birth and death rates had both tended to be high. In average years there may have been more births than deaths, while in bad years the reverse was likely. Death rates undoubtedly fluctuated more widely than did birth rates. High birth rates were deemed necessary in order to overcome the prevailing high death rates, though birth rates were generally not as high as they might have been because of a variety of social constraints.

During both the cultural and agricultural revolutions people increased their capacity to wrest a living from the earth, but it was not until the scientific-industrial revolution that, for the first time, they began to gain control over death. This control over death rates was a result of many changes, mainly changes that probably at first cut off the high peaks in the cyclical fluctuations of death rates. Better agricultural practices and improved distribution systems cut down on the localized effects of famines. Improved sanitary practices and facilities decreased the deaths from some diseases quite early. Then during the nineteenth century major medical advances accelerated the downward trend in death rates.

The innovations associated with each revolution were not contained in their area of origin but diffused outward, as is well illustrated by the spread of the Neolithic farming cultures of Europe. In 6,000 *B.C.* farming in Europe was primarily limited to a few sites near the Aegean Sea. During the next 1,000 years it spread northward into the Danubian Basin and by 4,000 *B.C.* to the North European Plain. An even more rapid diffusion of the industrial-scientific revolution has occurred. These innovations were carried by Europeans as they colonized new areas, and in the current century there are few places that have not been touched by industrialization to some degree.

The speed and direction of diffusion was governed by such things as distance, obstacles, nature of the environmental base, and receptivity of various social structures. One critical point is that the spread of new innovations was uneven. Whenever such innovations were introduced into a society there was a traumatic effect that necessitated new forms of organization, new patterns of leadership, and the acquisition of new skills. Rapid population growth often accompanied the changing socioeconomic conditions.

One further point to be emphasized about Deevey's interpretation of world population growth is the nature of the population growth curve for each revolution. After each rapid spurt in population growth, the growth rates slackened off—the numbers reached a plateau, and then further additions were slow to be achieved. Each revolution therefore removed, partially at least, some pre-existing constraint upon population growth, but it must also have set into motion forces that eventually brought growth under control. Obviously, these forces are of urgent concern in our present circumstance.

In response to the Industrial Revolution the world's population entered a period of rapid and sustained population growth. During the nineteenth century this growth was concentrated in the developed countries. By the middle of the twentieth century, however, population growth had subsided in the developed countries and was accelerating in the developing countries, setting the demographic stage for the new millennium.

Many would have (and some actually did!) predicted that the rapid population growth of the twentieth century (from about 1.6 billion in 1900 to 6.1 billion in 2000) would have resulted in a world of extreme poverty and economic deprivation as resource scarcities led to higher prices for basic commodities. Such was not the case, however, and we need to keep this in mind as we look ahead to discussions of population growth and economic well-being, food supplies, and environmental concerns.

Though no definitive yardstick is available for measuring such things, economists have suggested that the world's material standard of living increased perhaps nine-fold during the twentieth century— a considerable achievement. On the average people live longer, healthier lives now than they did 100 years ago. Geographically, however, the vast improvements in wealth during the twentieth century accrued mainly to the nations of Europe, the United States, and Japan. We enter the new century with vast differences in wealth among the world's nations—a person's place of birth largely determines his or her economic and demographic destiny.

THE HUMAN POPULATION TODAY

Today's population situation is unique in the world's history; not only is the current rate of increase still near the highest in human history, but the base population (6.1 billion) is also the largest ever. The historical record shows that the acceleration of world population growth started with the European countries and the lands that Europeans settled overseas (especially the United States, Canada, and Australia). However, the areas that are growing the fastest today are Africa, Asia, and Latin America, the so-called developing countries, where more than three-fourths of humankind dwell—more than 90 percent of the world's population growth occurs in these areas. Death rates in these countries have been falling during the last twenty-five years, whereas the birth rates have remained twice as high as they are in the developed nations (Goldstein and Schlag, 1999).

Between 1950 and 1987 the world's population doubled from around 2.5 billion to over 5 billion, an increase of over 2.5 billion people in less than forty years. The Cold War ignored growing populations and changing geographic concentrations of people. During those years population growth had been unequally distributed geographically; more than 85 percent of that growth occurred in the developing countries. In most of the developed countries today fertility hovers near or below replacement level (the United States is the major exception), so that an even higher percentage of population growth in coming decades will occur in the developing countries, those least able to absorb additional people. By the year 2000 80 percent of the world's population resided in the developing countries; nearly half of them are residents of either China or India (the world's second "demographic billionaire"). It is easy to see why, then, we can expect international migration to flourish in the decades ahead as globalization of the economy brings together capital, which is heavily concentrated in the rich countries, and young people, who are heavily concentrated in the poor ones.

Still another comparison between the developed and developing countries can be made by considering the differing age structures of their populations. A country that has a rapidly growing population has a large proportion of its residents in the younger age groups. In the rapidly growing areas of the world—Africa, Asia, and Latin America—high proportions (typically 30-45 percent) of the population are under 15 years of age, whereas in North America, Europe, and Oceania there are significantly lower proportions of young

Courtesy of Photo Disc

Africa has one of the fastest growing populations today. More than 90 percent of the world's population growth is made up of the developing countries of Africa, Asia, and Latin America.

people (20 percent or less). A population with a large proportion of old people will have different needs and requirements than a population with a great many young people.

POPULATION PROJECTIONS

So far we have viewed the present population situation mainly in the perspective of the past, but what does the future hold in store for the world's population? One answer to that question, based on a set of assumptions about the dynamics of a population over some time period, is the **population projection**. Young (1968, ix) cautioned us long ago, however, to remember that:

> The projection of a future population from a present growth rate is a hazardous undertaking at best. These rates contain many variables and are sensitive to small changes in these variables. Furthermore, since population growth is cumulative, very slight changes in present rates can make enormous differences when projected 100, 200, or more years into the future.

Demographers are careful to differentiate between projections and predictions. The *projection* for the size of a population at some future date is based on a set of *assumptions* about the demographic processes that will affect population growth over the time period. The simplest assumption is that the future rate of population growth will be the same as that of today. However, for most situations this is an unrealistic assumption. Typically, the projection is broken down into separate projections for the birth rate, death rate, and migration. These may then be combined into a single projection for the future population. What the projection shows is that, *if* the assumptions hold true for births, deaths, and migration, *then* the projection will be accurate. Demographers should not be held responsible if the assumptions are not fulfilled, you see. Often a projection may alter people's reproductive behavior and set into motion events that will assure that the projection will be off the mark. Usually more than one projection is made and quite often a series of projections is made, using different assumptions about future birth, death, and migration rates.

POPULATION PROJECTIONS: A BRIEF OVERVIEW

First, we should distinguish among the following three commonly encountered terms: projection, forecast, and prediction. A **population projection** is made on the basis of the population at some date and assumptions about births, deaths, and migration between that date and some future date. As Gibson (1977, 7) noted, "Population projections are 'correct' by definition (except for computational errors) because they indicate the population that would result if the base data (starting) population is correct and if the underlying assumptions about future change should turn out to be correct." The usefulness of projections, then, depends on what assumptions have been made and how well they accord with the actual events.

Demographers prefer to avoid the term *prediction* altogether because it suggests that only one projection has been made and it is considered as an ultimate truth, as occurring with a high degree of certainty. Past projections have often fallen so wide from their marks that the best most demographers will venture today is to choose one of a series of projections as a forecast, and that usually only for short distances into the future, and then only reluctantly.

Two broad classes of population projections exist: mathematical and component. **Mathematical methods** are easier to understand and to apply, but **component methods** are preferable for most projections, especially for those beyond the short term, which is usually taken to be five years or less. Whereas **mathematical methods** employ some mathematical formula to a base population using an assumed rate of growth over the projection interval, **component models** separately project births, deaths, and migration, then combine the "components" into an overall population projection. These latter projections, of course, are also done mathematically, and the terminology sometimes confuses people. For demographers and population geographers there is no escape from mathematics.

Within each class of projections different models exist, so a wide range of projection models can be called upon. The model that should be used, of course, depends upon several factors, such as the size of the area for which projections are being made, the assumptions that can or should be made, the types of data that are available, and the length of the projection interval. With respect to scale, for example, national projections require different considerations than do projections for local areas. Understanding and projecting migration are more critical for local area projections than is the projecting of births and deaths, perhaps; whereas projecting immigration at the national level may be much easier than projecting births, mainly because immigration is controlled, at least to some extent, by the national government. Also, data for local areas are not always available in sufficient detail for employing component models, thus making mathematical models more attractive, especially for short-term projections.

POPULATION PROJECTIONS: WORLD AND MAJOR REGIONS

Despite the difficulties, population projections are deemed essential and useful. They can stimulate our thinking about the consequences of population trends, for example. We need to keep in mind, however, what one perceptive United Nations study stated:

> The affairs of men are often subject to swings of the pendulum, erratic fluctuations, sometimes even dramatic upsets. That unsystematic variations will also occur in the future, particularly the long-range future, can be fairly taken for granted. But since these are intrinsically unforeseeable, only fairly smooth and systematic changes can be projected. In actual fact these may, at the most, constitute a long-term average trend around which there will occur unpredictable ups and downs. The calculated smooth changes should reflect those factors in the situation which are likely to prevail in the long run. (United Nations, 1974, 48).

Among the basic assumptions that the United Nations made for its population projections were the following:

1. At least a minimal degree of social order and control will be maintained.

2. Efforts at maintaining or improving the quality of life will continue and will not be totally frustrated.

3. Regional vital rates will move differently in terms of time, but eventually everywhere mortality and fertility will fall slightly below the lowest levels now observed.

The United Nations population projections are shown in Table 1-1. The potential for population growth is considerably higher in the developing areas of the world than in the developed ones, as has been suggested already.

After reaching 6 billion in 1999, the world's population is projected to increase by more than 50 percent over the next 50 years, to a population of 9.3 billion according to the medium variant of the United Nation's updated projections. This assumes that the world's total fertility rate drops to 2.15 births per woman by 2045 and that life expectancy increases from 65 years to 76 years. That would leave the world's population still increasing in 2050, but by less than 0.5 percent annually.

■■■ TABLE 1-1. Estimated and Projected Population of the World, Major Development Groups and Major Areas, 1950, 2000 and 2050 According to the Different Fertility Variants

Major Area	Estimated population (millions)			Population in 2050 (millions)		
	1950	2000	Low	Medium	High	Constant
World	2 519	6 057	7 866	9 322	10 934	13 049
More developed regions	814	1 191	1 075	1 181	1 309	11 62
Less developed regions	1 706	4 865	6 791	8 141	9 625	11 887
Least developed countries	197	658	1 545	1 830	2 130	3 150
Other less developed countries	1 508	4 207	5 246	6 312	7 495	8 738
Africa	221	794	1 694	2 000	2 320	3 566
Asia	1 399	3 672	4 527	5 428	6 430	7 376
Latin America and Caribbean	167	519	657	806	975	1,025
Europe	548	727	556	603	654	580
Northern America	172	314	389	438	502	446
Oceania	13	31	42	47	53	56

Source: United Nations Population Division (2001), p. 13.

POPULATION PROJECTIONS: THE UNITED STATES

For comparative purposes it is useful to look at what has happened to official population projections for the United States. These projections are extremely important because they serve as the basis for many other projections, including projections of the demand for housing, educational facilities, hospitals, and a myriad of other needs.

Population projections for the United States are shown by year in Table 1-2. Though these projections were published in early 2000, they do not include the 2000 census figures, which showed a population about 6 million larger than the 2000 figure used for projections. As a result, these projections are assuredly on the low side. The assumptions underlying the different variants are discussed in Hollmann, Mulder, and Kallan (2000). The middle series is considered the "most likely" variant, though with the usual caveat.

Given discussions in the United States about immigration rates, it is of particular interest to note the final column of projections because they assume no immigration. As a result, we see that the difference between the middle series variant and the zero immigration variant is about 76 million people by 2050 and more than 193 million by 2100. No matter how we view it, the United States will have a much larger population in the future if current high rates of immigration are sustained.

Table 1-3 compares the latest projections of the population of the United States with those made based on 1994 population estimates. The middle series projections clearly have been revised upward, reflecting mainly the expanding role of immigration in population growth. Keep in mind that not only are immigrants added to the population but also that their children will be added as well. It is likely that in the decades ahead the United States will grow more rapidly than either Japan or the nations of the European Union—our two major economic competitors in the world today.

In summary, selecting an appropriate growth rate for projecting a population requires numerous considerations. We must look not only at past trends and current patterns, but also try to look at ways in which these trends may be altered in the future. Our projections can be no better than the assumptions upon which they are based. In addition, studying projections forces us to confront different scenarios about the consequences of different growth patterns and the possibility of designing policies to affect those patterns (Lee, 2000).

SMALL AREA POPULATION PROJECTIONS

Though national population projections are of considerable importance, they are not the only ones that are of interest. Political units, from states to counties and cities, need to know something about their demographic futures, as do local school districts, highway planners, and urban and regional planning departments. In addition, numerous private corporations are interested in the changing demographics of local areas, so that they can better gauge local and regional changes in demand, marketing strategies, and even changing tastes and preferences. Thus the need for **small area,** that is, subnational, **population projections** exists and is growing.

At the same time, as you might expect, **small area population projections** are more difficult to make because local variations in fertility, mortality, and migration may be much wider than those at the national scale. Especially difficult to project are migration rates for local areas, because changes in the area's socioeconomic characteristics may quickly alter current migration patterns. For example, no one looking at the changes that had occurred in the Asian population in Long Beach, California, during the 1970s would have projected that that city and neighboring Lakewood would have a population of more than 20,000 Cambodians by the early 1990s. At best, small area projections are a tricky business, and changing migration patterns are the major culprit.

The United States Bureau of the Census does state projections, and even there the assumptions that must be made are difficult to choose. The different series result primarily from different assumptions about migration patterns. However, the Bureau wisely chooses not to do population projections for areas smaller than states, leaving that task to state and local agencies and to private firms such as Donnelley Marketing Information Systems. Detailed information on local area projections can be found in Irwin (1977).

CULTURE, POPULATION GROWTH, AND PLANNING

The United Nations designated 1974 as World Population Year and in August of that year convened the World Population Conference in Bucharest, Romania. The purpose of that conference was to

> Small area population **projections** are more difficult to make because local variations in fertility, mortality, and migration may be much wider than those at the national scale.

■■■ TABLE 1-2. Population Projections for the United States

	Middle Series	Lowest Series	Highest Series	Zero International Migration Series
1999	272,820	272,695	272,957	272,323
2000	275,306	274,853	275,816	273,818
2001	277,803	276,879	278,869	275,279
2002	280,306	278,801	282,087	276,709
2003	282,798	280,624	285,422	278,112
2004	285,266	282,352	288,841	279,493
2005	287,716	284,000	292,339	280,859
2006	290,153	285,581	295,911	282,219
2007	292,583	287,106	299,557	283,579
2008	295,009	288,583	303,274	284,945
2009	297,436	290,018	307,060	286,322
2010	299,862	291,413	310,910	287,710
2011	302,300	292,778	314,846	289,108
2012	304,764	294,120	318,893	290,514
2013	307,250	295,436	323,044	291,924
2014	309,753	296,723	327,293	293,334
2015	312,268	297,977	331,636	294,741
2016	314,793	299,197	336,069	296,144
2017	317,325	300,379	340,589	297,539
2018	319,860	301,521	345,192	298,921
2019	322,395	302,617	349,877	300,288
2020	324,927	303,664	354,642	301,636
2021	327,468	304,667	359,515	302,958
2022	330,028	305,628	364,524	304,251
2023	332,607	306,545	369,671	305,511
2024	335,202	307,412	374,960	306,735
2025	337,815	308,229	380,397	307,923
2026	340,441	308,999	385,971	309,070
2027	343,078	309,727	391,672	310,172
2028	345,727	310,413	397,507	311,230
2029	348,391	311,056	403,483	312,246
2030	351,070	311,656	409,604	313,219
2031	353,749	312,204	415,839	314,153
2032	356,411	312,692	422,154	315,049
2033	359,059	313,124	428,554	315,910
2034	361,695	313,499	435,041	316,737
2035	364,319	313,819	441,618	317,534
2036	366,934	314,086	448,287	318,304
2037	369,544	314,303	455,053	319,049
2038	372,148	314,472	461,917	319,773
2039	374,750	314,594	468,882	320,478
2040	377,350	314,673	475,949	321,167
2041	379,951	314,710	483,122	321,843
2042	382,555	314,707	490,401	322,506
2043	385,163	314,667	497,790	323,160
2044	387,776	314,591	505,290	323,807
2045	390,398	314,484	512,904	324,449
2046	393,029	314,346	520,633	325,087
2047	395,671	314,181	528,480	325,723
2048	398,326	313,990	536,447	326,359
2049	400,998	313,778	544,539	326,998
2050	403,687	313,546	552,757	327,641

	Middle Series	Lowest Series	Highest Series	Zero International Migration Series
2051	406,396	313,296	561,106	328,291
2052	409,127	313,030	569,589	328,949
2053	411,884	312,752	578,211	329,617
2054	414,667	312,461	586,975	330,297
2055	417,478	312,160	595,885	330,991
2056	420,318	311,850	604,943	331,700
2057	423,191	311,532	614,157	332,427
2058	426,097	311,206	623,527	333,172
2059	429,037	310,873	633,058	333,937
2060	432,011	310,533	642,752	334,724
2061	435,021	310,187	652,615	335,533
2062	438,067	309,833	662,648	336,365
2063	441,149	309,471	672,853	337,220
2064	444,265	309,098	683,233	338,098
2065	447,416	308,716	693,790	338,999
2066	450,600	308,321	704,524	339,922
2067	453,815	307,913	715,438	340,866
2068	457,061	307,488	726,530	341,830
2069	460,337	307,048	737,804	342,814
2070	463,639	306,589	749,257	343,815
2071	466,968	304,109	760,892	344,833
2072	470,319	305,608	772,707	345,865
2073	473,694	305,086	784,704	346,909
2074	477,090	304,540	796,883	347,966
2075	480,504	303,970	809,243	349,032
2076	483,937	303,375	821,785	350,107
2077	487,387	302,756	834,510	351,189
2078	490,853	302,111	847,420	352,278
2079	494,334	301,442	860,514	353,372
2080	497,830	300,747	873,794	354,471
2081	501,341	300,029	887,263	355,574
2082	504,866	299,286	900,922	356,681
2083	508,406	298,521	914,773	357,792
2084	511,959	297,732	928,818	358,907
2085	515,529	296,923	943,062	360,026
2086	519,113	296,093	957,506	361,149
2087	522,712	295,244	972,153	362,277
2088	526,327	294,375	987,006	363,409
2089	529,958	293,488	1,002,069	364,546
2090	533,605	292,584	1,017,344	365,689
2091	537,269	291,664	1,032,834	366,838
2092	540,948	290,727	1,048,542	367,992
2093	544,645	289,775	1,064,472	369,153
2094	548,357	288,808	1,080,626	370,319
2095	552,086	287,826	1,097,007	371,492
2096	555,830	286,830	1,113,615	372,672
2097	559,590	285,820	1,130,457	373,857
2098	563,365	284,796	1,147,532	375,048
2099	567,153	283,758	1,164,842	376,243
2100	570,954	282,706	1,182,390	377,444

Source: U.S. Census Bureau. www.census.gov/population/projections/nation/summary/np-t1.pdf. (Feb. 14, 2000)

■□■ TABLE 1-3. Comparison of Total Population, Present Series with 1994-Based Projections

(Numbers in thousands. As of July 1. Resident population.)

Year	Population			Average Annual Percent Change		
	Lowest Series	Middle Series	Highest Series	Lowest Series	Middle Series	Highest Series
1994-Based Projection[1]						
1998	268,396	270,002	271,647	–	–	–
2000	271,237	274,634	278,129	0.53	0.85	1.18
2025	290,789	335,050	380,781	0.28	0.80	1.26
2050	282,524	393,931	518,903	-0.12	0.65	1.24
2100	–	–	–	–	–	–
New Series (1998-Based)						
1998[2]	270,299	270,299	270,299	–	–	–
2000	274,853	275,306	275,816	0.84	0.92	1.01
2025	308,229	337,815	380,397	0.46	0.82	1.29
2050	313,546	403,687	552,757	0.07	0.71	1.49
2100	282,706	570,954	1,182,390	-0.21	0.69	1.52

Source: Hollmann, Mulder, and Kallan, 2000, p. 24.

focus world attention on problems associated with population growth. At the time it was the largest international population meeting ever held and had representatives from 136 governments around the world.

Though most governments recognized the existence of population problems in their own countries, as well as throughout the world, there was much disagreement and debate about the reasons for the problems and the types of solutions that should be implemented. A number of countries participating in the conference felt that the reason for high birth rates was the lack of social and economic development, so that the emphasis should not be put on population and family planning programs but rather on development. One of the frequently heard slogans at Bucharest was "Take care of the people, and the population will take care of itself." A different position, however, was taken by a number of other countries, including the Western European nations, the United States, and Canada, which felt that reductions in population growth rates would make a substantial contribution to the process of economic development and that what was needed first was a decrease in population growth to induce development.

A more lucid explanation of this relationship between population-family planning and development was expressed by Nortman and Hoffstater (1975, 3):

> Whatever the stance on the political stage, the most ardent family planning advocates recognize that contraception "alone" will not produce housing, schools, or steel mills; and among the staunchest supporters of the "new economic order," many appreciate the demographic value of legitimated and government-subsidized family planning services.

At least three different positions on the population problem can be identified. One position is that population growth is a crisis issue and the problem is so grave that catastrophe is near unless dramatic actions are followed in order to reduce the growth. A second position is held by those who feel that population growth will intensify and multiply other social problems, but that although population is important, it is not everything. A third position is held by those who feel that population is a nonproblem, or even a false problem, with the real problem being development or redistribution of income and power. In summary, the very nature of the population problem and the consequences of population growth are under closer scrutiny and examination today than at any time in the past (Hardin, 1999).

THE LAISSEZ-FAIRE POINT OF VIEW

The general argument stated by those in favor of some form of population control is that individual fertility decisions do not add up to what is socially optimal, or even desirable, hence such decisions cannot be left to individual families. Thus, parents intending to have children may impose a significant part of the cost and responsibility for those children on people other than themselves. These parents are therefore likely to have "too many" children.

On the other hand, however, there are those in favor of a laissez-faire solution. They feel that it is a question of individual choice, because it is the individual who bears the cost and receives the benefits of his own action. In general, the **laissez-faire argument** regarding population control is essentially the same as the laissez-faire argument in economics. Under proper functioning of the free market, without controls, the prices, both monetary and nonmonetary, that people pay for things reflect the real cost of production; and the prices that they receive reflect the real value of what they produce. Thus, when an individual makes an economic decision he or she bears all the costs and receives all the benefits. If the costs are less than the benefits, a positive decision is made; if the costs are greater than the benefits, a negative decision is made.

Laissez-faire population exponents feel that those best able to determine the costs and benefits of children are those who are contemplating having them. They are the ones who must assume the financial and social responsibility for that child and they are the ones who will benefit from that child. There are some, however, who believe that the costs and benefits for the family are not the same as the costs and benefits derived by the society for each additional child. For the individual family the ideal number of children may be five or six, whereas the ideal family size for the society may be only two.

> **Laissez-faire population exponents** feel that those best able to determine the costs and benefits of children are those who are contemplating having them.

■ ■

THE QUESTION OF CULTURAL GENOCIDE

Somewhat akin to the laissez-faire population exponents are those who feel that population control is a device proposed by more economically developed countries to control the less economically developed countries. Since the former group of countries are primarily the white non-poor nations and the latter group are the non-white poor nations, many believe that population control is a form of "genocide."

"Genocide" is a controversial concept with manifold emotional overtones. The United Nations Genocide Convention defined genocide as any of the following acts committed with intent to destroy, in whole, or in part, a national, ethnic, racial, or religious group:

- killing members of the group;

- causing serious bodily or mental harm to members of the group;

- deliberately inflicting on the group conditions to bring about its physical destruction in whole or in part;

- imposing measures intended to *prevent birth* within the group;

- forcibly transferring children of the group to another group.

The above definition was adopted unanimously by the General Assembly of the United Nations. According to the definition, mass sterilization of a compulsory nature would be considered genocide. Item (D) of the definition, "imposing measures intended to prevent birth within the group," is either directly or indirectly related to the family planning programs espoused by the United States and other developed countries. The important question thus becomes: Do family planning programs, as espoused by predominantly white, wealthy nations represent conscious, deliberate efforts to curtail nonwhite fertility, or do they reflect a genuine concern for the well-being and health of the rest of the world?

In an analysis of population control Darden concluded that, in his opinion:

> The poor and nonwhite should oppose any program which involves institutional limitation of population growth. Why? Because there is no guarantee that relative poverty would decline if the poor and nonwhites accepted such a fertility program. There might be fewer poor people in absolute numbers, but the gap between rich and poor would either get wider or remain constant.... In brief, institutionalized, coercive limitation of population growth is a policy aimed directly or indirectly at the poor and nonwhite and is therefore unacceptable as a solution to the problems of hunger, and other social ills in the United States and the world. (Darden, 1975, 51)

THE ETHICS OF POPULATION CONTROL

As previously mentioned, there are those who believe that any form of coercive population control is unethical. However, a significant number of people feel that in order to solve the population problem and limit population growth it will be necessary to induce people to limit the size of their families. They feel that the hazards of excessive population growth pose such critical dangers to the future of the species, the ecosystem, individual liberty and welfare, and the structure of social life, that there must be a reexamination and ultimately a revision of the traditional value assigned to unlimited procreation and to the increase in population size.

Callahan (1971, 2) outlined some general ethical guidelines for governmental action and presented them in a "rank order of preferences" from the most preferable to the least preferable. They are listed here and are still very much worth thinking about. The government has an obligation to do everything in its power to protect, enhance, and implement freedom of choice in family planning. This means the first requirement is to establish effective voluntary family planning programs.

If it turns out that the voluntary family planning programs do not curb excessive population growth, then the government has the right to go "beyond family planning." Callahan feels, however, that before governments take this second step they must justify the introduction of these new programs by showing that voluntary methods have been adequately and fairly tried. Callahan believes that the voluntary programs have not yet failed because they have not been tried in any massive and systematic way.

When the government has to choose among possible programs which go "beyond family planning," it has an obligation to first try those programs which, comparatively, are the least coercive. In other words, positive incentive programs and manipulation of social structures should be resorted to before "negative" incentive programs and involuntary fertility controls are applied. According to Callahan, if it appears that some degree of coercion is required, that policy or program should be chosen which:

- entails the least amount of coercion;
- limits the coercion to the fewest possible cases;
- is not problem-specific;
- allows the most room for dissent of conscience;
- limits the coercion to the narrowest possible range of human rights;
- least threatens human dignity;
- least establishes precedents for other forms of coercion;
- is most quickly reversible if conditions change;

In summary, the ethical considerations associated with population control are quite complex. Population policies, though they must take into account the interests and needs of particular regions and population groups, should have as their ultimate aim the best interests of the entire human species. Any plan to reduce world population growth to (some would argue even below) zero will have to carefully consider at least the following: economic development (including variations on the "Western model") and its role in reducing family size, the empowerment of women and gender equity, and the extension of family planning services to provide safe and efficient means of preventing unwanted births.

POPULATION DYNAMICS AND THE WORLD OF BUSINESS

Demographic considerations and their spatial or geographic components play an important role in today's business world. Understanding these issues is of vital importance to business executives, who are increasingly turning to demographic experts for answers to a variety of problems. As one writer noted, business executives need to ". . . understand population changes and their impact on basic corporate decisions such as labor supply, location of facilities, the changing nature of markets, and the age makeup of consumer groups" (Hyatt, 1979, 1). *American Demographics* has been one major response to such needs; this well-done monthly magazine presents succinct looks at the relationship between demographic trends and everything from where to market products to where to retire.

MARKETING

Marketing relies heavily on demographic statistics and their spatial or geographic aspects. Market segmentation and differentiation now play a key role in marketing strategies. According to Francese and Renaghan (1991, 50), ". . . many markets have become too complicated and too unforgiving to rely on just one or two demographic variables." A variety of variables, such as income, education, and age, have a symbiotic relationship and together form the basis of "database marketing." In order for marketers to succeed in the future they will have to understand the multi-dimensional demographic profiles of their market segment; in the United States multiculturalism itself is becoming a concern for marketing specialists, with growing Latino and Asian populations appealing to many product manufacturers and distributors.

The formation of households is an important variable to consider by planners for such utility firms as Pacific Gas and Electric and Chesapeake and Potomac Telephone—population gains and losses certainly shape the demand for power. Changing fertility rates and divorce rates, as well as the increase in the number of late marriages, have a tremendous impact on household formation and thus are important variables when predicting consumer demand for utilities.

Debates about affirmative action in the United States are shaped by demography as well. The design of equal opportunity employment policies by the General Motors Corporation involved considerable input of geographic population data. The company needed to know where to find female and minority workers. The American

Medical Association employs a population analyst who studies the mobility of doctors and the reasons that prompt doctors to move. These are just a few examples of the usefulness of population geography in the business world. Some population changes are obvious, whereas others are hidden. Some present important business opportunities, while others can cause setbacks. All population changes are important, however.

BUSINESS FORECASTING

Business people are always interested in the future. They are concerned about next week's sales, next year's profits, future changes in interest rates, and five-year capital investment schemes. Their natural tendency, however, has been to focus on the next fiscal quarter, instead of the next quarter-century. Short-term symptoms always seem to overshadow long-run causes. Worries about short-term problems tend to obscure the longer run, but, equally important, they obscure changes in population variables. For example, a home builder is usually concerned with changes in interest rates, but a shift in the divorce rate or a shift in migration patterns could be equally as important to that industry. Builders have only recently begun to appreciate the impact of the **"Baby Boom,"** and the boomers have yet to have their impact on Medicare and Social Security.

The impact of population change on business forecasting has recently taken on added importance. In a study on changing demographics and the future of business, James Hyatt observed that "... while only a few years ago the attention was on worldwide population growth rates, analysts are beginning to understand that for the United States a much more complex set of population shifts are at work. Fast, slow, and no growth are all occurring at the same time in different parts of the country" (Hyatt, 1979, 5). This important geographic dimension of population change is now receiving more attention from business forecasters.

Many of the present and future problems and opportunities to be faced by business have their roots in the most striking demographic phenomena of twentieth-century America: the unique high fertility period that followed the Second World War, the so-called "baby boom." During the **baby boom period,** between 1946 and 1964, there were nearly 80 million births in the United States, 50 percent more than during the preceding fifteen years. The baby boom or "bulge" has far-reaching effects in housing, employment, retail sales, education, and many other areas of concern to the business community. According to Hyatt (1979, 5), "Business managers would be well advised to keep in mind the location of that bulge from year to year, just as they pay attention to other economic indicators."

The first baby boomers began to turn 50 in 1996, and as Waldrop (1991, 24) suggested a few years earlier, "It will begin a population explosion among affluent, maturing householders. And it will turn the 1990s into peak years for consumer spending." These baby

During the **baby boom period,** between 1946 and 1964, there were nearly 80 million births in the United States, 50 percent more than during the preceding fifteen years.

■■■

boomers entering midlife brought about changes in a variety of business services. For example, the 1990s saw a rise in consumer demand for bifocal eyeglasses as those midlife boomers underwent inevitable changes due to the aging process. Visual and hearing impairments increase significantly after age forty-five. Increasing attention was paid to health concerns and boomers, in order to stay fit, had to be more selective about food choices. Talk of ice cream and twinkies gradually gave way to conversations about cholesterol, monounsaturated fats, and triglycerides. This brought about a rise in consumption of food items like fish, poultry, low-fat milk, whole grain cereals and fresh fruits and vegetables. The boomers are also relatively affluent, and in recent years they have been pouring money into mutual funds and other investments, a major factor in the strong stock market performance during the 1990s, despite an economy that has been struggling for the most part.

Although much of the attention of the business community has been focused on the young, there has been an increasing awareness of the aged and the aging of America (Solda and Agree, 1988). Indeed, adults over age 65 now outnumber teenagers and have become one of the fastest-growing population groups in the country. In 1900 only 3.1 million people in the United States were 65 years of age and older, but by 1985 that figure reached 28.5 million and projections for the year 2030 are that there will be 64.6 million. In 1985, one in nine Americans or 11.7 percent of the population was at least 65 years old, but by 2010, because of the maturation of the baby boomers, one in seven Americans will be at least 65 years old (Hooyman and Kiyak, 1988, 22).

As baby boomers become "senior" boomers, important ". . . public policy questions regarding how we will support the needs of a growing older population and how we will structure policies to assure a fair and equitable distribution of resources for all age groups" will have to be assured (Bouvier and DeVita, 1991, 27). Government bureaucracies, charged with planning the provision of services for the elderly, will be faced with a variety of decisions with both political and social consequences (Laws, 1991, 32). Some of these problems will be further complicated by the changing ethnic and racial composition of the United States.

Not only will there be more older Americans, but because of their relatively high disposable incomes, they will present a growing market for the business community. England (1987, 8) points out that "Americans over age 65 are the second-richest age group in U.S. society. Only those Americans in the next-oldest age bracket, from 55 to 64, are better off." The aged have assets nearly twice that of the median for the nation. "The spending power of the mature market may be one of the best-kept secrets left in the age of demographic scrutiny," suggests Lazer (1985, 23).

ZERO POPULATION GROWTH AND BUSINESS

To most entrepreneurs "growth" is a magic elixir; they are naturally attracted to growth, they love growth—be it in profits, sales, or incomes. Investors usually look for firms with solid growth records.

Also, to most business managers a growing gross national product (GNP) means a stronger economy with more consumer spending, more employment, and more sales and services. There is, therefore, understandable trepidation among those with businesses when they contemplate the prospects of a slowing, or even cessation, of population growth. They fear reduced demand, which in turn means less profit, smaller dividends, more unemployment, and general economic uncertainty (even though effective demand depends not just on numbers but also on affluence, the ability to pay).

The impact of population growth on the economy of the United States is assumed to be positive because it has continued for so long. Economic theorists have, however, found it difficult to find a direct correlation between population growth and economic well-being in highly industrialized nations. During parts of the nineteenth century population growth was high, yet per capita income growth was modest. On the other hand, in the twentieth century population growth slowed down while the GNP remained high. Among the three economic superpowers—Japan, the United States, and the European Union—only the United States has a significant rate of population growth, and a lack of growth does not seem to have diminished the affluence of its competitors.

LABOR FORCE

The composition and quality of the labor force are important variables for personnel managers to consider. During the next few decades the nature of the labor force will be considerably different from what it has been in the past. There will be more workers and they will need to be better educated. There will be more minorities, and more will be female and the mothers of young children. The number of workers over age 55 will be fewer in the next decade. Paradoxically, the average age of the labor force will be increasing. Many of these changes in the United States will be direct results of the baby boom. Millions of baby boom youth have been absorbed into the labor force.

The supply of workers available for businesses will be the result of two separate trends in the labor force. First is the size of the working age population, and second is the labor force participation rate. Among the most important demographic variables in determining the size of the work force is the age structure of the population. The bulge in the young working force cohort is now progressing through the population, so the number of young people entering the work force will be declining, though earlier projections are going to be off because of rising immigration and higher fertility in places such as California.

At the other end of the work force age spectrum, among those over 55 years of age, important changes will take place (aside from sagging middles and bifocals). Changing attitudes about retirement could have far-reaching impacts on the corporate world. Can business easily absorb those older workers who decide to continue their careers? Will companies find it harder to promote young workers if retirement ages advance? Will older workers continue to be productive? These are all important questions to personnel managers and are underlain by demographic changes in society. During the 1990s corporate downsizing resulted in the elimination of many jobs held by these workers, leaving them floundering to compete for lower-paying jobs with few or no benefits.

Labor migration in the new millennium promises to be at least as important as population growth, and it will affect the populations of numerous sending and receiving nations. As Wallerstein (1999, p. 17) observed, "We shall nonetheless see a rise in the real rate of migration, legal and illegal—in part because the cost of real barriers is too high, in part because of the extensive collusion of employers who wish to utilize such migrant labor."

THE EDUCATION ROLLER COASTER

The nation's education system, perhaps more than any other social institution, has been greatly affected by fertility variations. The impact of the postwar baby boom children on the schools was dramatic, costly, and painful. In the late 1950s and 1960s school enrollments soared. The elementary-school-age population (5-13 years) grew from 23 million in 1950 to 37 million in 1970, for example. Secondary schools faced similar problems by the early 1960s, and the high school population doubled in size between 1950 and 1975 (Bouvier, 1980, 21). High school enrollments rose 14 percent in 1957 alone, and a critical shortage in classrooms was evident. Administrators projected a need for an additional 750,000 teachers in three years. This demand caused a rapid increase in school budgets.

CHANGING ENROLLMENT TRENDS

School planners were totally unprepared for this rapid increase in enrollments. By the time colleges had geared up to graduate enough teachers and massive building programs had produced enough elementary and secondary classrooms, the crest of the baby boom wave was about to leave the K-12 school ages.

The mid-1960s saw enrollments in colleges and universities skyrocket. This increase was a result of both the sheer numbers of baby boomers and the increased proportion of young people going to college. In the 1950s only about 10 percent of those between ages 25-29 had completed at least four years of college. That proportion increased to 16 percent by 1970 and 23 percent by 1980. Between 1957 and 1975 college enrollments increased from 3 million to 11 million, creating a growth industry of its own. The baby boom generation thus became the most highly educated generation in American history (Bouvier and DeVita, 1991, 14).

The major change in school enrollment in the past two decades has been a significant increase in nursery and preschool attendance. The percentage enrolled in nursery schools, pre-kindergarten or kindergarten programs, or child-care centers with an "educational" curriculum, increased from 11 percent in 1965 to 38 percent in 1988 (Bianchi, 1990, 30).

Today, another school turnaround is underway. After years of stable or declining enrollments, the school-age population was projected to increase by about 8 percent during the decade of the 1990s. This increase was the result of the baby boom's children filling the elementary and secondary schools across the country (Bouvier and De Vita, 1991, 15); it was also very different from one region to another. For example, a 1997 study projected that Los Angeles County would have a countywide enrollment increase of 12.0 percent by 2005.

GEOGRAPHIC VARIATIONS

In order to understand these educational trends and their demographic causes, it is necessary to look beyond national statistics and to understand regional or more local circumstances. The geographic distribution of the population, as well as spatial differences in vital rates, presents a complicated picture. For example, the decline in births was not uniform across the United States in the 1960s and early 1970s. In some areas the peak in births was attained in 1957, whereas in other areas it was not reached until 1960 (Reinhardt, 1979, 10). Also, the decline in births within states was not uniform, and there were significant differences between metropolitan and nonmetropolitan areas.

These geographic differences, as well as other changes in population distribution, mean that some areas have had dramatic declines in the school-age population, whereas others have still been growing. Many rural areas and urban regions losing population have experienced dramatic declines, but many newly developing suburbs have enrollment increases, as do some inner-city areas. Nursery school attendance is more common in the Northeast than other U.S. regions, and among children who live outside the central cities of metropolitan areas (Bianchi, 1990, 30).

Migration has also had a significant impact on school-age populations. Increases can be seen throughout much of the Sunbelt and the West, with declines in the Midwest and the Northeast. An analysis of college-age students also points out significant spatial differences, with many Sunbelt and Western states being net importers of college students. Changes in demographic factors can have a significant impact on school planning policy decisions and should be incorporated into the decision-making process at all levels.

Population Data

A s you have undoubtedly surmised by now, population studies require an abundance and variety of data—we need numbers and rates in order to better understand both population changes and the ways in which those changes are in return related to socioeconomic and environmental variables. Demographic changes do not take place in a vacuum.

Populations and their associated characteristics are constantly in flux, so it is necessary to maintain current statistics to the extent that we can (though the costs of doing so are often high). Because of this dynamic nature of populations, however, it is also desirable to look at a full historic range of data (when possible) and not just at information for a short period of time. In turn, then, because of the large masses of statistical information required, coupled with the need for historical data, the population geographer usually obtains information from reports by government agencies or other large organizations. In most cases the individual scholar has neither the money nor the time to gather the required information.

Three broad types of uses for demographic data can be distinguished:

1. planning, policy guidelines, and projections ranging from five years to the long term

2. monitoring current demographic trends and applied programs

3. scientific study, at either the micro or macro level, of interrelationships between demographic phenomena and socioeconomic developments (Seltzer, 1973, 5)

Type of Use[a]	Accuracy	Timeliness	Detail Topical	Detail Geographical	User Confidence	Index of Total Burden[b]
Planning and projections	Medium	Medium	Low	Medium	Varies	8-10
Monitoring current trends	High	High	Medium	High	Varies	12-14
Scientific study	High	Low	High	Low	High	11

[a]The assessments and the use types are extension of distinctions developed by Tukey (1960). In terms of his dichotomy, "planning and projections" and "monitoring current trends" fall largely under the domain of decision theory, while "scientific study" would call for conclusion theory.
[b]Sum of scores where high = 3, medium =2, low = 1 and varies = 1-3.

Source: Reprinted with the permission of the Population Council from Demographic Data Collection: A Summary of Experience, by William Seltzer (New York: The Population Council. 1973), p. 6.

Each of these uses has specific requirements in terms of accuracy, timeliness, topical and geographical detail, and user confidence. The interrelationships of these variables for each data use are outlined in Table 2-1. The information in the table suggests that the greatest difficulties are encountered in monitoring current trends and programs; that the data used for longer-range planning purposes is the easiest to acquire and has the least stringent specifications; and that the information needed for scientific inquiry is immediate.

CONCEPTUAL DIFFICULTIES

At first glance many of the basic demographic measurement concepts seem to be concrete and unambiguous. The concepts of "age," "death," and "live birth" would seem to be straightforward and easy to define. The collection of accurate and reliable demographic data also involves some other concepts such as "city," "place of residence," and "household," which might appear to be relatively ambiguous. However, under close scrutiny, even the seemingly unambiguous terms (age, death, live birth) are difficult to conceptualize.

AGE DATA

Information on the age structure of a population is essential to many areas of population analysis. Ryder (1964, 449) noted that age is the central variable in the demographic model. It identifies birth cohort membership. It is a measure of the interval of time spent within the population, and thus of exposure to the risk of occurrence of the event of leaving the population, and more generally is

Courtesy of Photo Disc

In China, a person is considered one year at birth, and then ages by one year each New Year's Day.

a surrogate for the experience which causes changing probabilities of behavior of various kinds. Age as the passage of personal time is, in short, the link between the history of the individual and the history of the population.

Shryock and Siegel (1971, 201) also commented on the value of age data:

> Age is the most important variable in the study of mortality, fertility, nuptiality, and certain other areas of demographic analysis. Tabulations on age are essential in the computation of basic measures relating to the factors of population change, in the analysis of the factors of labor supply, and in the study of the problem of economic dependency. The importance of census data on age in studies of population growth is even greater when adequate vital statistics from a registration system are not available.

Gathering data on the age structure of a population would appear to be a simple process. The United Nations (1961, 18) defined age as the number of "completed solar years since birth." This type of information can be obtained by asking respondents either their ages as of their last birthdays or their dates of birth. Unfortunately, according to Seltzer (1973, 9), "these questions cannot be answered by populations that take no note of birthdays, and they will produce biased results when asked of those using a calendar other than the 'Western' solar calendar."

Where will these two graduates be counted in the next census?

Even though the "Western" method of determining age seems logical and straightforward to those who live in countries where it is used, other traditions for determining age prevail in other areas of the world, and if these traditions are not accounted for in the data-gathering process then substantial error and confusion can occur. For example, one of the most common non-Western methods of determining age is that used by the Chinese, who consider a person to be one year old at birth. Thereafter, a person ages by one year each New Year's Day. However, as noted by Hock (1967, 861):

> ... some of the Chinese respondents ... reckon their ages according to the Chinese system, some according to the Western system, and ... others might compute their ages according to Japanese reckoning, by adding one year to their age on the occasion of the Western New Year instead of the Chinese New Year.

Similar results were found in a study of South Korea. Thus, the answer to a seemingly simple question, "How old are you?", can have more than one interpretation.

TOTAL POPULATION

Do you assign people to the place that they customarily inhabit regardless of where they happen to be residing on the census day (**de jure**) or do you count them wherever they are physically present at census time regardless of their home place (**de facto**)?

Another seemingly simple question deals with the total population of a place. However, after careful scrutiny even this concept has its problems. For example, do you assign people to the place that they customarily inhabit regardless of where they happen to be residing on the census day (**de jure**) or do you count them wherever they are physically present at census time regardless of their home place (**de facto**)? Different countries make different choices.

In Great Britain the *de facto* census is favored. This method has the advantage of showing exactly where everyone was at a given moment (census day). Its obvious disadvantage is that certain population figures may be increased or decreased because of tourists, traveling salespersons, or other transients. The United States, on the other hand, has traditionally conducted a *de jure* census: people are tabulated according to their permanent places of residence. The *de jure* method provides information that is not affected by temporary or seasonal movements of people; however, it has two principal shortcomings: (1) it involves costly and time-consuming work to transfer people from their places of interview to their places of usual residence, and (2) it is sometimes difficult to be certain where a person's "usual" or "legal residence" is located. For the 1970 Census of the United States the "usual place of residence" was defined as follows:

> In accordance with census practice dating back to 1790, each person enumerated in the 1970 census was counted as an inhabitant of his usual place of residence, which is generally construed to mean the place where he lives and sleeps most of the time. This place is not necessarily the same as his legal residence, voting residence, or domicile. In the vast majority of cases, however, the use of these different bases of classification would produce substantially the same statistics, although there may be appreciable differences for a few areas. (United States Bureau of the Census, 1971, IV-V.)

No change in this definition was made for either the 1980 or 1990 censuses.

Definition of Urban

Even more troublesome than defining the total population is defining the term **"urban."** Almost every country has its own definition, and comparative studies of urbanization may be extremely difficult because of the variety of definitions. Also, within some countries the definition has changed over time. Most **urban** definitions fall into one of the following four categories: size, administrative function, legal identity, and site characteristics.

In 1990 the United States Bureau of the Census defined as "urban" all territory, population, and housing units in urbanized areas and in places of 2,500 or above outside urbanized areas. More specifically, the definition of "urban" included the following:

1. places of 2,500 or more that have been incorporated as cities, not including the rural portions of extended cities

2. census designated places of 2,500 or more

3. other territory, incorporated or not, that is included in urbanized areas

Other countries also use the size criterion, but with different sizes. In Denmark, for example, urban places are those with populations of

Most **urban** definitions fall into one of the following four categories: size, administrative function, legal identity, and site characteristics.

The **MSA** is a county or group of counties that has either a city of 50,000 or more people, using legal city boundaries as a definition of the city, or an Urbanized Area of 50,000 or more people with a combined metropolitan population of at least 100,000.

■▧■

The **PMSA** is a county or group of counties in which either local opinion supports separate recognition for a county or group of counties that demonstrate relative independence within the metropolitan complex or it had previously been recognized as a metropolitan area before the beginning of 1980 *and* local opinion supports its continued recognition as such.

■▧■

A county (or group of counties) may be recognized as a **CMSA** if it is a level A MSA *and* it includes a PMSA within it.

■▧■

at least 250, whereas in India urban places are those with populations of at least 5,000. Size definitions of urban places are often inadequate because the size of a minor civil division, such as a county, may be used in place of the size of an urban place. Exact definitions are necessary before comparative studies can be made.

Some countries prefer to use administrative units as designations of urban places, as is common in parts of Latin America. In Peru, for example, urban places are defined as ". . . capitals of all departmentos and distritos plus all towns larger than the average size of these administrative centers not possessing rural characteristics." The major problem with this kind of definition is that places with no administration functions will be considered rural, regardless of their population. Another possible criterion for defining urban places is legal identity—towns with legal identity are classified as urban and all others, regardless of size or function, are classified as rural. This definition traces its origin back to the chartered cities of the Middle Ages.

Finally, site characteristics may be used to define an urban place. A place may be considered to be urban if it has certain urban characteristics, such as piped-in water supplies, numbered streets, and lighting. Occasionally both site characteristics and the previously mentioned criteria are applied, and the resulting definition of an urban place can be quite complex.

The United States definition of urban has changed over time. Early American censuses created no distinction between rural and urban. A distinction was first made in the *Statistical Atlas of the United States* (1874), where urban was defined as towns greater than 8,000; subsequently, the Census of 1900 lowered the size limit to 2,500. Massive suburbanization, resulting in an increased number of people living in urban conditions, but not within incorporated city limits, rendered older definitions inadequate. In 1950 a density criterion for fringe areas was adopted, and an 11.5 percent increase in urban population was registered.

Subsequent to the 1980 Census, new definitions of larger urban regions were introduced to replace the previously used Standard Metropolitan Statistical Area (SMSA). These new areas were defined by the Office of Management and Budget with support from the Bureau of the Census in 1983. The new typography of metropolitan areas is as follows: **Metropolitan Statistical Area** (MSA), **Consolidated Metropolitan Statistical Area** (CMSA), **and Primary Metropolitan Statistical Area** (PMSA). These definitions remained in place for the 1990 census.

The **MSA** is a county or group of counties that has either a city of 50,000 or more people, using legal city boundaries as a definition of the city, or an Urbanized Area of 50,000 or more people with a combined metropolitan population of at least 100,000. MSAs are further categorized by size into four levels, A, B, C, and D. Whereas level A MSAs have at least 1 million people, those in level D have less than 100,000. The **PMSA** is a county or group of counties in which either local opinion supports separate recognition for a county or group of counties that demonstrate relative independence within the metropolitan complex or it had previously been recognized as a metropolitan area before the beginning of 1980 *and* local opinion supports its continued recognition as such. A county (or group of counties) may be recognized as a **CMSA** if it is a level A MSA *and* it includes a PMSA within it.

In addition, the Census Bureau recognizes the Urbanized Area (UA), which is defined by population density. Each Urbanized Area includes a central city and the closely settled urban fringe around it—together they must have at least 50,000 people and a population density that generally exceeds 1,000 people per square mile. Because Urbanized Areas include surrounding suburban developments, they often make the best units for comparative purposes in areas where MSAs have sizable rural or sparsely populated areas.

Several other problems occur when we attempt to define an urban place. One is with the process of annexation. For example, the 1950 population of the town of California, Pennsylvania, was 2,831 and the 1960 population was 5,978. Looking closely at this change, we see that most of the growth was due to annexation. Tracing population changes over time often is difficult because of changes due to annexation.

Another problem is reclassification. If a place had a population of 2,499 in 1980 and grew to 2,500 in 1990, then the net gain of one person would mean that 2,500 persons were subtracted from the rural population and added to the urban population. It has been estimated that one-third of the "gain" in urban population in El Salvador between 1950 and 1960, for example, was due to reclassification.

MIGRATION AND AREAL SUBDIVISIONS

Conceptual problems concerning migration have always been problematic. As Petersen (1975, 41) noted, "We know whether someone has been born or has died, but who shall say whether a person has migrated?" Everyone agrees that when we go to the neighborhood grocery store we have not migrated. Likewise, there is agreement that when we leave our home in one country and establish residence in another we have migrated. However, between these two extremes problems arise. How far does one have to move in order to be considered a migrant? How long must one stay at the new location? The questions of how far and how long, as well as other circumstances, play an important role in the definition of migration (discussed further in Chapter 8).

The geographic subdivisions that are used for collecting and mapping population data present further problems. As Zelinsky (1966, 22) pointed out, "Urban residence and function have generally spilled far beyond the political bounds of the metropolis; and if urban growth is rapid, no amount of technical alacrity can keep census divisions abreast of the actual urban-rural frontier."

Population on the Web

The World-Wide Web now contains numerous sites of interest to population geographers. For a complete overview the best place to look is in the Demography and Population Studies segment of the WWW Virtual Library. Maintained by Australian National University, this site provides numerous listings of URLs, data bases, and gopher servers of interest. The following is a partial list of sites that should be especially valuable to students and instructors in population geography classes.

- American Demographics
- Carolina Population Center (UNC, Chapel Hill)
- Census Information (EINet)
- Census Information (UC, Berkeley)
- Center for Demography and Ecology (UW, Madison)
- Center for Studies in Demography and Ecology (UW, Seattle)
  Centers for Disease Control (U.S.)
- Demographic and Health Surveys
- The Global Fund for Women
- Morbidity and Mortality Weekly Report (CDCP, Atlanta)
  Office for National Statistics (England)
- Office of Population Research (Princeton)
- PCdemographics
- Population Council (NY)
- Population Index (Princeton)
- Population Reference Bureau (Washington, D.C.)

Courtesy of Photo Disc

Population on the Web (Continued)

- Population Research Center (UT, Austin)
- Population Research Institute (PSU, University Park)
- Population Studies Center (UP, Philadelphia)
- RAND Institution (Santa Monica) Statistics Canada
- United Nations Information Services WWW Virtual Library
- United Nations Development Program
- United Nations Food and Agriculture Organization U.S. Bureau of the Census
- U.S. State Department, Country Background Notes (UM, St. Louis)
- The World Bank Server
- World Health Organization
- World Hunger Program

New sites on the Web emerge regularly, so keep your eyes open when you're surfing the net!

In summary, the collection and analysis of demographic data is wrought with conceptual problems, primarily problems of definition, and, unless these discrepancies are dealt with, gross inaccuracies can occur.

TYPES OF RECORDS

Information for demographic analysis generally comes from three sources:

1. census enumerations
2. vital registrations
3. sample surveys

These sources are the most comprehensive records available and are generally provided by national governments. A variety of fragmentary demographic data exists and may occasionally serve useful purposes, but such data are usually collected for selected small populations. Examples include the following: school censuses, church records, records of births and deaths recorded by hospitals, city directories, auto title transfers, and marriage licenses. They are often readily available, and their use should not be overlooked; however, they generally are inadequate substitutes for the data available from the major sources of demographic information.

A **census** is a total count of the population of a specified area, generally a nation, and ". . . a sort of social photograph of certain conditions of a population at a given moment which are expressible in numbers." (Wilcox, 1940, 195) In addition to enumerating the population, most censuses collect other data, such as age, sex, place of residence, place of birth, income, occupation, education, and religion. The quantity of information collected is mainly a function of the amount of money available for financing the census. A census is a major and expensive undertaking requiring considerable expertise and planning for it to be successful. Thus, censuses are usually conducted by national governments. The 1980 Census of the United States cost over a billion dollars, nearly five dollars per person enumerated, and the 1990 Census cost about 2.6 billion dollars, or over ten dollars per person enumerated.

Vital registrations are also major sources of demographic data. Vital statistics are recorded and compiled at or near their time of occurrence and usually include such events as deaths, fetal deaths, births, marriages, divorces, and at times disease and illness. Unlike the census, which is a static, cross-sectional view of a population at a specific moment of time, a **registration system** is a dynamic recording of events that can change rapidly.

The major responsibility for reporting vital events to civil registration authorities, depending upon the country, is given to a local registrar, parents or relatives, or to physicians, midwives, undertakers, religious officials; persons with special duties relative to births and deaths. Seltzer (1973, 35) suggested that "ideally, a national birth and death registration system obtains a report of each event shortly after it occurs, and statistical summaries of these reports, in conjunction with externally derived population estimates, can be used to compute vital rates." If the vital registration system is functioning effectively, it is capable of producing precise mortality and fertility estimates.

In some countries, such as the Netherlands, Finland, Belgium, and the Scandinavian countries, the vital registration system is so complete that a local registration bureau keeps a card for each individual. Major demographic events such as marriages, divorces, and changes in residence are noted on the card and this information can be available quite readily.

Because of the time and expense involved in a census, sample surveys are often used. Their advantages in terms of quality and cost are now well recognized. A major concern is assurance of a truly representative sample of the population. Many sample surveys dealing with demographic issues are conducted by census bureaus and other governmental and private agencies; sampling is also used within vital registration systems.

CENSUS ENUMERATIONS

By authority of the Constitution of the United States, the population of the United States must be enumerated every ten years. In other countries the time span between censuses may be as short as five

A **census** is a total count of the population of a specified area, generally a nation, and ". . . a sort of social photograph of certain conditions of a population at a given moment which are expressible in numbers."

Unlike the census, which is a static, cross-sectional view of a population at a specific moment of time, a **registration system** is a dynamic recording of events that can change rapidly.

years, or censuses may occur at infrequent intervals depending upon economic and political circumstances.

The United Nations (1954, 1) regards a census as ". . . the simultaneous recording of demographic data by the government, at a particular time, pertaining to all the persons who live in a particular territory." A census should also have the following characteristics:

- It should be of the population of a strictly defined territory.

- It should include everyone.

- It should be a personal enumeration.

- It should be conducted at one time, preferably on one day.

- It should be conducted at regular intervals.

A Brief History of Census Enumerations

The thought of counting an area's population occurred to people in ancient times. Perhaps the oldest account of census taking is that of the Incas, who organized their society according to a decimal system. Though this was not a census in any modern sense, it is at least a variation on the general theme.

Enumerations of various kinds were made in early times in Sumeria, Babylonia, and Egypt. A rather large amount of data, though quite fragmented, were recorded in ancient China. A census of sorts was taken in Rome during the sixth century, *B.C.*

Examples of early census counts appear in the Old Testament. Some passages illustrate both the purpose of the counts and the reasons for resistance to them. Many early population counts were for such purposes as conscription into the military service and the labor force. Other counts were taken for tax purposes. The word census comes from the Latin word, *censere*, which means "to value" or "to tax." Thus, it comes as no surprise to find that early attempts at enumeration were looked upon with considerable suspicion and resentment. Two different Biblical accounts are given of a census of Israel that was to be taken by David. The accounts differ both in the reasons for the census and the results. However, they agree that after David took the census he became remorseful, felt he had sinned by enumerating the population of Israel, and was finally punished for his deed.

According to Carr-Saunders (1936, 14), when the notion of taking a census was first proposed in England in the 1750s, ". . . the dire results of the census which David forced Joab to make were quoted by those who opposed the measure, and it was prophesized that some 'public misfortune' or an 'epidemical distemper' would follow if an enumeration were attempted." The result was a long delay in the first census of England, as well as continued anguish about censuses and census takers.

■■■ TABLE 2-2. Types of Censuses in the United States		
Census	**Periodicity**	**Recent Censuses**
Population	10 years	1960, 1970, 1980, 1990, 2000
Housing	10 years	1960, 1970, 1980, 1990, 2000
Agriculture	5 years	1979, 1984, 1989, 1994, 1999
Business	5 years	1982, 1987, 1992, 1997
Construction industries	5 years	1982, 1987, 1992, 1997
Governments	5 years	1982, 1987, 1992, 1997
Manufactures	5 years	1982, 1987, 1992, 1997
Mineral industries	5 years	1982, 1987, 1992, 1997
Transportation	5 years	1982, 1987, 1992, 1997

Source: United States Bureau of the Census.

MODERN CENSUSES

Modern census taking commenced in Scandinavia. Sweden began taking regular population counts in 1749; Norway and Denmark began in 1769. The first United States census was taken in 1790, and the United States was the first country to legislate a continuing, time-specific census. Thus, the practice of regular census taking originated in Western Europe around the middle of the eighteenth century and became widespread in Western Europe before the middle of the nineteenth century; during that interval enumeration efforts became more sophisticated.

It is not sufficient to know that a country has taken a census; we also need to know whether or not censuses are taken at regular intervals—many are not. For example, China took a census in 1953 and then not again until 1982.

Many improvements are still needed in census taking and demographic knowledge, in spite of the fact that only about two percent of the world's population lives in countries that have never taken censuses. The greatest recent gains in census taking have been made in sub-Saharan Africa.

THE CENSUS OF THE UNITED STATES

The United States Bureau of the Census is the record keeper for the Nation. The major output of the Bureau's data-collection activities are its statistical reports, which not only provide a record of the number of people in the United States but also document changes in characteristics of the people, manufacturing establishments, farms, retail stores, and a host of other business enterprises from which Americans derive their living. The Census Bureau publishes more statistics, encompasses a larger range of topics, and serves a greater variety of statistical needs than any other federal agency. According to U.S. law the Census Bureau is responsible for taking a variety of censuses; they are outlined in Table 2-2.

From the beginning, however, the primary task of the Census Bureau has been to provide the federal government with an accurate head count of the U.S. population. The count is to be conducted on the first day of April in every year that ends with a zero. The Bureau of the Census is required to notify the President by December of that year as to how many people he presided over eight months earlier.

The first census of the United States was strictly a head count and included only five questions, one of which was dropped in 1870—the one that asked the head of each household how many slaves he owned. It was conducted by 650 U.S. marshals and their deputies, and the results were issued in a single volume, fifty-six pages long. The entire operation cost $44,000, or approximately $.011 or 1.1 cents per person. However, the censuses have changed considerably since 1790. The volume of output has increased from a single short volume to some 15,000 separate publications.

Despite the fact that efficient, high-speed computers are now being used, census materials appear slowly, partly because of constant budget restrictions. The estimated cost of the 1990 census was $2.6 billion, and you can bet that the 2000 census cost more.

Though the United States was one of the first nations to establish a modern census, it lagged far behind in establishing a permanent bureaucracy to continue the task. Originally the State Department had the chore, but in 1830 it was transferred to the Department of the Interior.

A full-time bureau was not set up until 1902, when Congress decided to put the Census Bureau in the Department of Commerce and Labor. Later, when this department was split into two separate departments in 1913, the Census Bureau went into the Commerce Department, where it remains today. However, Congressional proposals to eliminate the Department of Commerce were made in late 1995. If Commerce is eliminated, the Census Bureau would remain, though it was not clear at the time where it would be housed.

Many changes in the type and number of census questions have been made since 1790. The largest number of questions ever asked was seventy in the 1970 census, although four-fifths of the people enumerated didn't have to respond to more than twenty. In 1980 and 1990 the number of questions declined slightly. Questions have been eliminated in the past because the information was no longer valid, for example, the number of slaves, or because the question was confusing. For instance, in 1960 enumerators were asked to rate the condition of the homes that they visited as either good, dilapidated, or deteriorating. This question was subsequently dropped because of the lack of agreement on the meaning of terms like dilapidated. Whatever problems and shortcomings the Census Bureau might have, however, we will continue to take a decennial census. As Anderson (1988, 240) concluded in her excellent history of census-taking in the United States:

> The census remains both an apportionment tool and a baseline measure for American social science. These dual functions have provided the impetus for technical innovation in the past and will

continue to do so in the future. . . Americans will undoubtedly continue to conceive of social issues in terms of census categories and to redistribute the benefits and burdens among the people—"according to their respective numbers."

VITAL REGISTRATION

Historically the collection of vital registration information has been the responsibility of ecclesiastical authorities. Most cultures have some form of religious ceremony attached to each of the major vital events: baptism and birth, wedding and marriage, funeral and death. The systematic recording of vital events has been traced back to *A.D.* 710 in parts of Japan, while in 1497 the Archbishop of Toledo instructed parish priests to maintain vital records. In England, the Vicar General under Henry VIII instructed the clergy to record the number of baptisms, burials, and marriages. In 1563 the Council of Trent made the keeping of burial and baptismal records compulsory in the Roman Catholic Church.

Most of the first annals were kept by parish priests, mainly to determine the severity of the plague. Many of those early records were destroyed by age, fire, or pests long ago. Gradually, toward the end of the eighteenth century, government agencies began taking over the vital registration function. For example, national civil registration began in Sweden in 1756, in France in 1792, in England in 1837, and in Ireland in 1864. Thus, vital registration data became available at about the same time that census enumerations started appearing.

VITAL REGISTRATION IN THE UNITED STATES

The United States was one of the first countries to begin taking a regular modern census, but one of the last countries to initiate the collection of vital statistics. Early attempts to collect vital statistics were sporadic. As early as 1639 Massachusetts had begun collecting and maintaining records of births and deaths, but this was done only locally. It was the government, however, and not the clergy, that began keeping the vital records, though the model was based on English precedent. Despite all attempts to enforce compliance with the registration system, early record keeping was not effectively accomplished.

During the nineteenth century there was an increased awareness of the importance of vital records. Shattuch, the founder of the American Statistical Association in 1839, used that association as a forum for lobbying for vital registration laws, and such laws were successfully established for Massachusetts in 1844. In 1850 the United States Census Office invited Shattuch to help develop plans for obtaining data on births and deaths, a venture that was unsuccessful.

In 1880 the Census Office created a Death Registration Area, which included all cities and states that registered 90 percent or more of their established deaths. At that time only Massachusetts and

Year	Birth Registration Area	Death Registration Area
1880	—	17
1890	—	31
1900	—	26
1910	—	51
1920	60	81
1930	95	95
1933 Both areas complete, with addition of Texas		

Washington, D.C., qualified. However, there was a gradual increase in death registrations as other areas improved their record collection techniques. It has been estimated that by 1890, 31 percent of the population in the United States was included in the Death Registration Area. Part of the reason for the slow expansion of the Death Registration Area was the lack of a central office to encourage states to register births and deaths. The Census Office was created each ten years for the purpose of taking the decennial census, but following each census it was disbanded. Thus, there was virtually no promotion of the Death Registration Area.

In 1902 the Permanent Census Act created the Bureau of the Census, which was charged with the responsibility of collecting vital statistics along with its other duties. Subsequently the entire record collection process was accelerated somewhat.

In 1915 the Birth Registration Area was established. At that time it contained approximately 33 percent of the population. Table 2-3 shows the additions to the Birth Registration and Death Registration Areas. By 1930 both Areas were 90 percent complete and, with the addition of Texas in 1933, both Areas were complete. In 1957 a Marriage Registration Area was established, and in 1958 a Divorce Registration Area was added. Neither of these areas is yet complete.

It was originally the task of the Bureau of the Census to collect vital registration information, however in 1946 the Public Health Service's National Office of Vital Statistics was authorized to collect the data. In 1960 another reorganization occurred and the National Center for Health Statistics was formed. Its National Vital Statistics Division currently collects, publishes, and distributes information on births, deaths, marriages, and divorces. The data are collected from county courthouses, registrars, and halls of records and tabulated and recorded on monthly and annual bases.

VITAL REGISTRATIONS THROUGHOUT THE WORLD

Birth and death registration is now virtually complete in most of North America, Europe, and Oceania, and in certain parts of the Caribbean and Latin America. However, as Seltzer (1973, 36) noted, "In view of the basic motivational and administrative obstacles to the

Courtesy of Photo Disc

Have these three children's births been recorded?

establishment of civil registration on a reliable basis, it is not surprising that official registration systems have been estimated as recording 40 percent or less of the live births occurring in Indonesia, Pakistan, India, the Republic of Korea, and the Philippines." Apparently death registrations are less complete than births in Asia also, and in most cases the coverage of tropical Africa is even more deficient than that of Asia. Considerable uncertainty exists in all the developing countries with respect to the completeness of their registration systems.

Two major problems in the developing countries relative to vital registrations are the following:

1. A considerable number of births and deaths are never reported.

2. Whole sections or areas of countries are not part of the data-collecting system.

The implications for demographic research are clear—a great deal of uncertainty surrounds the fertility and mortality data of developing countries.

Because of the expense and time involved in interviewing everyone in a country, demographers frequently use **sample surveys** to collect information.

SAMPLE SURVEYS

Because of the expense and time involved in interviewing everyone in a country, demographers frequently use **sample surveys** to collect information. The advantages of using a sample over a complete census are that

1. A sample can be taken much faster than a complete enumeration, and if timeliness is important to a particular study a sample may be the only means to collect the needed data.

2. A sample survey is much cheaper to administer than a complete enumeration.

3. The quality and accuracy of a sample can be greater than with A complete enumeration, particularly where there are a limited number of trained personnel to conduct a complete enumeration.

4. There is less paperwork involved, and data handling and processing can be easier.

In some nations, particularly the developing ones, a sample survey is the only economical method of collecting reliable demographic data. As Bogue (1969, 106) has commented:

> In some nations sample surveys are so widely used and have become so precise that a national census serves less to provide national totals of population composition than to provide data about smaller areas—states, cities, counties, townships, census tracts, and similar minor civil divisions—and to permit more detailed breakdown into refined categories and more elaborate cross-classifications of characteristics than do the sample surveys.

Requirements for an accurate and reliable sample survey include the following:

1. There must be an effective sampling design that will avoid biases in the selection process.

2. The sample design must lend itself to the estimation of sampling errors. It is difficult to develop and execute a sample design that is both economically feasible and scientifically correct without the help of a sampling expert.

Many variables affect the results of a sample survey, including the following:

1. characteristics of the person supplying the information

2. characteristics of the person collecting the information

3. salaries and benefits paid to those conducting the interviews

4. the amount of supervision

5. the collection procedure used

6. the type of questions used

7. the actual wording of the questions used

Multiple-round surveys are becoming more important in demographic data collection. In the multiple-round survey, or longitudinal survey, information about a specific group or problem is collected on a continuing basis. It is a useful method for collecting information, particularly in areas where there is poor census coverage. According to demographer Ralph Thomlinson (1976, 75), "Multiple-round surveys offer the triple advantages of cost reduction through sampling, keeping up-to-date by re-interviewing annually, or if desired, more frequently, and the opportunity to add and subtract questions as desired. In the United States the Bureau of the Census regularly conducts a variety of sample surveys. Two major categories of reports are published, *Current Population Reports* and *Current Housing Reports.*

QUALITY AND COMPLETENESS OF POPULATION DATA

Though there are exceptions, the quality and quantity of the collected demographic data are generally related to levels of economic development. In the poorest countries demographic data are usually either absent or incomplete and of poor quality, whereas in the wealthy countries demographic data are likely to be abundant and of good quality. Censuses and vital registration systems require a considerable outlay of money, and in many of the poor countries money is simply unavailable for such purposes. For this reason sample surveys, often done with the help of the United Nations, are being conducted in many of the developing countries. They cut down on the cost and, if carefully designed and executed, can provide useful information.

COMPLETENESS OF COVERAGE

Countries differ considerably in the number and types of questions they include on their census forms. Most censuses include questions on age and sex, and for perhaps half of the world's population censuses include questions on education and occupation. While about one-third of the world's people are asked questions on their language and religion, far fewer are asked about their ethnic identity. Though part of the reason for wide variation in census designs is financial, the variations are also affected by the political sensitivity of asking questions that some people may consider an invasion of privacy. Many people still view censuses with suspicion and resentment.

The list of questions asked in the 1990 United States Census was quite lengthy, and about the only topic not dealt with was religion. On the other hand, Gambia's 1963 census consisted of only five questions: place of enumeration, tribe, age, sex, and marital status. Beyond the problem of insufficient coverage, another problem is the lack of geographical detail. Though in the United States some data are available for geographic units as small as census tracts, block groups, and even blocks, in the developing countries very little data is published for comparable minor civil divisions. Even when such data are available in developing countries, their use is often limited by a lack of maps showing the exact boundaries of the divisions used.

Another problem of geographic coverage is the failure of census takers to visit isolated areas, either because of the physical difficulties involved in getting to such places or because of the fear of visiting such areas. The lack of adequate roads and communications systems in developing countries often means that some geographic areas will be either omitted or very poorly covered.

Even for countries that have detailed modern censuses, problems arise with historical studies. Questions often disappear from the census as we go back in time. A problem that often occurs in rapidly growing urban areas is the split tract problem; this happens when a census tract in a given time period is split into two tracts for the next census. Geographer Erick Howenstine (1993) provides help for people who run into this particular difficulty.

ERRORS AND OMISSIONS

Occasionally, census records are purposely falsified, as in the case of the 1940 Guatemalan Census. According to Whetten (1961, 20):

> The 1940 census might have been fairly adequate if the figures had not been deliberately tampered with. President Ubico, who wanted to show a large population for political purposes, issued orders to local authorities to alter the count after the census was taken. Relevant papers in the central files were carefully destroyed, but enough documentary evidence has been found in the offices of local authorities to indicate that the census results were inflated by at least 900,000 inhabitants.

Massive misreporting occurred in Nigeria in the 1962 census, as political parties in different regions attempted to increase their representation in the national government. The 1962 census was officially disavowed. Fortunately, such examples of deliberate falsification of census records are rare. For the most part people attempt to faithfully and honestly answer census questions, though they often give wrong answers when they misunderstand questions or suffer lapses in memory. Most census errors are a result of omission rather than commission.

ERRORS IN AGE AND MORTALITY REPORTING

Aside from the conceptual problems related to the collection of age data, misreporting of age is believed to be quite common, especially in the developing countries. Studies of African and South Asian populations confirm this skepticism about age data. In a study in Africa, Blacker (1969, 277) noted that, ". . . the widespread ignorance of age in the sense of numbers of completed years bedevils the collection of accurate data, and studies which have been made of reported African age distributions have revealed massive and deep-seated errors." The age situation in India has received a similar evaluation, and even in developed countries some data suggest that age is often misreported.

Problems similar to those of age misreporting can be found when examining mortality records. For example, because all mortality reports involve proxy respondents, a poor accounting of deaths is common for persons living alone or living with non-relatives prior to their death.

Mortality in the developed regions is concentrated among the elderly, whereas in the developing areas approximately half of the deaths occur to persons less than 15 years old; and many times more deaths occur in the first year of life than among all of those over 65. In the developed countries, where a larger proportion of older people are found, mortality surveys that omit the deaths of persons living alone or with non-relatives will have a larger impact. Underreporting of deaths of the very young is common in all areas, but it has a greater impact on the developing countries, where there is a larger proportion of young people.

ERRORS IN THE UNITED STATES CENSUS

In such a large undertaking as the United States census, with more than 480,000 workers in 1990, it is no wonder that errors have been made. Kahn (1974, 44) noted the following:

> While the general public is sometimes surprised to hear that census figures are not beyond question, this is no news to the nation's enumerators themselves. They are among the few bureaucrats in any federal establishment who have consistently and candidly and voluntarily conceded that they make mistakes.

Among the early critics who reminded the census takers of their fallibility, George Washington remarked that:

> Returns of the Census have already been made from several of the States and a tolerably just estimate has been formed now in others, by which it appears that we shall hardly reach four millions; but one thing is certain: our *real* numbers will exceed, greatly, the official returns of them; because the religious scruples of some would not allow them to give in their lists; the fears of others that it was intended as the foundation of a tax induced them to conceal or diminished theirs; and thro' the indolence of the people, and the negligence of many of the Officers, numbers are omitted. (Fitzpatrick, 1939, 329)

Fortunately, the Census Bureau realized the fallibility of its information and by 1950 started to conduct regular self-analyses called Post Enumeration Surveys. In 1970 they spent over $3 million for self-evaluation; more than the total amount spent on the census a century earlier. In 1980, and again in 1990, the task was both more complicated and more expensive.

In the four censuses immediately prior to 1980 in the United States there had been undercounts of 5.4 percent in 1940, 4.1 percent in 1950, 3.1 percent in 1960, and 2.7 percent in 1970 (Wolter, 1991). These figures coincide with undercounts in other developed countries. At first glance a 2.7 percent error may seem insignificant, but with a base population of over 200 million people in 1970 it meant that about 5.4 million Americans were not counted. Perhaps even more important is to consider *who* was not counted. The undercount was not distributed equally among all groups in the population. In 1970, according to the Census Bureau, there was a net under-enumeration of 6.5 percent for African-Americans and 2.2 percent for whites, similar to the 1960 undercounts of 8.3 percent for blacks and 2.7 percent for whites.

The 1980 Census

For the 1980 census, accuracy was improved. However, the most surprising result of the 1980 census was apparent by the end of 1980. The *error of closure*, the difference between the precensus estimate of the population and the actual count, was nearly 5,000,000 persons. In other words, the Census Bureau's count of 226, 504, 825, not counting the United States citizens that were living abroad, was far beyond the 1980 estimated population.

Ironically, the magnitude of the error of closure, which resulted from a combination of factors, occurred primarily because of the Census Bureau's concerted effort to provide as complete a count as possible. For the 1980 census the Census Bureau, learning from previous experiences, worked to improve such efforts as the following:

1. checking all dwellings reported as vacant

2. verifying addresses

3. extensive cross-checking of census rolls against other lists, such as driver's licenses

4. more intensive canvassing of pool halls and similar areas for people who had no permanent address

5. more precensus advertising, much of which was aimed directly at various minority groups

The estimated number of illegal aliens counted turned out to be a substantial 4.8 million, about 2.3 million of whom were counted in two states, Texas and California (Hauser, 1981). However, undercounts still occurred, especially for minority groups. The Census Bureau's preliminary measurements suggested an undercount of under 2.0 percent for the entire population, a figure that was gradually believed to be near 1.2 percent. Again undercounts were higher for some segments of the population; African-Americans were undercounted proportionately more than whites, for example, with the undercount for Hispanics somewhere in between. The estimated undercount for black males was 6.7 percent, compared to 1.5 percent for white males. An error of over 15 percent was found for African American males between ages 20 and 55. Because the allocation of federal funds often depends on census enumerations, errors of any magnitude can have considerable political and socioeconomic consequences.

Undercounts of African American males occur for several reasons. Many African American men in some age categories are highly mobile, hence they are easily missed by census takers. Some other possible reasons for undercounting in certain neighborhoods were outlined in a report by an enumeration supervisor who stated:

> One of the addresses to which I obtained access had been marked in the listing book as follows: "Will not return—something thrown at me." The occupants of the house with the "vacant" apartments were a sullen group. The bleary-eyed household head of the first floor apartment had a bottle on the table and glared silently at me while his wife answered the questions. . . . My experiences apparently were prosaic compared to some I learned about from discussion with crew leaders and enumerators; e.g., the lady who blundered into the house full of dope addicts; another lady who had been chased out of a house by a threatening occupant. . . . Although this sort of experience is not a regular occurrence by any means, it need happen to an enumerator only once to be unnerving. Situations in which an enumerator is in real danger probably are rare—but cases of suspicion, sullenness, and vituperative hostility in this area are far above our expected norm. It makes the recruitment and retention of good enumerators exceedingly difficult. . . . The inevitable concomitant of these conditions is under-enumeration—both of persons and housing units (Pritzker and Rothwell, 1967, 14).

The 1990 Census

The 1990 Census once again undercounted the population, this time by a figure estimated by the Census Bureau to be between 4.3 and 6.3 million. This estimate, based on a postcensus survey of some 165,000 households, translates into an undercount of between 1.7 and 2.5 percent, with a figure around 1.8 percent considered likely, somewhat higher than the error in 1980. As in previous censuses, the highest undercounts occurred for minority populations in metropolitan areas. African American males were undercounted by an estimated 8.0 percent. Despite considerable pressure, especially from large cities, Robert A. Mosbacher, then Secretary of Commerce, refused to make official adjustments of the census counts.

The 2000 Census

Planning for the 2000 Census began early in the 1990s, indicating the immensity of the task—accurate enumeration of a population of more than 270 million. As journalist Faye Fiore (1997, A5) put it, "This is the biggest federal operation short of going to war, the only single thing Washington does that sets out to touch every single soul in the United States." The census questionnaire appears in Appendix A. Attempts have been made to shorten the form and to make it simpler for people to read.

The 2000 Census was full of surprises. For example, the census counted 281,421,906 residents, nearly 7 million more people than the Census Bureau had estimated it would find on April 1, 2000. It also showed the decade of the 1990s to be the greatest period of population growth in the nation's history—population grew by nearly 33 million people in a decade (nearly the equivalent of the 2000 population of California). Furthermore, population growth occurred in all 50 states, though hardly at the same rates. Twelve seats in the House of Representatives changed from one state to another, more than had been expected. The big winners (two new seats each) were Texas, Florida, Georgia, and Arizona—New York and Pennsylvania each lost two seats.

After a three-decade decline, the response rate for the 2000 census pegged upward, with an initial return of 67 percent of mailed-out questionnaires. The count was also thought to be the most accurate one that the Census Bureau had produced for some time, an important consideration considering that nearly $200 billion in federal funds are allocated annually on the basis of census figures (Anderson and Fienberg, 1999).

After considerable argument and deliberation, the Census Bureau declined to statistically adjust the new census data in order to adjust for undercounts, which may have been around 3 million people or just over 1 percent (Prewitt, 2000). The comparability of racial and ethnic data has been confounded by the decision to allow people to choose more than one race for the first time ever. Nearly 7 million people took advantage of the idea and identified with more than one race (Skerry, 2000).

Census 2000 also made clear that rapid changes are occurring in the ethnic/racial composition of the United States, a trend that seems likely to continue. The Hispanic population grew much faster than expected (from 22 to 35 million in a decade) and is now virtually equal nationally to the African American population. The Asian population grew rapidly also. As an example of what is probably to come for the nation, the non-Hispanic white population in California slipped below 50 percent, leaving the union's most populated state with a population that has no racial-ethnic majority.

SUMMARY

The collection and use of all types of demographic information is important for governments, social scientists, businesses, and others. Censuses, sample surveys, and vital statistics are essential to the functioning of modern societies. As you begin to use demographic statistics, certain volumes are well worth knowing about. The United States Bureau of the Census publishes a detailed catalog of census materials. Also of considerable interest to those seeking population data for the United States is the Census Bureau's *Statistical Abstract of the United States*, which is published annually and contains a wealth of information; it is the first volume that reference librarians look to for questions about this nation's data.

Demographic data outside of the United States are available in a variety of different sources including the United Nations' *Demographic Yearbook*, which is published annually, *The Universal Almanac*, edited by John W. Wright, *World Resources*, an annual volume edited by the World Resources Institute and published by Oxford University Press, and, of course, the censuses of individual countries.

Population Distribution and Composition

3

The description and analysis of spatial distributions have been recurrent themes in geography; where phenomena are located, and the reasons for their particular locations, have traditionally been a central part of the discipline. In that vein, then, in this chapter we look at the "where" of population and analyze some of the reasons behind the spatial distribution of the world's population, the age and sex structure of populations, and examples of the distributions of both the elderly and major racial and ethnic populations in the United States.

It is obvious that there have been (and continue to be) temporal changes in the world's population. Though the overall trend in population growth had been one of gradual increase until approximately 300 years ago, when a more rapid rate of increase gradually became perceptible, there have been large fluctuations of numbers in specific areas (China and Europe, for example) throughout most of human history. Uneven growth over time has been matched by a similarly uneven growth over space; that is, growth is asymmetrical geographically because the populations of some countries and continents have grown more rapidly than others.

The number of inhabitants in a given area—be it a city, county, state, nation, or any other political unit—is usually one of the first population facts called for before any analysis of an area is undertaken. The numbers of people in specific geographical units are common elements that are needed for various administrative and research purposes; in many cases first judgments about a country or region are based upon a combination of the extent of its geographical area and the size of its population. More importantly, this information is not just of geographic importance. According to Smith and Zopf (1976, 39):

> Sociologists must have the facts relative to the numbers of inhabitants in various areas before they can compute indexes of criminality, juvenile delinquency, marriage, and so on; administrators must have them in order to determine how state and federal funds for education, agriculture, road construction, and so on, are to be apportioned among the counties, states, or other political divisions with which they are concerned.

Many other disciplines, including economics and political science, as well as many private and public agencies, including city and regional planners, also need accurate population counts.

GLOBAL DISTRIBUTION PATTERNS

An analysis of the distribution of the world's population, shown in Figure 3-1, reveals some fascinating patterns:

- First, the great majority of the earth's inhabitants live on a small part of the land area. Approximately 90 percent of the people live on about 10 percent of the land.

- Second, over 90 percent of the world's population live north of the equator and less than 10 percent live to the south; however, it must be remembered that over 80 percent of the total land area is located in the Northern Hemisphere as well.

- Third, the earth's population is concentrated on the margins of the continents. Estimates are that approximately 70 percent of the world's population live within 1000 kilometers (600 miles) of the sea and about 67 percent live within 500 kilometers (300 miles) of it (in the United States an increasing proportion of the population lives in coastal areas as well).

- Fourth, population numbers generally decline with altitude. Approximately 56 percent of the world's population live below 200 meters (656 feet), whereas close to 80 percent live below 500 meters (1640 feet) (Clarke, 1972). Except for the attraction of minable natural resources (minerals and fuels, especially) and recreational opportunities (skiing, for example), mountainous areas are limited in their range of economic activities.

There are four areas in the world with large population clusters: southern Asia, eastern Asia, western Europe, and east-central North America (the smallest). The southern Asian cluster includes India, Pakistan, Sri Lanka, Burma, Cambodia, and Thailand, along with related areas. The eastern Asian cluster includes China, Korea, Japan, and the Philippine Islands. Together, these two clusters include over one-half of the world's population. The western European cluster

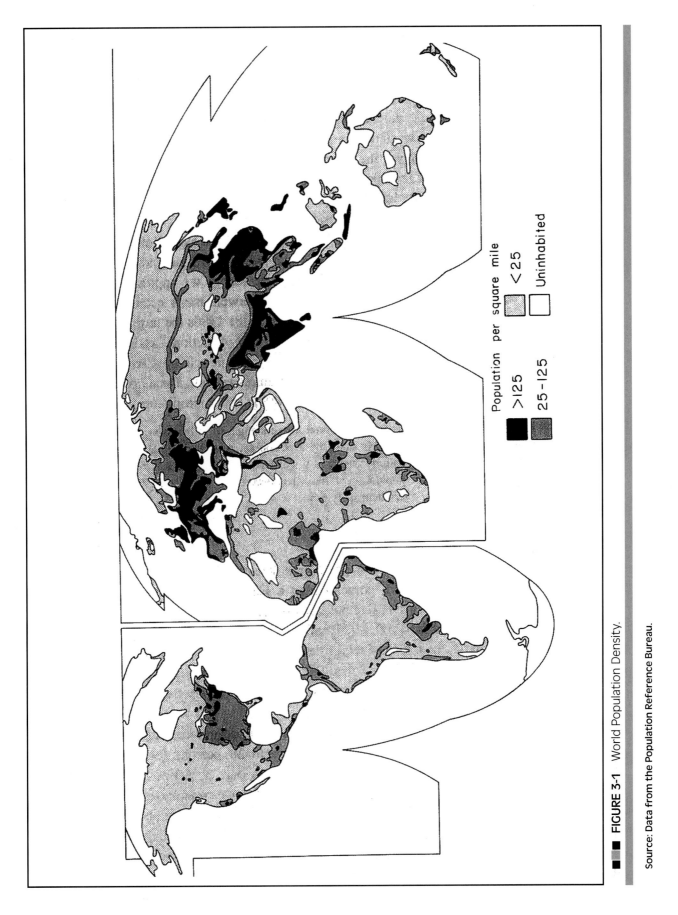

FIGURE 3-1 World Population Density.

Population per square mile

■ >125
▨ 25-125
▨ <25
□ Uninhabited

Source: Data from the Population Reference Bureau.

includes most of the Commonwealth of Independent States, the United Kingdom, Germany, and the other western European nations. The last and smallest cluster includes urbanized eastern parts of both Canada and the United States.

ECUMENE AND NONECUMENE

One manner of describing the distribution of the world's population is to divide the world into two groups, the **ecumene** and the **nonecumene.** The ecumene is the permanently inhabited portion of the world, whereas the nonecumene is the uninhabited or virtually uninhabited portion. Though at first glance it may seem relatively simple to divide the world into ecumene and nonecumene, it is actually quite difficult because neither the inhabited nor the uninhabited areas of the world are represented by a continuous distribution.

Climate plays an important role in determining these broad spatial patterns. The permanent icecaps are relatively easy to identify as part of the nonecumene, as are the world's most arid realms, but much of the rest of the nonecumene is isolated (some mountain ranges, for example) and either unoccupied or sparsely occupied (Trewartha, 1969). Of the total land area of the earth, approximately 30 percent is barren waste, without any significant human occupance; large areas of forest are also nearly uninhabited by people. The nonecumene, then, consists of perhaps 35-40 percent of the earth's total land area; a substantial portion of the earth's surface is without permanent human occupance. Of course the nonecumene is smaller today than it has been in the past because an established feature of human history has been a progressively expanding ecumene—population growth, technological advances (air conditioning, for example), and the discovery of minable minerals in isolated locations all contribute to that augmentation.

Further investigation of the ecumene reveals some absorbing regional characteristics. The Anglo-American and western European population concentrations share numerous similar attributes. For example, the Anglo-American cluster was a direct result of European colonization (Native Americans comprise less than one percent of the region's population today) and thus bears the stamp of European culture. Both of these clusters have advanced technology, high per capita incomes, well-developed regional specialization of economic activities, low percentages of their labor forces employed in agriculture, and a large proportion of residents living in cities.

By contrast, the two major Asiatic concentrations, with the exception of Japan, still exist mainly in the developing world, though rapid industrialization and other signs of economic development are changing the picture. Asian NICs (newly industrialized countries), sometimes dubbed "Fighting Tigers," include South Korea, Hong Kong, Taiwan, and Singapore. Since the 1980s the People's Republic of China has been undergoing numerous economic reforms aimed at bettering the Chinese economic system. Soon the Chinese economy

The **ecumene** is the permanently inhabited portion of the world, whereas the **nonecumene** is the uninhabited or virtually uninhabited portion.

will have the planet's largest total GNP (though it will remain much lower in terms of per capita GNP); its growing role in world trade is already apparent. Changing economic circumstances in these countries are resulting in changing demographic pictures as well.

POPULATION DENSITY

There are several ways to measure population density. The most common measure compares the total population of a place to its total area, a measure referred to as **arithmetic density.** Table 3-1 depicts the population density for the principal regions of the world for 1960, 1975, and 1995. Though Europe and Asia are the most densely settled continents, the average density of the developing countries is more than triple that of the developed countries (compared to less than double in 1960). During the 1980s the population density in Asia surpassed that in Europe, which previously had been the most densely settled continent.

Although density figures are convenient for analyzing differences in population distribution, they are often misleading. In the mid-1990s there were about 42 persons for every square kilometer on the earth's land surface, but Oceania has only 3.3, Africa 23.7, and North America 13.7, whereas the more densely settled continents have considerably higher densities, with 103.1 for Europe and 125 for Asia. Densities obviously vary on a continental scale, but they vary to an even greater degree within smaller areas. For example, Australia has a density of only 2.3 inhabitants per square kilometer, whereas the Netherlands has 455. Other examples are evident in Table 3-2. Smaller nation-states, and particularly island nation-states, often have dense populations, in contrast with larger nation-states like the United States and Russia, which have densities of 28.6 and 8.5 respectively. Population density values, however, have the least significance when large areas with marked environmental differences are involved. Though the average density for the former U.S.S.R. was 13 people per square kilometer, it is important to note that the density in the Asiatic portion of that country was only about 10 percent of the density in the European portion. Arithmetic density is widely used because the necessary data are easily obtained; keep in mind, however, that it tells us very little about the relationship between people and resources in a nation. We can find rich nations with high densities (Germany, Japan), rich nations with low densities (U.S., Australia), poor nations with high densities (Pakistan, Bangladesh), and poor nations with low densities (Sudan, Chad).

A somewhat more refined measure of density is the ratio of population to arable land, a measure referred to as **physiological** or **nutritional density.** Arable land is that portion of the earth's land surface that is suitable for tillage, a definition which obviously leaves out some productive areas of nonarable land, such as mining land, natural pasture, forests, and scenic regions. Nutritional density cannot

There are several ways to measure population density. The most common measure compares the total population of a place to its total area, a measure referred to as **arithmetic density.**

A somewhat more refined measure of density is the ratio of population to arable land, a measure referred to as **physiological** or **nutritional density.**

■■■ TABLE 3-1. Land Area and Population Density in the World and Major Areas, 1960, 1975, and 1995

Area	Land Area (1,000 km²)	1960	1975	1995
World total	135,779*	22.1	29.4	42.0
More developed regions	60,907	16.0	18.6	19.2
Less developed regions	74,872	27.0	39.5	60.5
Europe	4,936	86.1	96.0	103.1
U.S. and Canada	21,515	9.2	11.0	13.7
Oceania	8,509	1.9	2.5	3.3
South Asia	15,775	54.9	80.4	116.6
East Asia	11,756	67.0	85.5	122.7
Africa	30,320	9.0	13.2	23.7
Latin America	20,568	10.5	15.8	23.4

*Not including the Antarctic continent.
Source: United Nations.

■■■ TABLE 3-2. Population Density, 1995

Area	Per. Sq. Kilometer	Per. Sq. Mile
Afghanistan	28.2	73
Egypt	62.2	161
Chad	5.0	13
Mexico	49.0	127
Brazil	18.5	48
Honduras	48.6	126
U.S.	28.6	74
Japan	332.4	861
Australia	2.1	6
Germany	234.0	606
Netherlands	55.6	1180

Source: United Nations

account for the great variations in productivity within arable areas that are the result of such environmental characteristics as drainage, soil, and climate. According to Trewartha (1969, 74), nutritional density provides a better indicator than does arithmetic density of the degree of crowding in a region compared with its physical potential for producing food and agricultural raw materials. Japan's arithmetic density in 1960, for example, was 655 per square mile; its nutritional density was an unbelievable 4,680, a fact which suggests the necessity for significant imports of food and vegetable raw material.

Several other density measures have been suggested, including the "standard land unit," which includes varying measures for different modes of land use, and various indices based on the calories produced by a country, the values of national income, and

Courtesy of Photo Disc

With economic development, employment structures shift from mainly employment in agriculture to increasing proportions of employment in manufacturing and services. Both of these tend to concentrate populations in large cities, such as Toronto.

productivity. Some density maps give the ratio of the agricultural population to the amount of arable land, a measure that eliminates nonfarmers. As Beaujeu-Garnier (1966, 32) commented, "All of these methods have their interest in showing up this or that aspect of population density, but none is of undisputed absolute value; and, above all, none escapes from the criticism that, being applied to large and complex areas, it distorts a phenomenon which is essentially localized and variable."

FACTORS IN POPULATION DISTRIBUTION

An analysis of the distribution of the world's population suggests the importance of certain fundamental factors:

1. the historical evolution and duration of a population in a certain area

2. the natural environment and its associated aspects such as landforms, soils, precipitation, temperature, and natural resources

3. the socioeconomic and technical development of a region

As we shall see, the latter factors have become increasingly important as the Industrial Revolution and modernization in its broadest sense have spread throughout the world. Given an increased technological base, natural environments have less influence on a nation's population distribution. Storm windows and natural gas help mitigate the harshness of cold winters, whereas air-conditioning has turned many desert areas into such delightfully desirable places as Palm Springs, California.

With economic development, employment structures shift from mainly employment in agriculture to increasing proportions of employment in manufacturing and services, both of which tend to concentrate populations in cities. Thus, a major redistribution of population from rural to urban areas usually accompanies modernization.

SEX AND AGE STRUCTURE

Two of the basic characteristics of any population are sex and age. Both of these factors are important in relation to other population variables such as mortality, morbidity, marriage, fertility, and migration.

SEX STRUCTURE

> The **sex ratio** is defined as the number of males per 100 females.

The sex ratio is one of the simplest measures of population structure, and statistics on sex are usually easy to get and accurate. There is no ambiguity about the meaning of female and male and there is little need to misrepresent such information. The **sex ratio** is defined as the number of males per 100 females. For example, the sex ratio for the population of the United States in 2000 was

$$\text{Sex ratio} = \frac{\text{number of males}}{\text{number of females}} \times 100$$

$$= \frac{138,053,563}{143,368,343} \times 100 = 96.29.$$

A sex ratio of 100 indicates that there are equal numbers of males and females in a population; a ratio above 100 indicates that there are more males than females; and a ratio below 100 indicates that there are more females than males. The basic factors that determine the sex ratio are

1. death rate differences between the sexes

2. net migration rate differences between the sexes

3. the sex ratio of newborn infants.

All modern censuses calculate the proportions and numbers of females and males in the population. The first census in the United States determined the sex structure of the white population, and since 1820 the sex structure of all segments of the population has been tabulated. Because of the relative simplicity of the question pertaining to sex, census officials since as early as 1930 have concluded that the data on sex structure is the most reliable of all population data included in their tabulations.

Courtesy of Photo Disc

The rush of males to the mining areas of California during the Gold Rush of 1849 resulted in an extremely high sex ratio for that state.

Examining sex ratios and correlating them with other demographic factors such as births, deaths, and migration reveal some basic facts:

1. The sex ratio at birth is relatively high (105) for the United States and for most of the rest of the world.

2. In general, females have lower death rates than males at all ages.

3. Females outnumber males in migration from rural to urban areas and in other short-distance migrations.

4. Males outnumber females in long-distance moves.

The sex ratio of the population of the United States has varied greatly from place to place, time to time, and group to group. Figure 3-2 depicts the changes in the sex ratio of the United States from 1840 to 2000. The sex ratio of 94.2 in 1980 was the lowest in American history. According to Guttentag and Secord (1983, 32) "Low sex ratios, signifying a shortage of men, provide impetus for feminist movements. This is not to say that low sex ratios can bring movements about by themselves, but rather that they provide some facilitating conditions." Their work raised some challenging questions about sex ratios and society, though no geographers to our knowledge have followed up on their ideas.

Changes in the sex ratio over time in the United States are primarily due to changing immigration patterns. The all-time high

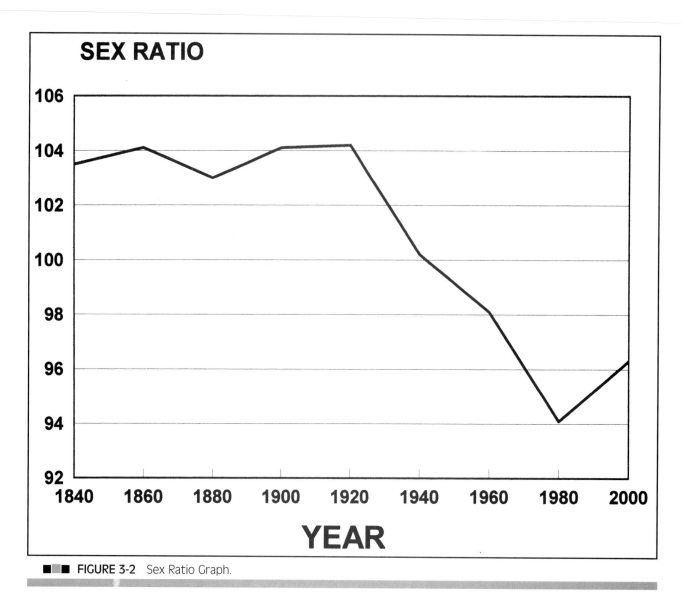

FIGURE 3-2 Sex Ratio Graph.

Source: Data from *Statistical Abstract of the United States and Census 2000*.

of 106, which occurred in 1910, was the result of large numbers of immigrants, primarily males. The low sex ratio of 1870 was primarily due to the decline in immigration during the Civil War and the large numbers of males killed in battle.

A state-by-state analysis of the sex ratio of the United States reveals some interesting changes in spatial patterns over time. Figure 3-3 shows sex ratios for each state in 1790. These sex ratios, however, represent only the white population. The sex ratio for all of the states was 103.7. A state-by-state analysis reveals that higher sex ratios were found in what were then considered to be frontier areas, whereas the lower sex ratios were found in the more settled areas of New England. This same pattern also appears in Figure 3-4 for 1850, though these data include both white and nonwhite residents. Once again, the frontier areas are male-dominated, whereas the more

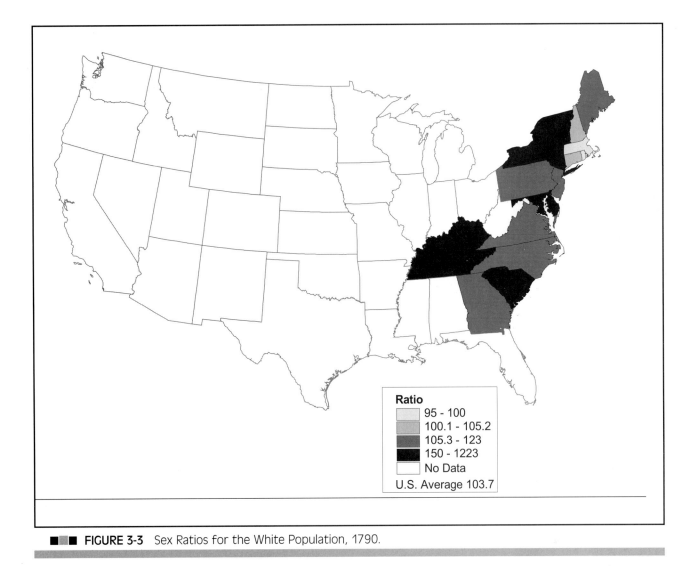

■■■ FIGURE 3-3 Sex Ratios for the White Population, 1790.

densely settled and urbanized areas have a lower sex ratio. The rush of males to the mining areas of California during the Gold Rush of 1849 is apparent in the extremely high sex ratio of that state (1222.9). It is also interesting to note that the lowest sex ratio, 95.0, is found in Washington, D.C., though in general urban areas have lower sex ratios than rural areas.

Figure 3-5 shows sex ratios at the turn of the century. The sex ratio of the United States was still above 100, mainly because of large numbers of immigrants. The opening up of the western frontier brought about large imbalances between the proportions of the sexes. The westward movement was largely dominated by males, making the sex ratios in several western states extremely high, whereas large numbers of females were left behind in the South and East. Washington, D.C. once again had the lowest sex ratio. Utah, although a western state, had a much lower sex ratio than the other western states, undoubtedly because of the unique nature of the settlement of

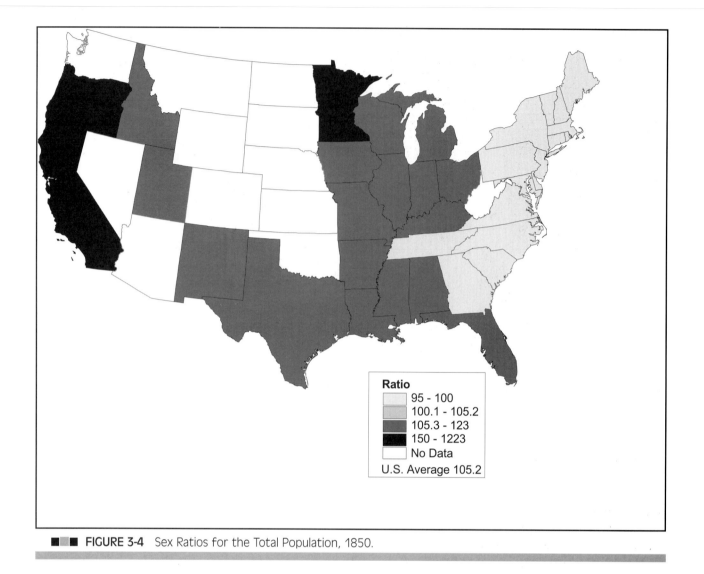

Ratio
95 - 100
100.1 - 105.2
105.3 - 123
150 - 1223
No Data
U.S. Average 105.2

■■■ **FIGURE 3-4** Sex Ratios for the Total Population, 1850.

Utah and the role of the Mormon Church. Migration to Utah often involved families, whereas single males were most evident in early migrations to the other western states.

With the passing of the frontier came a reduction in the sex ratio for the United States, as well as a lowering of sex ratios for western states, as is apparent in Figure 3-6 for 1940. Not only is there a lowering of the sex ratios in the West, but also the sex ratios for the various parts of the country tend to cluster close to the national average.

Figure 3-7 illustrates sex ratios in 1970. It is apparent that all sections of the United States are well settled and that only relatively small regional differences exist within the country. The sex ratio for all sections of the country is close to the national average of 94.8. The only exceptions are Alaska and Hawaii, with Alaska having the highest sex ratio of all states. Alaska can be considered the last frontier and we would expect it to have a higher proportion of males, again

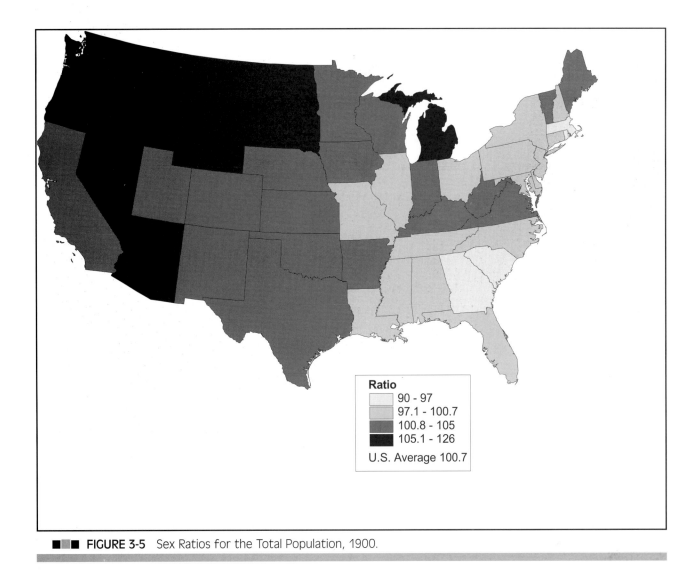

Ratio
	90 - 97
	97.1 - 100.7
	100.8 - 105
	105.1 - 126

U.S. Average 100.7

■■■ **FIGURE 3-5** Sex Ratios for the Total Population, 1900.

because of the dominance of males in migration to frontier-type areas. Though the sex ratio increased slightly between 1970 and 2000, the spatial pattern of sex ratios in 2000, as shown in Figure 3-8, differed little from the pattern in 1970.

INTERNATIONAL VARIATIONS IN SEX STRUCTURE

Though there are more males at birth, females tend to outlive males, especially in regions where there is good medical care. Thus, we would expect to find lower sex ratios in the developed countries and higher ones in the developing areas, where maternal mortality and fertility are still high.

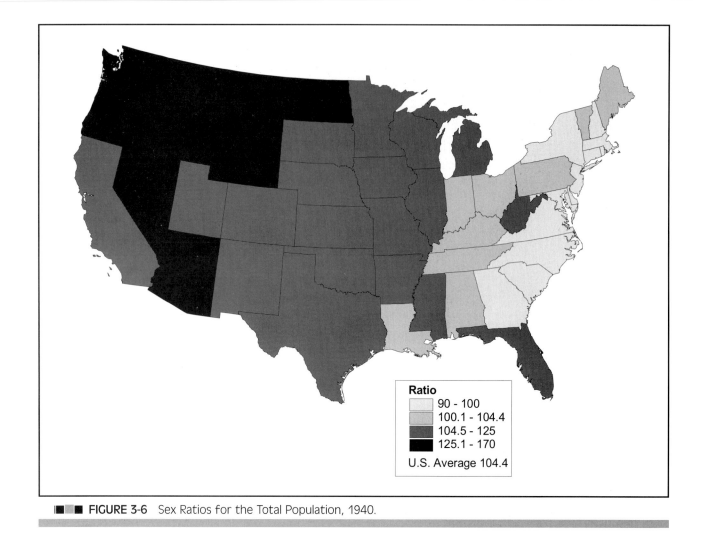

Ratio
	90 - 100
	100.1 - 104.4
	104.5 - 125
	125.1 - 170

U.S. Average 104.4

FIGURE 3-6 Sex Ratios for the Total Population, 1940.

Very few countries in the world have sex ratios higher than 105 or lower than 90, as shown in Table 3-3. The reasons for high sex ratios include such variables as in-migration of males, high female mortality due to poor health facilities, undercounts of females, and high fertility, with its resulting young age composition. In some areas, such as West Malaysia, Hong Kong, and Cuba, migration and the influx of laborers have been important variables determining sex ratios. However, in other areas, such as Libya, Pakistan, India, and Sri Lanka, migration was not important; high sex ratios in these areas are probably due to higher mortality rates for females than for males.

The countries with low sex ratios are primarily found in Europe and their low sex ratios can partly be traced to large numbers of emigrants, mainly males. Another reason is the loss of males during World War II.

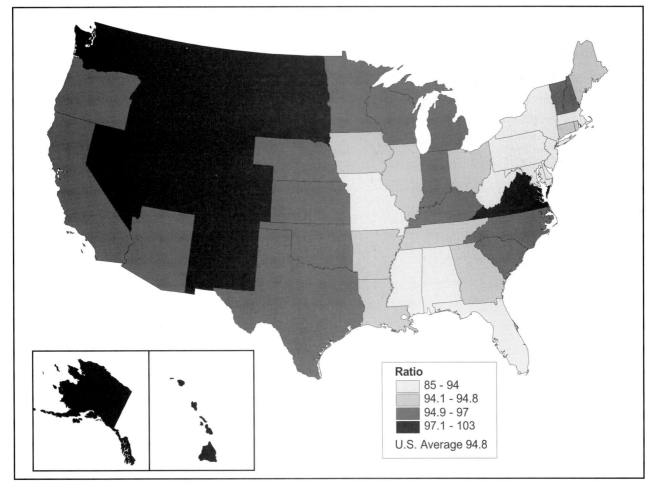

■■■ FIGURE 3-7 Sex Ratios for the Total Population, 1970.

Ratio
- 85 - 94
- 94.1 - 94.8
- 94.9 - 97
- 97.1 - 103

U.S. Average 94.8

■■■ TABLE 3-3. Sex Ratios of 105 or More

Country	Sex Ratio
Libyan Arab Republic	108
Cuba	105
Sri Lanka	108
Hong Kong	106
India	106
Iran	107
West Malaysia	106
Pakistan	111
Albania	105
Sex Ratios of 90 or less	
Malawi	90
Austria	87
German Federal Republic	89
German Democratic Republic	84
U.S.S.R	85

Source: United Nations, *Demographic Yearbook*, 27th Issue (New York: United Nations).

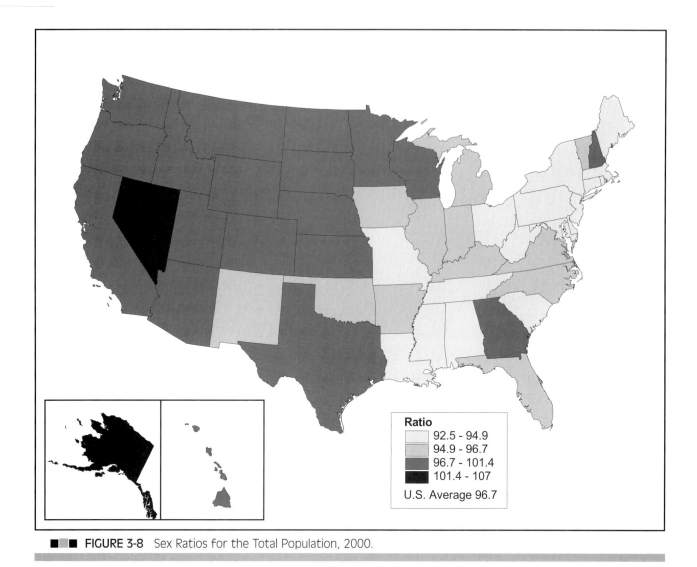

Ratio

	92.5 - 94.9
	94.9 - 96.7
	96.7 - 101.4
	101.4 - 107

U.S. Average 96.7

■■■ **FIGURE 3-8** Sex Ratios for the Total Population, 2000.

SEX AND GENDER

Though demographers and geographers have long recognized sex ratios as a measurement of the relative distributions of males and females in a population, there has been a growing interest in gender studies as well during the past two decades. Gender studies among geographers and other social scientists are concerned with socially created differences between the sexes (in contrast to biological distinctions). Such social differences between the sexes (gender roles) typically arise from the differential ways in which roles in a society are assigned to the two sexes; from early educational opportunities in developing countries to the "glass ceiling" in America, from women as secretaries and nurses to men as construction workers and doctors.

Because such social-typing can vary considerably from place to place, geographers have become interested in spatial manifestations of gender differences. Common grounds for feminist geography (and the geographical study of gender) include at least the following: the portrayal of women's experiences, concern with the quality of life for women, and a political vision of gender equity (Monk, 1994). Additional ideas about geography and gender are included in Monk and Hanson (1982), Bondi (1992), and Bondi (1993); recent books focusing on various aspects of gender studies include Adepoju and Oppong (1994), Peters and Wolper (1995), and Jones, Nast, and Roberts (1997). Two brief essays by Rodger Doyle, each accompanied by a world map, have examined female illiteracy (Doyle, 1997) and women in politics (Doyle, 1998). Low rates of female participation in politics and high levels of female illiteracy are characteristic of many nations in Africa, Asia, and Latin America.

AGE STRUCTURE

Much of a population's economic and social behavior is determined by the proportion of people found in various age groups. Smith and Zopf (1976, 149) suggested the following three reasons why a knowledge of the age distribution of a population is basic in nearly all population analyses:

1. Age is one of the most fundamental of one's own personal characteristics; what one is, thinks, does, and needs is closely related to the number of years since one was born.

2. Groups are determinants of primary social and economic importance in any society.

3. The qualified student of population must possess the technical skills needed to bring out the significant features of the age composition of a population with which he is concerned and also those required to make the proper allowances or corrections he may attempt.

Because the age structure is so closely related to other demographic characteristics, such as the birth rate, death rate, migration rate, and marriage rate, caution must be used when comparisons and analyses involving these characteristics are undertaken.

FACTORS AFFECTING AGE STRUCTURE

Many factors determine the age structure of a national population, but the primary variable is the birth rate. A population with a high birth rate will have a large proportion of young people, whereas a population with a low birth rate will have a smaller proportion of young people. The role mortality plays in determining the age structure of a national population is best described by Coale (1964, 49) as follows:

> Most of us would probably guess that populations have become older because the death rate has been reduced, and hence people live longer on the average. Just what is the role of mortality in determining the age distribution of a population? The answer is surprising—mortality affects the age distribution much less than

does fertility, and in the opposite direction from what most of us would think. Prolongation of life by reducing death rates has the perverse effect of making the population somewhat younger.

Migration generally has little or no impact on the sex structure of national populations because most countries have restrictions on international migrations. However, for sub-national populations, for example the populations of states, counties, cities, or even particular districts within cities, migration often has a considerable impact on age structure. Migration is generally age-selective. That is, almost all types of migrations involve a large proportion of young people. An area that has experienced a great deal of out-migration usually has fewer young people in the age group 20-40, whereas a region with a significant amount of in-migration may show an excess of young adults.

Catastrophes such as famine, pestilence, or war may also influence the age structure of a region. For example, war has a twofold effect on the age structure. First, many young adults are combatants in a war and are killed; and second, there is usually a sharp decline in the birth rate during a war.

POPULATION PYRAMIDS

A **population pyramid** is a useful aid in examining the age structure of a population and it also provides information about the social attributes of the population concerned. As illustrated in Figure 3-9, a population pyramid is really two bar graphs placed back-to-back, with the vertical center line representing zero. The number or percentage of people in each age group, by sex, is indicated with horizontal bars; and the vertical axis gives the different age groups, usually in five-year intervals. The shapes of population pyramids vary according to the country or community and the time period. A pyramid for a particular country represents the demographic history of that country over the past two or three generations.

A national population with a large number of children, as the result of a high birth rate, would have a pyramid with a wide base, like that shown in Figure 3-10; whereas the pyramid for a population with a low birth rate would have a small base, like the pyramid for the United States in Figure 3-11. The pyramid is also affected by death rates, though usually to a lesser extent. In wartime the death rate for certain age groups may rise considerably, and the bands representing those age groups will be shorter. For example, the population pyramid for Berlin (Figure 3-12) shows a noticeable lack of men in the 20-29 age brackets, most likely due to World War II and the mortality among young men who fought in the war. Finally, selective migration can also change the shape of the pyramid.

It is possible to recognize certain broad categories of population structure by comparing the shapes of pyramids for different countries. Four types of population structures can be recognized, each coinciding with a particular cultural development. The first type

---

A **population pyramid** is a useful aid in examining the age structure of a population and it also provides information about the social attributes of the population concerned.

■■■

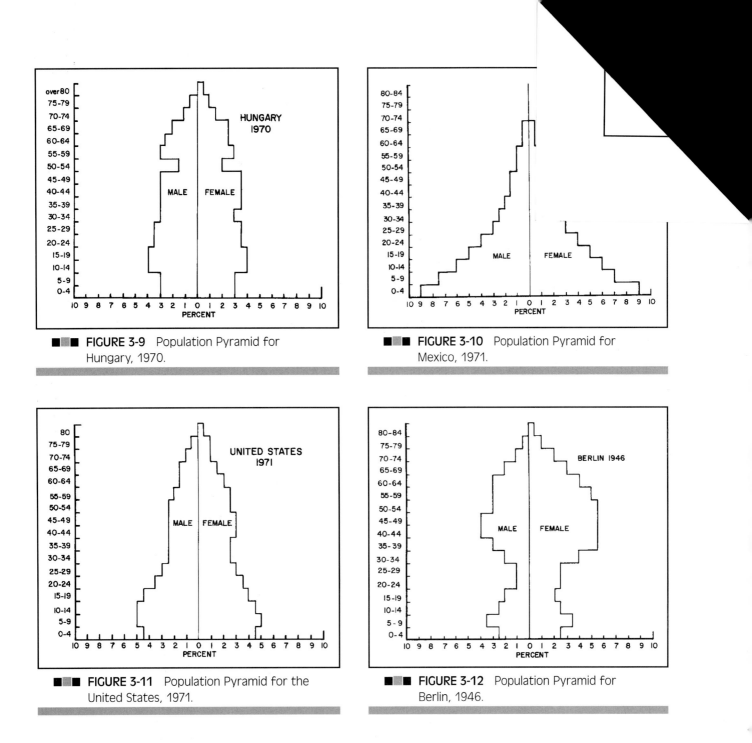

■■■ **FIGURE 3-9** Population Pyramid for Hungary, 1970.

■■■ **FIGURE 3-10** Population Pyramid for Mexico, 1971.

■■■ **FIGURE 3-11** Population Pyramid for the United States, 1971.

■■■ **FIGURE 3-12** Population Pyramid for Berlin, 1946.

(Figure 3-13) is a pyramid with a regular triangular shape, typical of a country with high birth and death rates. Few countries today have this type of pyramid, but it was common for most countries during the seventeenth and eighteenth centuries. The second type of population pyramid has a narrow top, wide base, and concave sides (Figure 3-14); it represents a country with a falling death rate, particularly in the youngest age category, and a high birth rate. This pyramid is typical for most of the developing countries in Africa, Asia, and Latin America. A beehive shape characterizes a third type of population pyramid (Figure 3-15) and indicates a relatively stable

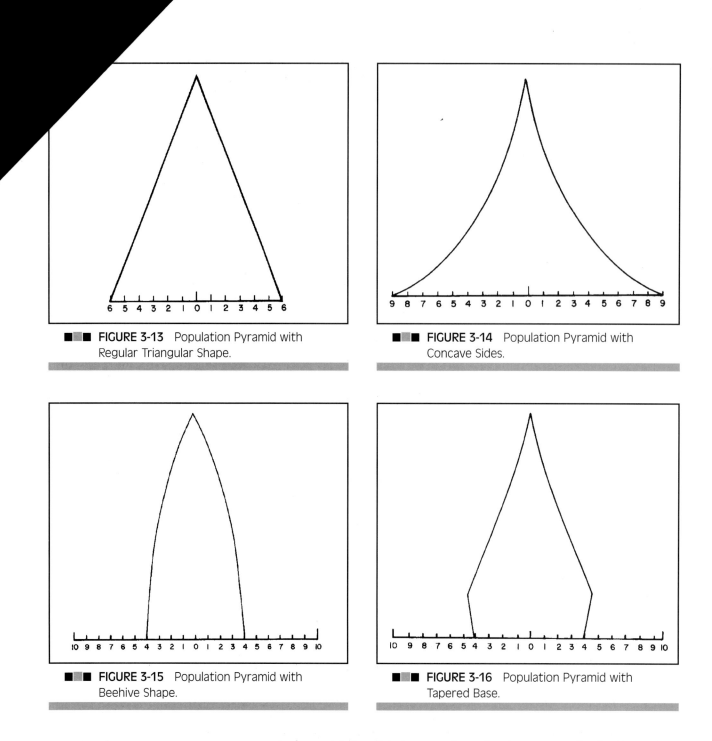

FIGURE 3-13 Population Pyramid with Regular Triangular Shape.

FIGURE 3-14 Population Pyramid with Concave Sides.

FIGURE 3-15 Population Pyramid with Beehive Shape.

FIGURE 3-16 Population Pyramid with Tapered Base.

population with low birth rates, low death rates, and a high median age. Most European countries would have pyramids of this type. England, Wales, and Sweden are good examples. The final pyramid type (illustrated in Figure 3-16) represents a country that has had a rapid decrease in fertility, thus giving a tapered shape to the bottom of the pyramid.

Changes in birth and death rates are reflected in the shape of a population pyramid. Deviations from the normal Christmas-tree shapes can be found for areas that have few women and/or children, as shown in Figure 3-17.

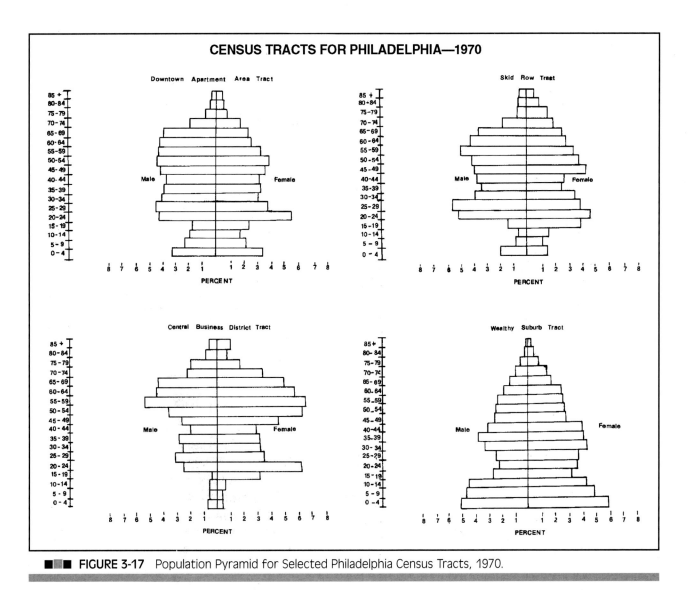

■■■ FIGURE 3-17 Population Pyramid for Selected Philadelphia Census Tracts, 1970.

DEPENDENCY RATIO

The **dependency ratio** is a simple statistic that measures the role of age composition on the productive activity of a population by comparing the proportion of the population in the nonproductive ages with those in the working ages. It assumes that people between ages 20 and 64 are the productive segment of the population, whereas those under 20 and over 64 years are the dependent segment of the population. It may be calculated as follows:

$$\text{dependency ratio} = \frac{\text{population under 20 plus population 65 years and over}}{\text{population 20-64 years}} \times 100$$

> The **dependency ratio** is a simple statistic that measures the role of age composition on the productive activity of a population by comparing the proportion of the population in the nonproductive ages with those in the working ages.

■■■

The purpose of the dependency ratio is to measure the number of dependents that each 100 people in the productive years must support. Thus, some nations have a high dependency ratio because of a large number of elderly, whereas other nations have high ratios because of a large number of children.

In the absence of comparable international data on economic activity, the dependency ratio is often used to reflect variations in economic dependency, even though some people in the dependent age range are producers, whereas many persons in the productive age range, particularly in Western societies, are economically dependent.

According to an analysis by the Population Reference Bureau (1977, 6):

> In general, dependency ratios are much less favorable in developing nations, due to their relatively high fertility levels; thus large population proportions are under 15 years of age. In Ghana, for example, where 47 percent of the population is under 15 and only 4 percent of the population is over 64, there are 104 persons in the "dependent ages" for each 100 persons in the "working ages." On the other hand, in the highly industrialized country of Japan, where 24 percent of the population is under 15 and 8 percent is over 64, there are only 47 persons in the "dependent ages" for each 100 persons in the "working ages."

WORLD PATTERNS OF AGE STRUCTURE

A world view of age structure reveals some interesting patterns, as we see in Figures 3-18 and 3-19. The world's youngest populations are found in the developing nations of Africa, Asia, and Latin America, where almost half of the population in many countries is under 15 years of age. The elderly comprise only 4 percent of the entire populations of Africa, Asia, and Latin America. For example, in Tanzania, only 3 percent of the population is over 64, while 45 percent is under 15. In 2000 the following countries were among those with the highest proportions of persons under 15: Gaza (50 percent), Benin (49 percent), Ghana (43 percent), Syria (45 percent), Niger (48 percent), Swaziland (47 percent), Honduras (42 percent), and Nicaragua (44 percent). High birth rates in these countries ensure a steady supply of new parents for the next generation, even if those parents in turn decide to have smaller families.

Europe ranks as the world's most elderly continent with 14 percent of its population aged 65 and over. The proportion of elderly is even higher for several individual nations of Europe. For example, in 2000 the countries with the highest proportion of persons 65 and over included: Germany (16 percent), Sweden (17 percent), Austria (15 percent), Norway (15 percent), United Kingdom (16 percent), Belgium (17 percent), France (16 percent), Switzerland (15 percent), and Denmark (15 percent). As you might imagine, birth rates (and percentages of young people) are quite low in these countries. Aging populations in European countries are forcing many governments to reconsider the generosity of medical and other benefits for the elderly.

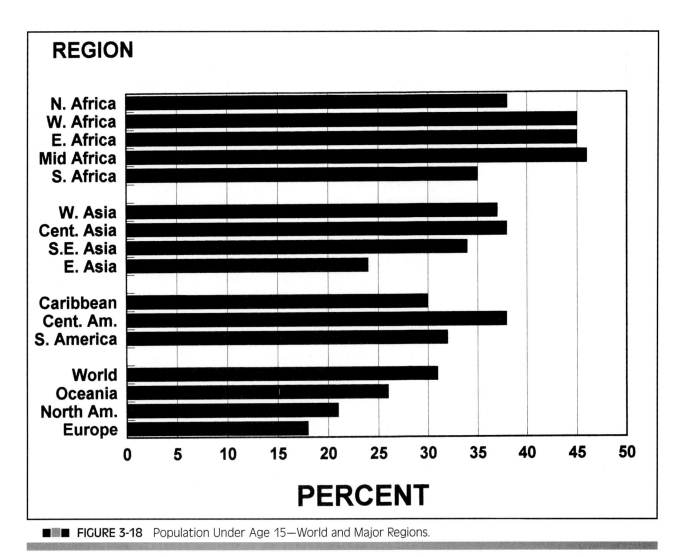

■■■ FIGURE 3-18 Population Under Age 15—World and Major Regions.

Source: Data from 2000 World Population Data Sheet. (Washington, D.C.: Population Reference Bureau, Inc.)

BABY BOOMERS IN THE UNITED STATES

Following World War II the United States, as well as most European and many other countries, experienced an increase in fertility. In the United States the cohort that is generally recognized as the "Baby Boomers" includes children born between 1946 and 1964 (there are about 77 million boomers, the first of whom—among them such noted celebrities as Kareem Abdul-Jabbar, Diane Keaton, and Bruce Springsteen—began to turn 50 in 1996, which should substantially impact membership in the American Association of Retired Persons). In 1946 there were 20 percent more babies born in the United States than in the previous year, a not unexpected result of the return home of missing husbands and lovers. However, one year was apparently not enough to make up for the war years, and Americans went on something of a fertility binge, totally unexpected by demographers and other social scientists (keep in mind that the best of economic, demographic, and other projections have been known to fall far from the

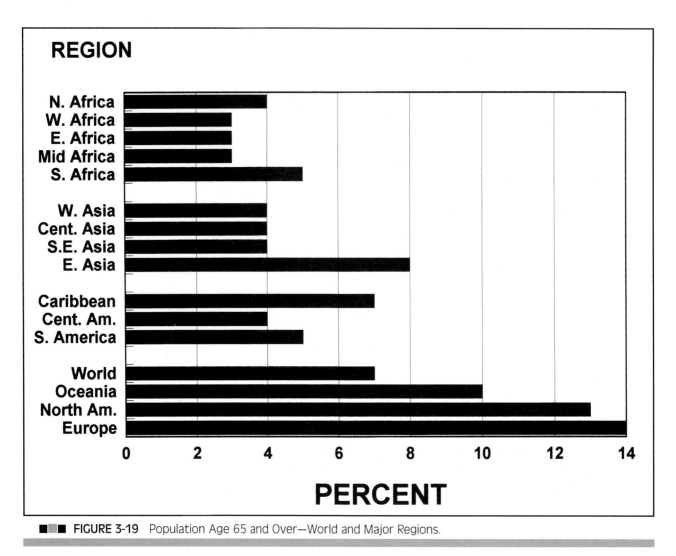

FIGURE 3-19 Population Age 65 and Over—World and Major Regions.

Source: Data from 2000 World Population Data Sheet. (Washington, D.C.: Population Reference Bureau, Inc.)

mark). At the same time, the baby boom was not a return to the high fertility of earlier periods; rather, it was a result of a significantly larger percentage of women marrying and having two or three children.

The sustained period of earlier marriages and higher fertility, so different from what occurred in the depression years of the 1930s, was at least partly a response to better economic conditions after World War II. Bouvier and De Vita (1991, 7) also noted that "Because men's wages were rising during this period, women felt little economic pressure to enter a male-dominated work world and thus were likely to be ambivalent about the desirability of 'yet another child' and apt to 'relax contraceptive vigilance.'" Interestingly, the end of the baby boom coincided rather well with the introduction of oral contraceptives in the mid-1960s.

By the time the leading edge of the baby boom cohort began to reach public schools in the early 1950s it was apparent that this cohort was going to create unprecedented problems, from the need for more schools as they entered kindergarten to their retirement needs, which we can expect to become extremely problematic after the turn of the century. Though the first baby boomers will not turn 65 until 2011, early retirements should begin to show up in larger numbers a decade earlier.

Having been supplemented over the years by immigration, as if their numbers were not sufficient already, in 1990 the Baby Boomers numbered around 80 million, or nearly one-third of the population of the United States. At that time they ranged in age from 26 to 44 and formed a prime component of everything from the workforce to the housing market. In commenting on those Boomers, now ensconced in middle age, essayist Roger Rosenblatt (1996, 33) noted:

> Some of the same kids who wanted to "turn on" as they tuned in are, as mature adults, alarmedly opposed to drugs. They who mocked and derided conservative religion now practice it. Many of the celebrants who ran naked at Woodstock in the name of sexual freedom are today those who are most steamed up about sexual scenes in movies, rap lyrics, and pornography over the Internet. Most "political correctness"—a reactionary movement, however liberally imposed—is the work of Boomer professors and is carried on by younger heirs. The Baby Boomers have evolved as both values-driven moralists and the deconstructionists of institutions.

Though we don't want to completely endorse the idea that "demography is destiny," the Baby Boomers and their impact on American society are clear evidence that demographics cannot be set aside. Changing age, sex, and ethnic compositions can, and probably will, have impacts on the economic and social systems within which they are found. Demographers don't think Boomers are preparing adequately for retirement (Russell, 1998), for example.

THE ELDERLY IN THE UNITED STATES

As we approach the twenty-first century most Americans can expect to attain at least the Biblical age of three score years and ten; life expectancy today is over 76 years, with males nearing an average of 73 and females approaching an average of 80. However, an analysis of the elderly in the United States over the past 90 years shows that only recently have most Americans been able to reach such advanced ages, as is illustrated in Figure 3-20.

In the United States, being **elderly,** or old, is usually defined according to the Federal social security program as having reached 65 years of age. The elderly population of the United States has risen consistently since 1900, when slightly more than 3 million Americans were aged 65 and over. The elderly population increased nearly threefold to 9 million by 1940 and then to nearly 30 million in 1990 (Table 3-4). In 90 years there was almost a 1,000 percent increase in the elderly population, compared with just over a 325 percent increase in the total population, from 76 million to 250 million, during the same time period.

In the United States, being **elderly,** or old, is usually defined according to the Federal social security program as having reached 65 years of age.

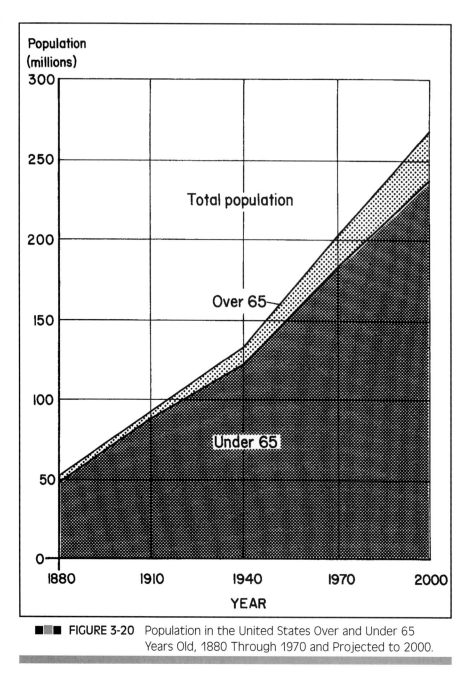

■ ■ ■ FIGURE 3-20 Population in the United States Over and Under 65 Years Old, 1880 Through 1970 and Projected to 2000.

Source: The Commission on Population Growth and the American Future, *Demographic and Social Aspects of Population Growth* (U.S. Government Printing Office, 1972), pp. 52-53.

The rapid increase in the elderly population in the United States is the result of three major factors.

1. High fertility rates occurred during the late nineteenth and early twentieth centuries, and because of this large birth cohort there has been an increased number of people in the 65 to 75-year age bracket.

■■■ TABLE 3-4. Population 65 and Older in the United States: 1900-2000

Year	Population (thousands)	Population Increase from the Preceding Decade (thousands)°	Percent of Population Increase from the Preceding Decade
1900	3,099		
1910	3,986	887	28.6
1920	4,929	943	23.7
1930	6,705	1,776	36.0
1940	9,031	2,326	34.7
1950	12,397	3,366	37.3
1960	16,659	4,262	34.4
1970	20,156	3,497	20.4
1980	24,830	4,674	23.2
1990	29,835	5,005	20.2
2000	34,991	5,156	17.3

Source: Leon Bouvier, Elinore Atlee and Frank McVeigh. (1975) "The Elderly in America," *Population Bulletin,* 30(3):4, courtesy of the Population Reference Bureau, Inc., Washington, D.C. Newest figures from the United States Bureau of the Census.

2. There has been a marked decline in mortality over the past 75 years, mainly because of advances in sanitation and medicine, which allowed more people to live to age 65. Presently, around 75 percent of newborn children are expected to reach age 65, whereas in 1900 only 39 percent of newborns were expected to reach that milestone.

3. Large increases in the elderly population in recent years were due to the high level of immigration prior to World War I. These migrants were primarily young adults, and thus a large cohort of that age group was added to the population.

Population projections by the Bureau of the Census forecast a continuing increase in the numbers of elderly in the population until at least the year 2030. There are expected to be 31.7 million by 2000, 34.9 million by 2010, 39.2 million by 2020, and 51.4 million elderly by the year 2030. These figures should be reasonably accurate projections because the future elderly have already been born, though immigration may have a significant impact on the later figures.

Not only has the total number of elderly been increasing rapidly, but the proportion of the elderly in the United States has also increased. In 1900 about one out of every 25 Americans was 65 or older, whereas in 1990 about 1 out of 8 Americans was elderly. The proportion of the elderly in a population is affected by variations in the numbers of people in other age categories. The increase in the proportion of the elderly is primarily attributed to declining fertility rather than to the lowering of the death rate. Changes in the proportion of the elderly over the past 60 years are illustrated in Figure 3-21.

The end of the Cold War, major economic and demographic trends, and global communications systems are converging at the end of the millennium to produce unprecedented human population movements. In turn, these movements are going to reshape demographic trends within both sending and receiving countries; they are already reshaping American demography, for example, though not without considerable debate, much of it emotionally charged (Isbister, 1996; Brimelow, 1995).

As the Cold War reached its sudden and unexpected end, euphoria swept the nations of Europe and elsewhere. On November 9, 1989, celebrants drank Champagne atop the Berlin Wall, hugged and danced in the streets, and cheered as the Wall came tumbling down. Then-President George Bush proclaimed the coming of a "New World Order." However, fading euphoria and reality checks on several continents suggest that we may be much closer to a new world "disorder."

Regional conflicts, often strongly nationalistic in origin, have flared up from Bosnia and Chechnya to Rwanda and Haiti, generating streams of frightened and battered refugees in their wake. Women, children, and the elderly are suffering disproportionately as uncertainty and instability replace the old order—any new order seems distant and indefinable, especially while the major powers resist interfering in places such as Bosnia. From Vietnam, Laos, and Kampuchea in the 1970s, Afghanistan in the 1980s, and Colombia, Rwanda, Burundi, Bosnia, and Chechnya in the 1990s streams of refugees

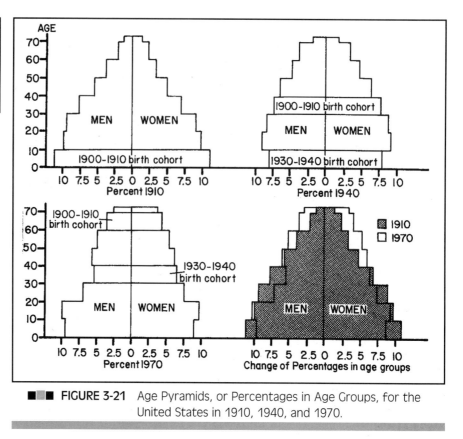

■■■ FIGURE 3-21   Age Pyramids, or Percentages in Age Groups, for the United States in 1910, 1940, and 1970.

Source: U.S. Bureau of the Census, *General Population Characteristics*, Final Report PC(1)-B1 United States Summary (U.S. Government Printing Office, 1972), p. 1-276.

One major reason why Americans are living longer now than ever before has been the substantial decline in heart attack deaths that has occurred since about 1950. However, this greater longevity is to some extent a mixed blessing at best. For example, Crimmins (2001, 5) pointed out that "Improvement in health does not necessarily accompany an increase in life expectancy since only about half of the disability and functioning loss at older ages is caused by lethal diseases. The other half is caused by conditions that are not linked to mortality trends, including arthritis, vision loss, and Alzheimer's disease." More attention needs to be focused on the quality of life for elderly folks, rather than simply the extension of life. There is hope, of course, that stem cell research and other biotechnology innovations may aid in healthier lives for the elderly, but their promise remains in the future.

SPATIAL DISTRIBUTION OF THE ELDERLY

Though the elderly population in the United States has increased dramatically within the past century, these increases have not been evenly distributed within the country. The North Central states and the Northeast have larger proportions of the elderly than the West and South. States with smaller proportions of the elderly either have populations with relatively high fertility (South Carolina, Georgia,

Courtesy of Photo Disc

This infant has his whole future ahead of him. Present statistics indicate that 75 percent of newborn children are expected to reach age 65.

International Migration and Demographic Change (Continued)

have sought safe havens elsewhere, generally in different parts of their own country or in neighboring countries.

At the same time, worldwide demographic and economic changes are generating ever larger streams of international migration, most of which are focused on the developed countries of North America and the European Union. At least three important trends are converging as the millennium ends: (1) demographic—rich countries have slow-growing or even stagnant populations, whereas poor countries have been growing rapidly, (2) economic—the global economy (victory in the Cold War has been accepted by many as "proof" that unfettered capitalism must prevail) is bringing capital and labor together in unprecedented ways and amounts, and (3) communications—global networks (examples include CNN and the Internet) are tying the world together in such a way that nearly everyone recognizes the great disparities that exist between rich and poor nations (the richest 20 percent of the world's population generates about 85 percent of the world's GNP).

As a consequence of these converging trends, both capital and people are going to be on the move as never before. Capital, concentrated in the developed countries, will go in larger amounts to poor countries, but not fast enough to stem the flow of great numbers of people who will move to rich countries, where the capital is. Both of these trends will accelerate, as we are seeing from California to Belgium and Germany. In turn, these trends are disturbing the status quo, creating social and political backlashes that must be dealt with in rational ways if

Louisiana, New Mexico, and Utah) or have received large numbers of relatively young in-migrants from other areas of the country (Maryland, Colorado, and Nevada). States with over 12 percent of their populations classified as elderly include Arkansas, Nebraska, Iowa, South Dakota, Missouri, and Florida. Many similar states have lost a considerable number of young people due to out-migration, whereas others have an influx of retirees, which increases the proportion of the elderly. In both cases the age distribution becomes older. Nebraska and Florida are states that have similar proportions of the aged but for very different reasons. Bouvier, Atlee, and McVeigh (1975, 8) noted that in 1970, 14.6 percent of the population of Florida and 12.4 percent of Nebraska were 65 years or over. Between 1960 and 1970, however, the aged population increased by almost 80 percent in Florida but by only 12 percent in Nebraska. The high proportion of the elderly in Florida is explained by in-migration of the old; in Nebraska it is caused by the exodus of the young in addition to the longevity of the state's population (Figure 3-22).

Within America's metropolitan areas the distribution of the elderly is also of interest. Cowgill (1978), for example, discussed "gray ghettos" in many of the older core areas of major cities. Goodman (1987) used Lorenz curves and Gini coefficients to study the distribution of the elderly in three American cities: Baltimore, Philadelphia, and Pittsburgh. Though Pittsburgh had the largest elderly population, he found that the elderly were more concentrated in

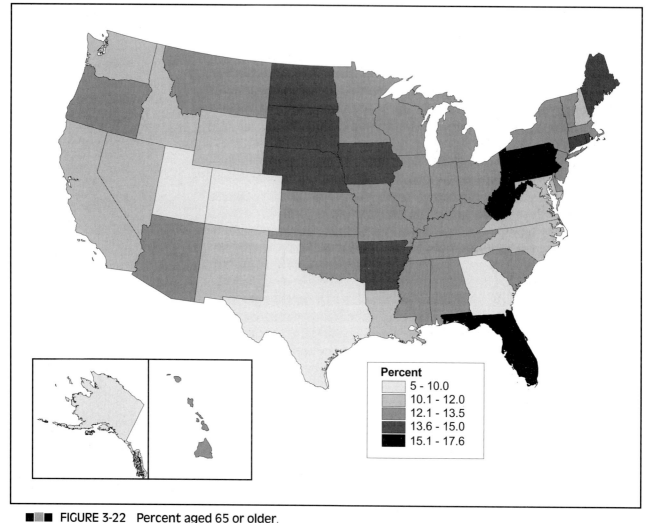

■□■ FIGURE 3-22   Percent aged 65 or older.

Source: U.S. Census Bureau, 2000 Decennial Census.

Percent

░	5 - 10.0
▒	10.1 - 12.0
▓	12.1 - 13.5
▓	13.6 - 15.0
■	15.1 - 17.6

International Migration and Demographic Change (Continued)

societies everywhere are to provide better lives for future generations.

International migration and the changing racial and ethnic compositions of individual nations are inextricably linked together. A few examples in the United States serve to focus attention on internal changes that cannot be separated from increasing flows of immigrants.

Since the 1930s immigration to the United States has been the central cities of Philadelphia and Baltimore than they were in Pittsburgh. They were most concentrated in Baltimore.

However, as Soldo (1980, 13) earlier noted, "This does not reflect intentional discrimination; rather, while their children have moved to the suburbs, the elderly have stayed behind." In small towns and rural areas the elderly are also often present in relatively high proportions for the same reasons; those who left such places are typically young. "Contrary to popular opinion," Soldo (1980, 13) went on to say, "relatively few Americans move during their retirement years." Most who do move do not move very far. Among the longer-distance movers the primary destinations are in the Sunbelt, especially in California, Arizona, and Florida. Even these migration streams are selective, however, because most of the elderly people involved are the "young" and more affluent elderly, those most able to make such moves (Soldo, 1980). General problems of an aging population are discussed in Peterson (1999).

RACE AND ETHNICITY IN THE UNITED STATES

Race and ethnicity are important aspects of population composition, especially in societies that are large, diverse, and undergoing differential demographic changes. Though the United States is hardly the only such country, it serves as a useful example. Geographers Allen and Turner (1988) provided the most ambitious look at ethnicity in America that has ever been undertaken. They noted (Allen and Turner, 1988, p. 205), importantly, that "Differences in values, occupational directions, lifestyles, religious identities, and patterns of socializing based on ethnic background may sometimes be obvious but are more likely to be subtle and hidden from public view . . . ethnicity has played a major role in shaping social structure, patterns of political and economic competition, religion, lifestyles, food preferences, and the processes of cultural change in America." Furthermore, such changes can be observed at a variety of scales, from the national to the local. Though here we treat examples only at the national level (observing variations among states), geographers have been interested in other scales (and other nations) as well; good examples are the study of the social geography of Canadian cities by Bourne and Ley (1993) and the study of rapid changes in a New York neighborhood by Alba (1995).

The United States has always been shaped by changing racial and ethnic patterns, and its future will be shaped by them as well. Though less detailed than the work of Allen and Turner, McKee (1985) also provided some useful geographic perspectives on ethnic patterns in modern America. Aside from these recent works on racial and ethnic groups in the United States by geographers, sociologists and others have much to offer as well. Some recent examples include Fuchs (1990), Gonzales (1990), Gonzales (1991), Lamphere (1992), Roberts (1993), Murdock (1995), Ingoldsby and Smith (1995), and Gonzalez, (2000).

Any discussion of ethnicity brings with it problems of definition and measurement, many of which are beyond the scope of this brief geographic survey. For example, Glazer (1983, p. 234) recognized many of the definitional problems and noted:

> The term 'ethnic' refers to a social group that consciously shares some aspects of a common culture and is defined primarily by descent. It is part of a family of terms of similar or related meaning, such as 'minority group,' 'race,' and 'nation,' and it is not often easy to make sharp distinctions between these terms. 'Race' of course refers to a group that is defined by common descent and has some typical physical characteristics. Where one decides that a 'race' ends and an 'ethnic group' begins is not easy.

The issue of racial identity became even more complex when the 2000 census for the first time allowed respondents to check more than one box when they identified their racial identity. About 7 million people, half of them under age 18, chose more than one racial category, leaving demographers and other social scientists with some new things to think about. As Gregory Rodriguez (2001, M1) noted,

International Migration and Demographic Change (Continued)

increasing steadily (Figure 8-6), and since the 1960s the origin of these immigrants has shifted increasingly from Europe to Asia and Latin America. Today there are vast differences in rates of growth for major racial and ethnic groups within the United States. Rapid growth in the Asian and Hispanic populations results from both increased immigration rates (including many refugees from Southeast Asia) and the increased fertility of many recent arrivals, especially among the Latino population. In California—the new Ellis Island and recipient of more immigrants since 1980 than any other state—for example, the total fertility rate for Latinas is slightly more than twice that for Anglo women.

At every level, from the national to the local, rapid changes in the racial and ethnic compositions of populations are occurring. For example, the composition of California's population has changed considerably in the past fifteen years. For the first time in recent decades Asians outnumber African Americans, though the two groups together are about 40 percent smaller than the Latino population. The state no longer has a majority population; Anglos are the largest minority still, but if current trends continue, then Latinos will become the largest minority in two or three decades (this already is the case in Los Angeles, both within the city and within the county). In the city of Long Beach, with a population of just over 400,000 according to the 1990 Census, 65,000 immigrants took up residence just during the decade of the 1980s.

Thus, at every geographic scale the impact of increased immigration is becoming apparent— the nature of places is being altered. Without doubt, growing numbers of immigrants are generating concern, both in North America and in the European Union. Less apparent, however, is how political units—from cities to nations—are going to cope with such rapid changes in population composition. Immigrant bashing has become more common, as have attacks on immigrants and even citizens who might look like immigrants—civil society cannot accept such acts.

Yet, here and elsewhere, the volume of both international migration and refugee movements will continue to increase in the decades ahead. Most countries, especially those in the developed world, are going to experience growing ethnic diversity. With luck, stronger international linkages will develop between immigrant nations and those that are providing large numbers of immigrants. Successful policy changes are going to have to be based on the assumption that more immigrants are going to be seeking permanent residence. Finally, social and political adjustments are likely to lag behind economic and demographic changes well into the next century.

"While most Americans still chose only one box last year, the government's official recognition of hybridity has not only muddled the statistical portrait of the nation, it has also undermined the popular belief that race and culture are immutable."

Between 1990 and 2000 high immigration rates and relatively high birth rates resulted in Latinos and Asians being the two fastest-growing racial/ethnic groups in the nation. Both of these groups tend to have high rates of intermarriage with other groups, including non-Hispanic whites, resulting in ever more blending of future generations. In fact intermarriage is on the increase for all groups in the United States. As Rodriguez (2001, M6) also noted, "The newly released census data will not so much resurrect the melting-pot concept as broaden it to include cross-racial and not only white-ethnic mixing."

Though distinctions can be made at many scales, the Census Bureau data and classifications are used for the discussions that follow, though we need to keep in mind that imperfections exist in both the data and the categories. Though we wish to focus on geographic aspects of ethnicity in this section, you should also keep in mind the following comment by Weeks (1992, p. 280) that "To be a member of a subordinate racial, ethnic, or religious group in any society is to be at jeopardy of impaired life chances."

Distributions of racial and ethnic groups are important and need to be studied at various scales, from national to intraurban, but space constraints allow us here to consider only the national distributions of the three largest such groups in the United States: African-Americans, Hispanics, and Asians and Pacific Islanders. The overall **racial and ethnic composition** of the United States in 2000 is shown in Figure 3-23; Jaffe (1992) has recently written a detailed account of the North American Indian population.

African-Americans

According to the 2000 Census there were almost 34.7 million African-Americans in the United States, though as a group they were certainly undercounted. They comprised 12.3 percent of the nation's population. Their share of the United States population has changed over time, from a high of 19.3 percent in 1790 to a low of 9.7 percent in 1930. The Hispanics have now surpassed African-Americans as the nation's largest minority. In all likelihood by the middle of this century Anglos will become the largest minority in a nation that will no longer have a majority racial or ethnic group.

Figure 3-24 shows the distribution of African-Americans in 1990. Slightly over half of all African-Americans, almost 53 percent, still live in the South, around 57 percent live in the central cities of metropolitan areas, and their struggle for equality in America has still not been fulfilled. As Weeks (1992, p. 280) noted, "Being of African origin in the United States is associated with higher probabilities of death, lower levels of education, lower levels of occupational status, lower incomes, and higher levels of marital disruption than for the white population."

Thomas Robert Malthus (1766-1834).

It was partially in response to such a utopian dream that Malthus wrote his famous essay on population.

MALTHUS

Perhaps more than any other population theorist, Thomas Robert Malthus (1766-1834), both an economist and a clergyman, has had a profound impact on attitudes and ideas concerning population growth. His name still appears in most discussions of population prospects and seems ineradicably associated with the topic of population. However, demographers do not all agree that Malthus made a significant contribution to the scientific study of population; at the very least he elevated it to a higher level as a topic for serious discussion (Petersen, 1999).

Malthus was born in England in 1766 into a rather well-to-do family and lived during a time of revolution, both political and industrial. For most of their history Europeans had been conditioned to expect, and to accept, a bitter lot and a short life. This fatalism was conditioned by a belief that there were not enough material resources to go around. The major economic thought of Malthus's day was Mercantilism, an economic system with strong government policies directed toward the accumulation of wealth. Poverty was associated with the merits of man and the will of God, and the earth was viewed as a place of testing and punishment. Mercantilist population policies were mainly pronatalist because more births meant more workers, hence more aggregate wealth. At the same time, as Wrigley (1989, 31) noted, Malthus and the two other great classical economists, Adam Smith and David Ricardo, were in complete agreement that, "The secular tendency of real wages was likely to be flat, if not tending downward, because any increase in the funds available to pay wages would be matched by a proportional rise in the number of wage earners."

However, during Malthus's time industrialization was accelerating and ideas were rapidly changing; the acceptance of deprivation was increasingly being challenged. Liberty, equality, and fraternity became the new watchwords. Rapidly increasing production and increased trade suggested that resources might not be so limited after all. Rather, abundance might be possible. Out of this background a rash of utopian schemes, wild dreams, and visions of the future were forthcoming. The future was seen as a time when poverty, misery, vice, want, greed, and even death itself might be eliminated. With a little bit of luck, and a few changes, the perfect society could evolve. Malthus's father was caught up in these new ideas concerning the perfectability of man and society. He invited many utopians to his home for discussions, but within these discussions the young Malthus took a position opposite to that of his father.

The first edition of Malthus's famous population essay, published in 1798, was entitled: *An Essay on the Principle of Population as it affects the future improvement of society; With remarks on the speculations of Mr. Godwin, M. Condorcet, and other writers*. It was written neither as a text in demography nor as an exposition of some new law of population growth. Rather, its intent was to refute some of the utopian ideas that were then gaining currency. His chief targets were the authors mentioned in his title. Commenting upon Condorcet, Malthus (1798, 3) wrote: "I have read some of the speculations on the perfectability of man and of society with great pleasure. I have been warmed and delighted with the enchanting pictures which they hold forth. I ardently wish for such happy improvements. . . ." Unfortunately, Malthus then argued, the road to utopia would always be blocked because he believed that man would always press up against the limit of subsistence.

Malthus's first edition of the *Essay* was rather simple. His "principle" of population was the result of two postulates and one assumption. The first postulate was that "food is necessary to the existence of man," and the second was that "the passion between the sexes is necessary and will remain in its present state." Malthus assumed that population would always tend to increase at a geometric rate, whereas food production could only be increased at an arithmetic rate. In his own words:

> Population, when unchecked, increases in geometric ratio. Subsistence increases only in an arithmetical ratio. A slight acquaintance with numbers will show the immensity of the first power in comparison of the second . . . the human species would

increase as the numbers 1, 2, 4, 8, 16, 32, 64, 128, 256, and subsistence as 1, 2, 3, 4, 5, 6, 7, 8, 9. In two centuries the population would be to the means of subsistence as 256 to 9; in three centuries as 4,096 to 13; and in two thousand years the difference would almost be incalculable.

According to **Malthus,** population growth would always press against the means of subsistence, unless it was prevented by some very powerful and obvious checks. These checks, according to Malthus, could be categorized as either *positive* or *preventive*.

Preventive checks were those that affected the birth rate, and in his view moral restraint was the only acceptable one, though he recognized that "vices" such as homosexuality and birth control could also have an impact. The positive checks were those that affected the death rate, including misery, disease, famine, and war. Thus, Malthus felt that there was no way to escape the positive checks and that Godwin's argument that things would become perfect if we could just get rid of government and private property was wrong. The "Principle of Population" was inflexible, inexorable, and inescapable. No exercise of reason could remove its effect, he argued, producing his own dismal view of the future of humankind and helping to saddle economists forever with the description of their discipline as the "dismal science."

From today's perspective a number of shortcomings can be identified in Malthus's thinking. He emphasized land as the major limiting variable on expanding food supplies, yet other inputs such as improved crop rotation patterns, fertilizers, and new hybrid strains of seeds have been, and continue to be, major factors in increasing food supplies. Furthermore, food is not our only necessity. Clothing, shelter, and other items are essential to at least some degree, and industrialization, well underway in his own era, certainly increased the availability of many of these items more rapidly than the rate of population growth. He also failed to consider the major changes that transportation systems would undergo, as well as the expansion of trade that would accompany such changes. Because of his clergical leaning, he also failed to foresee the possibility that contraceptives would be widely accepted, or to even imagine that couples would decide to limit the sizes of their families in response to changing socioeconomic conditions.

Though considerable discussion and controversy appeared after the publication of Malthus's book in 1798, his theory dropped from favor by the middle of the nineteenth century, as the Industrial Revolution found ways to reduce the pressure of population, including emigration to the New World, and to raise wages above the subsistence level for more workers. However, the rapid increase in population during the twentieth century, coupled with considerable malnourishment in many areas, has revived an interest in Malthus's ideas. Many "Neo-Malthusians" still feel that population growth will still outrun the food supply and that the world will not be able to

According to **Malthus,** population growth would always press against the means of subsistence, unless it was prevented by some very powerful and obvious checks.

■▩■

continue supporting a growing population. Strangeland (1904, 356) summarized Malthus's contribution as follows:

> Malthus's work was a great one, written in an opportune time, and though it cannot lay claim to any considerable originality as far as theories presented are concerned, it was successful in that it showed more fully, perhaps more clearly, and certainly more effectively than had any previous attempt, that population depends on subsistence and that its increase is checked by want, vice, and disease as well as by moral restraint or prudence.

The indelible mark that Malthus has left on population studies has been largely negative and pessimistic. Yet, as geographer Vaclav Smil (2000, p. xxvii) perceptively noted with respect to Malthus, ". . . the final message of that famous English cleric was not one of despair, but one of cautious hope." Smil (2000, p. xxvii) went on to say that "When a man, shortly before his death, deliberately changes the conclusion and the emphasis of his previously much-publicized work, then it is this final version, rather than the initial product, that should be seen as the intellectual bequest to be associated with his name." We agree, and appreciate Smil's calling attention to this. At the end of his 1803 book, Malthus wrote:

> On the whole, therefore, though our future prospects respecting the mitigation of the evils arising from the principle of population may not be so bright as we could wish, yet they are far from being entirely disheartening, and by no means preclude that gradual and progressive improvement in human society. . . . And although we cannot expect that the virtue and happiness of mankind will keep pace with the brilliant career of physical discovery; yet, if we are not wanting to ourselves, we may confidently indulge the hope that, to no unimportant extent, they will be influenced by its progress and will partake in its success.

Of course is was not this cautiously hopeful, and often ignored, conclusion that led to the rise of neoMalthusianism, but rather the more despairing earlier work of Malthus discussed above.

THE NEOMALTHUSIANS

Neomalthusians continue to argue that population growth is problematic. Not only do they argue, as Malthus did, that population growth creates problems relative to food supplies, but they go beyond that to relate population growth to environmental problems as well. They see the consequences of rapid population growth not just in mass poverty but also in the deteriorating quality of the earth itself as a home for humans. At the same time they go far beyond Malthus's "moral restraint" as a means of controlling population growth, supporting family planning, contraceptives of every sort, and even abortion.

Two major voices for neomalthusianism are biologists Paul Ehrlich and Garrett Hardin. Both generated considerable public interest in the problems of population growth in the late 1960s, and both

Neomalthusians see the consequences of rapid population growth not just in mass poverty but also in the deteriorating quality of the earth itself as a home for humans. At the same time they go far beyond Malthus's "moral restraint" as a means of controlling population growth, supporting family planning, contraceptives of every sort, and even abortion.

■ ■ ■

have pursued those themes since then. Ehrlich's 1968 book, *The Population Bomb*, set forth the central themes of rapid population growth, uncertain and inadequate food supplies, and environmental degradation. Ehrlich and Ehrlich (1990, 225) revisited the population issue and its relationship to environmental and other problems, and concluded that "A central problem facing us now is finding ways to convince national and international leaders and the world's people that opportunities for action to assure global environmental security are fast slipping away...it's the top of the ninth and humanity has been hitting nature hard...we must always remember that nature bats last!" The writings of Ehrlich, Hardin, and other neomalthusians are discussed in detail in Chapter 9.

BOSERUP

In 1965 Danish economist Ester Boserup proposed an argument about the relationship between population growth and food supply that was in many ways directly the opposite from Malthus's. Rather than arguing that population growth "depends" on agriculture, treating it as a dependent variable as Malthus did, she began her investigation by stating that, "Population growth is here regarded as the independent variable which in its turn is a major factor determining agricultural developments." (Boserup, 1965, 11) In essence, she argued that population growth and critical population densities could stimulate agricultural innovation and change. As a starting point, she suggested replacing the distinction between cultivated and uncultivated land with the concept of frequency of cropping. Given definitions of land use and patterns of tool adoptions in agrarian societies that she outlined, she was able "...to define the concept of intensification in agriculture in a new way, namely as the gradual change towards patterns of land use which make it possible to crop a given area of land more frequently than before." (Boserup, 1965, 43)

Faced with population growth in an agrarian society, people would be confronted with a threat to their standard of living. Whereas **Malthus** would argue that death rates would rise to curtail the population growth and restore equilibrium, **Boserup** would see another possibility, namely an intensification of the agricultural system. She argued that by working more hours and adopting more intensive ways of growing crops, people could cope with a growing population. The primary trade-off that they would be forced to make is one of leisure for work. She envisioned a succession of increasingly more intensive agricultural practices as population growth continued, ranging from forest-fallow to bush-fallow, through short-fallow and annual cropping, to multi-cropping, the most intensive of the five systems, and noted that "Under the pressure of increasing population there has been a shift in recent decades from more extensive to more intensive systems of land use in virtually every part of the underdeveloped regions." (Boserup, 1965, 16)

> Whereas **Malthus** would argue that death rates would rise to curtail the population growth and restore equilibrium, **Boserup** would see another possibility, namely an intensification of the agricultural system.

In a subsequent work Boserup (1981) broadened her arguments somewhat but continued to argue that population growth acted as a stimulant to technological change, at least under some conditions. However, she recognized also that high rates of population growth could overload systems. Agricultural intensification has continued to occur throughout the underdeveloped world, but in some cases, despite total growth in agricultural production, per capita growth has remained constant or even declined as populations grew at unprecedented rates. For example, such conditions have occurred in various parts of Africa in recent decades.

Though Boserup provided some compelling arguments for treating population growth as an independent variable that in turn drives agricultural intensity and technological innovation, she is not without her critics as well. One often made argument against her work focuses on population growth itself and asks why, if food was scarce to begin with, a population would begin to grow, especially since slow or no growth has been typical for most of human history. Another criticism is that she treats agrarian societies as closed systems, whereas migration may act as a safety valve for population growth in many cases, though that role has certainly become more limited with the rapid expansion of population in the twentieth century.

MARX

Foremost among Malthus's critics was **Karl Marx,** who argued that no such thing as overpopulation existed.

Down through the years several theorists have disagreed with the basic Malthusian ideas. Foremost among Malthus's critics was **Karl Marx,** who argued that no such thing as overpopulation existed. Marx believed that Malthus was a bourgeois, chauvinist clergyman who upheld the established order of social inequality. Furthermore, Malthus was accused of copying the writings and ideas of the classical school of economic theorists, who proposed unrestricted competition, self-interest, private property, and individualism as solutions to Europe's population explosion. Malthus was thus branded as an easy target for revolutionary criticism and an archenemy of the worldwide socialist movement, since he was considered a literary oppressor of the underprivileged and an apologist for the exploiting class.

In the *Communist Manifesto* and in other works, Marx expounded the familiar class struggle theory that an uprising by the proletariat would eventually lead to a classless society without private property, and that an egalitarian ethic would prevail. He believed that the output of food and resources was outstripped by population growth not because there was a natural law that subsistence could not keep pace with population growth, but because poor people, under the capitalist system, were denied access to the control of food and other resources. Furthermore, he argued, if poor people were given control over the means of subsistence—equipment, knowledge, land, and an adequate share of the wealth—their production of goods and services would far surpass the growth of population. Exploitation by the rich and private ownership of the means of production blocked this solution. In essence, Marx felt that there was no population problem; the problem was a maldistribution of resources.

Communists and socialists have generally agreed with the concept that it is not population growth but the maldistribution of resources that is the basis of the population problem. However, there is a basic difference between the socialists and the communists, as demographer Ralph Thomlinson (1976, 44) noted in his statement that "whereas communists prefer to deny the existence of population problems, socialists acknowledge the problems but turn to revision of the social order for a solution." Even in the few remaining communist countries today, however, birth control is often widely practiced. Of course, in the remaining communist countries, such as the People's Republic of China, the reasons given for birth planning are often improved health, better education for the next generation, and improved conditions for women, not direct control of population size (Jaffee and Oakley, 1978).

THE DEMOGRAPHIC TRANSITION: AN OVERVIEW

The original demographic transition theory, or model, was based on historical observations of demographic changes in western European countries. The gap between the birth rate and death rate in Europe was closed by a declining birth rate, not by a rising death rate as Malthus had predicted. A new kind of demographic stability was reached, first in northwestern Europe and then elsewhere. This transition from a relatively stable demographic regime with high birth rates and death rates to a similarly stable regime with low birth rates and death rates is known as the **demographic transition.**

The basic *demographic transition model* postulates a deterministic causal link between modernization on the one hand and both mortality and fertility reduction on the other. Conceptually it can be viewed as an idealized sequence of stages through which a given population passes, and the end result is a stable demographic regime with low birth and death rates. The data upon which the transition model is based came primarily from the demographic experience of northwestern Europe, principally England and Wales. Figure 4-1 illustrates the changes in vital rates that have taken place in England since 1700 and provides the idealized view of this descriptive model.

Prior to its industrialization in the eighteenth century, England generally experienced high birth and death rates, though not as high as those currently being experienced in many Third World countries. The result was a relatively stable population in which births were approximately balanced by deaths. A decisive break occurred around the middle of the eighteenth century, when the death rate began a long-term decline. The reasons for the initial decline are still not completely understood, though most experts agree that it was probably due to unusually good harvests at the time. The good harvests were

This transition from a relatively stable demographic regime with high birth rates and death rates to a similarly stable regime with low birth rates and death rates is known as the **demographic transition.**

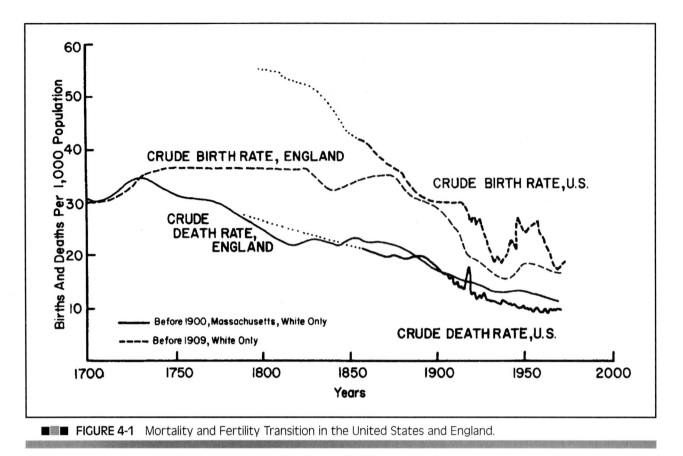

■■■ FIGURE 4-1   Mortality and Fertility Transition in the United States and England.

Source: Abdel R. Omran, "Epidemiologic Transition in the United States," *Population Bulletin*, Vol. 32, No. 2 (May, 1977), p. 15. Courtesy of the Population Reference Bureau, Inc., Washington, D.C.

a result of such agricultural improvements as crop rotation and advances in animal husbandry. There was also a slight rise in the death rate in the early nineteenth century, probably associated with the growth of factory towns and their squalid, unhealthy living conditions. However, after a few decades the death rate resumed its decline, which has continued until today.

In England following 1700 the birth rate initially increased slightly as a result of the Industrial Revolution. With industrialization and the disappearance of the apprentice system, unskilled young people could get factory jobs and did not have to endure long apprenticeships before they received compensation. Thus, young people could marry earlier and start families at younger ages. Improved living conditions may also have added slightly to the birth rate—healthier mothers would be more likely to experience full-term pregnancies that resulted in proportionately more live births and produced healthier babies.

Fertility in England stayed relatively constant for the next one hundred years, then started its major decline around 1870. The primary reason for the decline was not a change in marriage patterns, but a redefinition of the ideal family size within marriage. Several reasons for this decline in ideal family size have been suggested. First, the contribution of children to family welfare was declining. In the preindustrial period children contributed a great deal to family welfare. Because the overwhelming majority of people were involved in agricultural pursuits, children were very useful in weeding crops, fetching water, and gathering wood. In the early textile factories and mines children served an important economic function because they worked at a much cheaper rate than adults. Children were also a form of social security; they were often the only means of support for their parents when the parents became unable to work. However, as more restrictions were put on child labor practices in the middle and late nineteenth century, and as populations generally became more urbanized, children became more of a burden to parents. The state began to take on a welfare function and to provide for older citizens. With increasing industrialization the number of people engaged in farming dropped significantly and the role of children as producers was further diminished.

A second reason for the redefinition of ideal family size had to do with rising expectations. The Industrial Revolution produced a tremendous volume of goods and services, so people could increasingly shift from satisfying needs to satisfying wants. Luxuries sometimes became viewed as necessities; social mobility and the attainment of riches were easier if one had fewer children. With the growth of knowledge came more favorable attitudes toward family planning, resulting in a significant and sustained decline in the birth rate.

Though Figure 4-1 illustrates the demographic experience of England, general trends appeared to be similar for most of the Western world, with the exception of France, where fertility declines began much earlier, especially within some social classes (van de Walle and Muhsam, 1995). The timing of the transition varied from place to place, but the results were essentially the same, at least when viewed at the national scale. The new balance of vital rates represented an improved condition of human efficiency and health, as well as a level of material well-being that had never before been achieved.

In the typical experience of demographic transition the decline in birth rates lagged behind the decline in death rates and occurred after industrialization was well along. Obviously the demographic transition was not foreseen by Malthus, who believed that a population equilibrium would be reached by a rise in the death rate rather than a lowering of the birth rate.

The general pattern of the demographic transition can be associated with different areas of the world today. Figure 4-2 represents an idealized demographic transition model. Stage A indicates an area with a high birth rate and a high death rate, equivalent to the preindustrial era in Europe. No major world region remains in this stage, though a few African countries, such as those of the Sahel, probably still do. Stage B represents a population with a high birth rate and a declining death rate. Many of the African nations are now in this stage. Parts of Asia, excluding at least Japan, China, Hong Kong, and Taiwan, may also fit into this category. Stage C includes nations with

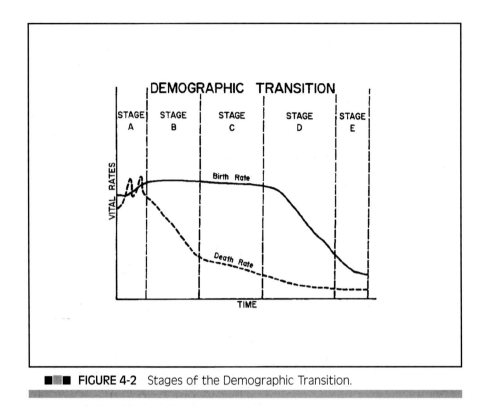

DEMOGRAPHIC TRANSITION

STAGE A | STAGE B | STAGE C | STAGE D | STAGE E

VITAL RATES

Birth Rate

Death Rate

TIME

■■■ **FIGURE 4-2** Stages of the Demographic Transition.

high birth rates and low death rates, such as many of those in tropical Latin America and some African nations, where populations are growing at some of the fastest rates in the world. Kenya, for example, had a crude birth rate of 45 and a crude death rate of 12 in 1995, yielding an annual population growth rate of 3.3 percent. Countries in stage D have declining birth rates and low death rates. Examples include countries in temperate Latin America, such as Argentina and Uruguay, as well as the People's Republic of China, Taiwan, and South Korea. In the decades following 1970 birth rate declines in developing countries became more common, somewhat slowing the world's overall growth rate. The final category, Stage E, depicts a population with low birth and death rates, and includes the countries of the Western world and Japan.

If we accept the demographic transition model and believe that there is a causal link between modernization and a decline in both vital rates, then the obvious solution to rapid population growth is to modernize the world as rapidly as possible. However, if there is progressive step-wise movement from Stage A to Stage E, then the speed at which nations move through these stages is critical. The nations in Stage D are heading in the right direction and in a short time should reach an equilibrium between births and deaths. Though the nations in Stage C are experiencing a rapid increase in population, they are going in the right direction and should soon experience a decline in birth rates and a concomitant decrease in population growth. The critical nations are those found in Stages A and B. They have yet to reach their greatest rates of demographic increase and still must pass through Stage C, the period of most rapid population growth. Africa and Asia contain the largest populations, and if they follow the European experience then they will experience a tremendous increase in numbers. However, the People's Republic of China appears to be experiencing a rapid decline in fertility. As early as 1982 the Chinese were officially encouraging the one-child family, a drastic demographic measure (and one with serious consequences, including female infanticide and distorted population pyramids).

Clearly the demographic transition model is not a simple one which, once stated, is applicable to all places or all times. As Beaver (1975, 9) commented:

> In general, the theory of the demographic transition is not well explicated, but is weighted down with numerous supplementary arguments drawn from different disciplines. The only clear implication of this theory is the prediction that socioeconomic development will lead to natality decline, and that this will tend to occur sometime after a major decline in mortality.

THE RELEVANCE OF THE TRANSITION MODEL

Though the demographic transition model has received considerable attention in recent years, its relevance seems increasingly in doubt. Whether or not the transition model is useful in explaining the population changes that are occurring in the developing countries is an important issue. Conditions in today's developing countries differ considerably from those that prevailed in the Western nations as they moved through the demographic transition. Policy formulation in developing countries often depends on the concepts of the transition model, and to the degree that these concepts are validated they will be useful in guiding policy development. However, questions need to be raised about the degree to which the developing countries will duplicate the experience of the developed countries upon which the demographic transition model is based. Some time ago Teitelbaum (1975, 422) noted that:

> The situation is a mixed one; in some respects the different circumstances in developing countries suggest great obstacles to the timely completion of the transition by means of a "natural" decline in fertility along the European pattern. In other respects these differences provide reasons to anticipate an unusually rapid completion of the transition in these countries.

It is important to see that differences between the European experience and the likely experience of today's developing countries may both favor and operate against rapid movement through the demographic transition. Some of the differences and their possible influences on demographic change in the developing countries are discussed below.

POPULATION GROWTH

As we have already seen, population growth in the developing countries is occurring at unprecedented rates. Africa is the world's fastest growing major region with a 2.4 percent rate of natural increase in 2000. By comparison, when most European countries were undergoing their demographic transitions very few countries experienced annual growth rates of more than 1.5 percent.

One of the obvious problems associated with rapid population growth is the population-related growth in demand for goods and services and the growth in demand for social investment, especially in education. Whereas the European countries were not too much affected by these difficulties, most of the developing countries today face severe problems and find that population growth increases the difficulty of raising the levels of living (Kindleberger and Herrick, 1977).

To make matters even worse, it probably will be much more difficult to halt rapid growth in the developing countries than it was to slow down the growth in European countries during their demographic transition. The younger age structures in the modern developing countries, the result of much higher fertility, add a great momentum for further growth, a compelling *demographic inertia*. Even if the fertility rate in the developing countries declined to replacement level within the next decade, a highly unlikely event, they would still continue to grow for another 60 years or more and would nearly double in size. For a country such as India, even a doubling of its current population (nearly one billion) would present considerable problems.

MORTALITY DECLINES

One of the major reasons for rapid population growth in the developing countries is the rapid decline in mortality that has occurred as modern medicines and public health programs have been introduced. The European countries experienced a gradual decline in mortality that was closely related to the economic and social forces of industrialization and economic development. In many developing countries today mortality declines have been much more rapid. As a consequence, the developing countries have mortality levels that are considerably below those that prevailed in Europe during early stages of industrialization. Further mortality declines seem likely, especially in Africa; however, such declines may be slowed (or even reversed) as a result of the widespread effects of AIDS on that continent and increasing social instability (such as was seen in Rwanda).

FERTILITY LEVELS

The fertility levels of most of the developing countries today are higher than those that prevailed in Europe before the European countries began their demographic transitions. Many African countries had crude birth rates of 40 or more in 2000, for example. By comparison, the crude birth rate in early nineteenth century Britain was estimated to be around 35. The major reason for the difference is marriage patterns. Extensive nonmarriage and late marriage were typical in nineteenth century Europe, whereas the practice in most developing countries today is early and nearly universal marriage.

Although high birth rates prevail in the developing countries as a result of the prevailing marriage pattern, coupled with low rates of contraceptive usage, hopeful signs may be seen as well. For one thing, it may be possible to reduce fertility in the developing countries today by altering the marriage patterns as well as by introducing more effective contraceptive practices. Part of the recent fertility declines in the People's Republic of China have resulted from such efforts. Raising the legal marriage age decreases the number of years females are exposed to the possibility of conception, for example.

Of course, motivation toward smaller families is going to be necessary before significant declines in fertility are recorded in most of the developing countries. The "demonstration effect" of the European transition and the predominance of small families in Europe and the United States has recently diffused to most of the developing world as a result of expanded trade and better communication systems.

The role of governments and various international agencies may also aid in bringing about rapid fertility declines. Family planning programs are most likely to be effective if they are supported by the government. Governments today are more likely to be involved in many forms of planning than were their nineteenth-century counterparts. Furthermore, many developing countries today have social scientists and planners who actively participate in the formulation and execution of national policies for economic development and social change. Such planning was almost nonexistent in the European countries during their periods of demographic transition and industrialization. Also, financial and technical assistance for both development and family planning programs is available from such organizations as the United Nations and the United States Agency for International Development.

Encouraging signs of fertility declines in many Third World countries have been noted in the 1990s. Family planning programs, communications about family planning, and the widespread availability of modern, effective contraceptives are having significant effects throughout the Third World. According to Robey, et al (1993):

> The developing world is undergoing a reproductive revolution....Contrary to the expectations of many observers, developing nations are not experiencing the classical demographic transition that took place in many industrialized countries over the past century...recent evidence suggests that birth rates in the developing world have fallen even in the absence of improved living conditions. The decrease has also proceeded with remarkable speed.

MIGRATION

Migration undoubtedly played a significant role in the stabilization of population in parts of nineteenth-century Europe. International migration operated as a "safety valve" and softened the effect of rapid population growth. Millions of European citizens moved to Oceania and the Americas during the nineteenth century. However, this potential "safety valve" for releasing population pressure is no longer available for large numbers of people because of current economic and political realities. During the European demographic transition the growing rural population was able to find opportunities for gainful employment in urban areas as industry rapidly expanded. New skills and occupations were acquired as people moved from rural to urban areas. In the developing countries today, however, industrialization is unable to create jobs in the cities fast enough to absorb the rapidly growing labor force—thus encouraging fertility to remain high. In some countries attempts are being made to slow down the rate of rural-urban migration.

EDUCATION AND ECONOMIC DEVELOPMENT

Economic development and modernization for many developing countries has proceeded more rapidly than in nineteenth-century Europe. To the extent that a direct correlation exists between modernization and fertility decline, it suggests a possibility for a more rapid fertility decline in the developing countries and a quicker completion of the demographic transition.

However, education also has an impact on fertility; in general, lower fertility is associated with higher educational levels (this is especially true for females, who are at a considerable disadvantage for educational opportunities throughout much of the Third World). Rapid population growth in the developing countries, together with the large proportions of people in the younger age groups, means a rapidly increasing demand for educational facilities. The developing countries usually cannot keep up with this demand, and the result is most likely to be deferment of the goal of universal education along with its potential effect of lowering fertility.

Of course, care must be taken in trying to forecast the demographic consequences of modernization. Caldwell (1976, 358) commented that:

> The major implication of this analysis is that fertility decline in the Third World is not dependent on the spread of industrialization or even on the rate of economic development. It will of course be affected by such development in that modernization produces more money for schools, newspapers, and so on; indeed, the whole question of family nucleation cannot arise in the nonmonetized economy. But fertility decline is more likely to precede industrialization and to help bring it about than to follow it.

Courtesy of Photo Disc

Since females from third world countries are at considerable disadvantage for education opportunities, third world countries are at a disadvantage when it comes to lowering fertility rates.

We are still a long way from understanding the specific interactions between socioeconomic and demographic changes. Undoubtedly other differences between the transition experience of the European countries and the likely course of demographic events in the developing countries could also be identified. Changes in the composition of the labor force in the developing countries would have a downward effect on fertility if more women found occupations outside the home, for example. Once females have opportunities for fulfillment outside of the traditional role of housewife and mother, they will be more motivated to have smaller families. Once females have outside incomes, a part of the cost of having children will be the income foregone (opportunity cost) during and following pregnancy. This opportunity cost may be enough to discourage having as many children.

SOME FINAL COMMENTS

The demographic transition model has been in a respectable position for many years. Recently its validity has been increasingly questioned, especially with respect to its applicability to developing countries. From around the middle of the eighteenth century onward in Europe, marital fertility underwent a sustained decline;

it was mainly associated with the modernization that was sweeping the area. At a general level the transition model appears compatible with the events that occurred. However, when causal links are established in a model and that model is then used to predict the course of fertility decline in today's countries, its success may be quite limited. Most likely the course of events will differ significantly from the earlier patterns.

As Wilson (2001, pp. 67-68) commented, "If we consider demographic transition in the light of broader modernization theory, it is clear that social and demographic change has progressed far more rapidly than economic development. . . . while huge economic gaps remain between rich and poor countries, we are moving into a world in which that distinction is of diminishing demographic relevance."

Furthermore, the validity of the demographic transition model for explaining the demographic changes that occurred in Europe during the transition has been questioned as well. Freedman (1979, 64) wrote that, "detailed empirical work has been unable to establish combinations of development variables at specific levels which were systematically related to the European fertility declines." In some countries fertility declines began in less advanced regions, rather than in the more advanced ones. Also, Freedman (1979, 64) noted that "detailed study of the European fertility transition has shown that many areas that were culturally similar, for example, in language or ethnicity, also demonstrated similar fertility patterns, without prime reference to socioeconomic developmental indices critical to transition theory." Though it appears that spatial and temporal variations in fertility decline were probably more complicated than we have believed, the new evidence does not necessarily invalidate the demographic transition model. It does suggest, however, that we need to look more closely than ever at the model, especially with respect to its predictive ability. We cannot specify neat causal relationships. For example, we cannot state that a 10 percent rise in a nation's level of urbanization will result in a corresponding specified decrease in fertility.

Recent studies of the demographic transition in Europe have highlighted another problem, one associated with the way that fertility is usually measured. The typical way to chart the transition is to look at crude birth and death rates. However, as van de Walle and Knodel (1980, 15) pointed out, "France and Ireland around 1900 had almost the same crude birth rate." But, they went on to say, "this result was obtained in France by family limitation (within marriage) and in Ireland by high proportions of single persons in the reproductive ages." Thus, historical demographers are now decomposing overall fertility into the various components that comprise it. The major component, of course, is marital fertility, for which an index has been developed. Related data for selected countries are shown in Table 4-1.

TABLE 4-1. Index Numbers of Marital Fertility: Selected European Countries, 1850-1910

Country	\ Date \ 1850	1860	1870	1880	1890	1900	1910
Belgium	109	109		100	89	71	59
Denmark	99	94	96	100	97	89	77
England-Wales	100	99	102	100	92	82	69
Germany			103	100	96	90	73
Ireland			99	100	101	100	101
Italy		105	100	100	99	98	95
Netherlands	100	98	102	100	97	90	78
Norway			93	100	98	93	90
Scotland		101	103	100	95	86	—
Sweden	96	101	101	100	96	91	80
Switzerland		107	102	100	96	91	76

Source: Etienne van de Walle and John Knodel. "Europe's Fertility Transition. New Evidence and Lessons for Today's Developing World." *Population Bulletin* 34, No. 6 (February, 1980), p. 18.

After reviewing several studies of European fertility decline, van de Walle and Knodel (1980, 20-21) provided the following useful summary:

1. The past was largely characterized by natural fertility, that is, the deliberate practice of family limitation was mostly absent (and probably unknown) among most of the population prior to a fairly recent time. This was true, even though a substantial proportion of births may have been unwanted. High infant and child mortality may have reflected unconscious efforts to eliminate unwanted children.

2. The transition from high to low fertility represented a shift from natural fertility to family limitation, occurred rapidly, and was an irreversible process once underway.

3. The onset of long-term fertility decline was remarkably concentrated in time and took place under a wide variety of socioeconomic and demographic conditions.

4. Differences in the start and in the speed of the fertility decline seems to have been determined more by the cultural setting than by socioeconomic conditions.

Before leaving the subject of demographic change, two more theories need to be discussed. These theories enhance our understanding of past demographic changes and suggest some additional factors to consider when projecting future demographic change.

During the 1980s considerable attention was given to refining and reinterpreting the European experience of the demographic transition. One shift in focus has been from the role of individual choice in fertility decline to that played by relatives, friends, and neighbors in influencing local fertility behavior (Coale and Watkins, 1986). Watkins (1990, 242) argued the need for paying more attention to the role played by "...members of the community with whom individuals interact on a day-to-day basis, as well as the members of...'imagined communities'...I assume that in the end it is individuals who act in the privacy of their bedroom; I propose, however, that even when the couple is literally alone in the bedroom, the echoes of conversations with kin and neighbors influence their actions."

Watkins (1990) focused on the diversity of marriage and marital fertility both within and among the countries of western Europe during two different time periods, 1870 and 1960, and found that between the two time periods diversity increased to a maximum and then decreased to below what it had been in 1870. She also discovered that variations in fertility during the former period were closely related to linguistic diversity. She found that national social integration ultimately diminished demographic diversity within nations, though at the same time demographic diversity among nations became more pronounced. She identified the following variables as most important in reducing within-nation demographic diversity:

1. the integration of national markets

2. the expansion of state functions

3. nation-building

Within western Europe today fertility rates are the lowest in the world; they are close to (or even below) replacement level in most European nations. Interestingly, in nations such as Italy and Spain fertility levels have dropped so low that a few demographers are starting to talk about a "second" demographic transition, one that leads to sustained negative population growth rates. Whether such trends materialize or not, time will tell. It is clear, however, that young European women (with a few exceptions, such as the Irish) are not currently inclined toward having more than one or two children. Young Japanese women are also opting for very low reproductive rates.

MULTIPHASIC RESPONSE

As we have seen, countries that have achieved a reasonable modernization level have also experienced a decline in birth rates and death rates; and the decline in birth rates, with the possible exception of France, has lagged behind the decrease in death rates. What is not made explicit is the causal mechanism, the direct linkage between modernization and the declines in birth and death rates.

Modernization is a complex process involving sweeping changes in the socioeconomic fabric of societies as they are transformed from primarily rural-agricultural societies to primarily urban-industrial societies. Somewhere in this complex of changes there were reasons for the observed declines in vital rates, but their specification is elusive. The search for better explanations continues, even as new patterns may be emerging in the Third World.

Demographer Kingsley Davis argued that the demographic transition model was perhaps too simplistic, mainly because it considered only changes in vital (birth and death) rates. In so doing, the transition model ignores other possible demographic responses that a society could make when confronted with population growth. Davis (1963, 345) stated that "The process of demographic change and response is not only continuous but also reflexive and behavioral—reflexive in the sense that a change in one component is eventually altered by the change it has induced in other components; behavioral in the sense that the process involves human decisions in the pursuit of goals with varying means and conditions." Thus, the subject of demographic change is complex. Davis attempted to incorporate the complexities in an analysis of demographic change in the industrialized, or developed, countries.

He entered the subject with a discussion of demographic change in Japan, the only non-Western country so far to industrialize. One of the major demographic responses to modernization in Japan was abortion. Both abortion rates and birth rates are well documented and appear in Table 4-2. As Davis argued, abortions have also been common in many other areas of the world, but their numbers have not been so well documented. Other demographic responses in Japan included the increased use of contraceptives, sterilization, emigration, and postponement of marriage. For example, in 1920, 17.7 percent of females aged 15-19 were married. By 1955, however, this figure had dropped to 1.8 percent. Of the Japanese situation Davis (1963, 349) stated:

> It is the picture of a people responding in almost every demographic manner then known to some powerful stimulus. Within a brief period they quickly postponed marriage, embraced contraception, began sterilization, utilized abortions, and migrated outward. It was a determined multiphasic response, and it was extremely effective with respect to fertility.

The stimulus to this **multiphasic response** was not population pressure and the prevalence or threat of dire poverty; Japanese industrial output grew more rapidly between 1913 and 1958 than did that of Germany, Italy, and the United States. Rather, according to Davis (1963, 352), the stimulus was "in a sense the rising prosperity itself, viewed from the standpoint of the individual's desire to get ahead and appear respectable, that forced a modification of his reproductive behavior."

Davis went on to further test these ideas in agricultural areas of northwestern Europe. With sustained population growth in these areas it was observed that families tended to remain large. Birth rates were not decreasing. However, another response was occurring; migration. High birth rates could be sustained only if there were sufficient outflows from the population. Presently, however, with international migration restricted as it is, emigration is unlikely to solve the pressures generated by population growth.

		Annual Totals		Sum per 1,000
Year	Births	(000's) Abortions	Sum	Population
1949	2,697	102	2,798	34.4
1950	2,338	320	2,658	32.1
1951	2,138	459	2,596	30.8
1952	2,005	798	2,803	32.8
1953	1,868	1,067	2,935	33.9
1954	1,770	1,143	2,913	33.1
1955	1,727	1,170	2,897	32.6
1956	1,665	1,159	2,825	31.4
1957	1,563	1,122	2,686	29.6
1958	1,653	1,128	2,781	30.4
1959	1,626	1,099	2,725	29.5

■■■ **TABLE 4-2.** Births and Abortions in Japan

Source: Kingsley Davis, "The Theory of Change and Response in Modern Demographic History," *Population Index,* No. 4 (October, 1963), p. 347. Used by permission.

REVOLUTIONARY CHANGE: THE CASE OF CUBA

Cuba has received more attention from policymakers, the mass media, and scholars than most developing countries. Although it only ranks eighth in size among Latin American nations, it is the most populous nation in the Caribbean. In area, it is about the size of Pennsylvania, and its 2000 population of 11.1 million is nearly as large as that of Pennsylvania.

The importance of Cuba, and the reason for its receiving so much attention, is clearly not because of its size; it is due to its unique history of social and economic change. Cuba was one of the last American nations to sever ties with Spain, and it had a long and close relationship with the United States. When the revolutionary movement led by Fidel Castro overthrew the government of Fulgencio Batista in 1959, it became the first socialist nation in the Western Hemisphere, only 83 miles from Florida.

The social and economic revolution has had a profound impact on the socioeconomic fabric of the nation. Along with, and partly because of, this change in the social and economic order, the past few decades have brought about significant demographic changes. Among today's developing countries, Cuba probably has one of the lowest fertility rates (below replacement level) and one of the highest life expectancies.

Demographic changes in Cuba have been significantly different from those in most developing countries. It underwent its demographic transition before World War II, a time when most other developing countries had not started their transitions (Collver, 1965). The death rate started to decline in the early part of the twentieth century and birth rates started to fall in the 1920s. Migration also played an

■ ■ ■ TABLE 4-3. Population Growth in Cuba: 1899-2000

Census Year	Enumerated Population	Growth Between Censuses (Percent)	Net International Migration
1899	1,572,797[a]	33.9	127,257(1900-1909)
1907	2,048,980	29.1	233,535(1910-1919)
1919	2,889,004	26.6	268,062(1920-1929)
1931	3,962,344	15.9	-147,963(1930-1944)
1943	4,778,583	21.1	-21,920(1945/49-1953)
1953	5,829,029	22.1	
1970	8,569,121		-582,742(1959-1978)
1979[b]	9,772,855		
1986	10,221,000		
2000	11,100,000		

Source: Sergio Diaz-Briquets and Lisandro Perez. (1981) "Cuba: The Demography of Revolution," *Population Bulletin*, 36(1):5 and *World Population Data Sheet 2000.*

important role in Cuba's demographic development, with a large influx of African slaves and Chinese indentured servants in the nineteenth century, along with hundreds of thousands of Spaniards and other Europeans. As illustrated in Table 4-3, and quite unlike other developing countries, Cuba experienced a decline in its population growth rate from the beginning of the century to the immediate post-World War II period, and then, as with other developing countries, growth accelerated because of a notable decline in mortality.

MORTALITY CHANGES

Demographic trends in mortality prior to the revolution were primarily shaped by economic circumstances. The introduction of sanitary reforms during the United States' occupation of 1899-1902 and a vigorous economy fueled by large-scale foreign investments in the sugar industry during the first quarter of the twentieth century brought about a significant decline in mortality (Diaz-Briquets, 1977). With the depression of the 1920s and 1930s, and its disastrous consequences for the Cuban economy, mortality rates stabilized. With the advent of modern drugs and insecticides following World War II, Cuba's mortality rates, like those of most other developing countries, experienced an accelerated decline.

On the eve of the communist revolution in 1958, Cuba was already one of the most demographically advanced developing nations, with a life expectancy of over 60 years. Among developing countries of the Western Hemisphere, this life expectancy was surpassed only by those of Jamaica, Puerto Rico, Argentina, and Uruguay. Since the revolution, an enormous effort has been put into the development of medical facilities, and currently, virtually free or very low-cost medical coverage is available across Cuba through a system of polyclinics and rural, regional, provincial, and national hospitals (Danielson, 1979).

As the result of the development of medical programs, Cuba has now achieved excellent health standards with life expectancy at 75 years (2000), nearly equal to that of the United States. Infant mortality was down to 7.0 per thousand live births by 2000, compared to 7.0 per thousand in the United States. Cuba's crude death rate in 2000 was 7, below that of the United States, which was 9. This figure for Cuba reflects both the young age of the population and significant health improvements. As in developed countries, the leading causes of death have shifted primarily from infectious to degenerative diseases.

FERTILITY CHANGES

Information about pre-revolution fertility is scarce, but the limited data suggests a decline in births during the depression years of the 1920s and the 1930s. This was the result of delayed marriages and more widespread adoption of abortion as a fertility limitation method (Gonzalez et al, 1978). Following World War II the crude birth rate stabilized in the low 30s and by the eve of the revolution it had dropped to the mid- to upper-20s.

Though fertility has fluctuated since the revolution, the general trend has been significantly downward (Figure 4-3). A few years before the revolution the birth rate was around 26 per thousand. With the advent of the revolution in 1959 the birth rate rose from 26.1 in 1958 to 35.1 in 1963. This was the highest birth rate in the post-revolutionary period and similar to the level of the 1920s. After 1963 the birth rate fell, with only a brief rise in 1971, a rise attributed to the disruption of normal activities in 1970 in an all-out effort to produce 10 million tons of sugar cane. Thousands of workers were moved from the city to rural areas to help with the harvesting, and the rise in 1971 might have been a reflection of the births averted in 1970 during the harvest. Also, the late 1960s and early 1970s saw a marked increase in marriages.

A dramatic decline occurred in the birth rate from 1973 to 1979, a decline of 40 percent. Overall, there was a 58 percent decline, 20.3 points in the birth rate, between 1963 and 1979; in 2000 the crude birth rate was 14. Cuba's 2000 rate of natural increase, 0.7 percent, equals that for the United States and makes Cuba's rate among the lowest in the developing world.

Though the age structure of the Cuban population tends to exaggerate this fertility decline, as depicted in the crude birth rate, other fertility measures also point toward a significant decline. Between 1970 and 1978 the total fertility rate dropped by nearly half, from 3.7 to an estimated 1.9 births per woman; by 2000 it was down to 1.6, well below replacement level.

The increase in births in the early 1960s can be directly attributed to the social, political, and economic changes brought about by the

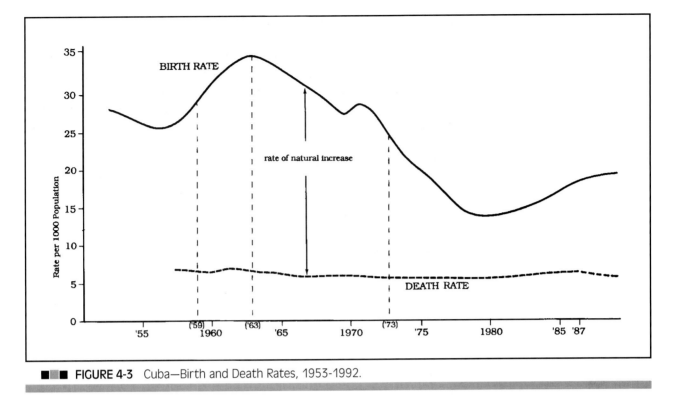

■■■ FIGURE 4-3 Cuba—Birth and Death Rates, 1953-1992.

Source: Sergio Diaz-Briquets and Lisandro Perez. (1981) "Cuba: The Demography of Revolution," *Population Bulletin* 36(1):13 and *1992 World Population Data Sheet.*

revolution. The primary factor behind the rise was probably the growth in real income among the poorer groups, who then felt that the future was more promising and that they could afford more children. At the same time, marriage rates went up and the average age at marriage declined. There was also a shortage of fertility limitation options—partly the result of the economic blockade of Cuba by the United States—and the disruption of contraceptive supplies as well as governmental restrictions on abortion.

An analysis of this birth rate surge shows distinct geographical differences (Figure 4-4) within Cuba. Birth rates increased most in urbanized provinces such as Havana, where the rate went up almost 60 percent between 1958 and 1963. In the least urbanized or modernized provinces, like Oriente, the rate rose only 17 percent.

The causes of fertility decline since 1963 are many and varied. The proximate determinants of the decline—changes in trends of abortion, marriage, divorce, and contraception use—are relatively easy to trace. Restrictions on abortion were eased in 1964 as part of changes in health laws. Since that time free hospital abortions have become available on request to women aged 18 and over during the first ten weeks of pregnancy. Later abortions are also available with a physician's permission. There has been a drop in abortion rates since Cuba's high in 1974 of 69.5 abortions per 1000 women aged 15-44. The drop suggests an increase in the use of contraceptives.

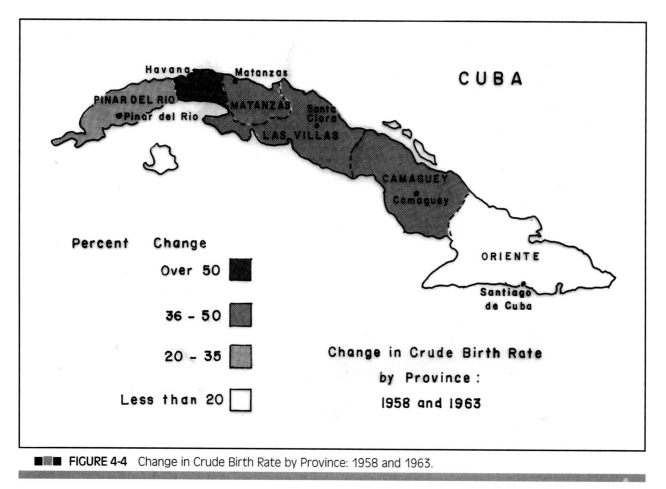

FIGURE 4-4 Change in Crude Birth Rate by Province: 1958 and 1963.

Source: Compiled by authors from statistical data published by the Republic of Cuba.

Though national statistics on contraceptive use are not available, evidence reviewed by Hollerbach (1980) suggested that contraceptive use has been widespread. The IUD is the most prevalent form of contraception, with oral contraceptives second in usage.

The marriage rate has fluctuated since the revolution, the rate of 6.2 per thousand population in 1978 was considerably less than the 10.8 per thousand marriage rate in the United States. This decline could have contributed to the fertility decline, as could the fivefold increase in divorce rates since the revolution. Many socioeconomic changes since the revolution have also played a key role in fertility decline. One thing is certain: the decline was not the result of official antinatalist policies. The government's ideological position is that overpopulation is not one of the causes of poverty. The prevailing view is that changes brought about by the revolution have eroded societal norms favoring childbearing. A good summary of this viewpoint is presented in Hollerbach (1980, 100):

> The decline in fertility, especially rapid since 1973, has not been achieved through antinatalist policies (such as those of China), nor through the creation of demographic targets, which are characteristics of policies in some developing nations. Rather a vari-

ety of economic and political factors are responsible, the most significant of which have been increased educational levels . . . adult education programs and expanded enrollment in higher education; the urbanization of rural areas through the concentration of social services and development projects there; construction of small urban communities, and reduction of disparities between urban and rural income levels; and more recently, the incorporation of women into the labor force.

Hollerbach's main emphasis, as with those of other demographic analyses of Cuba, is that modernization has been primarily responsible for these demographic changes. Cuba's experience, therefore, has valuable lessons to be exported to other developing nations throughout the world (Harrison, 1980). Other demographic analysts, however, believe that this view—that Cuba's population has been affected uniformly and almost exclusively by modernization sparked by the revolution—needs to be complemented by looking at other factors. They believe that modernization explains the demographic changes for the most disadvantaged groups in Cuba, but that economic setbacks, particularly in recent years, have also had an impact. Their central argument is that different sectors of Cuban society have limited their fertility for different, though overlapping, reasons. According to a study by Diaz-Briquets and Perez (1981, 22):

> Cuba's fertility decline since the mid-1960s has been a response to difficult economic conditions as well as to the undoubted progress made in many social areas. This more comprehensive explanation makes it questionable that poor, high-fertility countries around the world might draw a lesson from Cuba's experience, as has often been asserted. The political, historical, and social context that produced the Cuban revolution is unique in many ways, and the fertility response appears to be just as unique.

CONSEQUENCES OF FERTILITY CHANGES

Unlike other developing countries, Cuba experienced a baby boom and bust similar to those in many developed nations such as Canada and the United States. The large bulge in Cuba's population pyramid (Figure 4-5) illustrates the large number of people in the 10-24 year age group. The 10-19 age group bulge is a result of the baby boom in the 1960s and the marked decline in infant mortality rates at that time. The narrow base of the pyramid reflects the more recent fertility decline.

Like other countries that have experienced large changes in fertility rates, Cuba has had, and will continue to have, problems trying to adjust to the varying sizes in age cohorts. Cuba has had problems similar to those in the United States. Perez (1977) has pointed out the problems of suddenly needing to accommodate large numbers of children in the school system and of having large numbers of people attempting to enter the labor force at a time when job prospects are dim. Reports of increases in Cuba's crime rate can be associated with

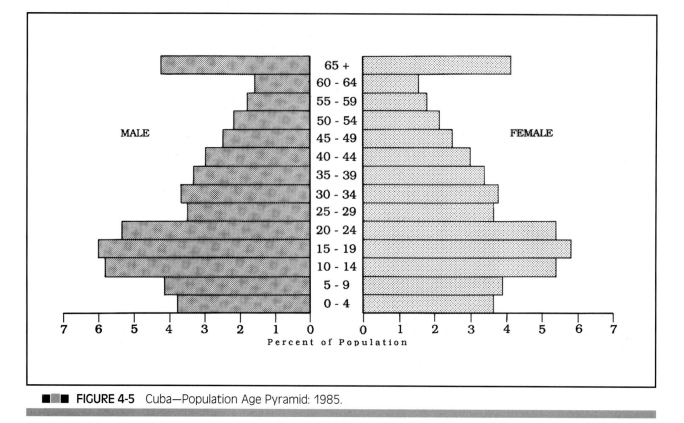

■■■ **FIGURE 4-5** Cuba—Population Age Pyramid: 1985.

Source: Compiled by authors from statistical data published by the Republic of Cuba.

age structure. An increase in people 15-24 years of age is probably part of the reason for the rise in crime because, like in many other countries, crime rates are highest in this age group (Salas, 1979, 195).

Further evidence of pressures from the changing age structure and its effect on Cuba's economy was apparent in the 1980 Mariel sealift. Though there is not conclusive evidence of government orchestration, emigration from the port of El Mariel to the United States certainly was of benefit to the Cuban economy: "As if by magic, thousands of housing units became available, unemployment pressures were somewhat reduced, and many young people, among whom the crime rates were highest, left the country." (Diaz-Briquets and Perez, 1981, 24)

EMIGRATION

Migration has always played an important role in Cuba's population dynamics. During the nineteenth century large influxes of African slaves, Chinese indentured servants, Spaniards, and other Europeans affected Cuba. During the first three decades of the twentieth century, migrants from Spain predominated, though there were also significant numbers of migrants from Haiti, Jamaica, and other Latin American countries. Migration dwindled as an important factor by

■▣■ TABLE 4-4. Cuban Migration to the United States, 1959-1990

Year	Number
1959 (Jan. 1-June 30)	26,527
Year Ending June 30	
1960	60,224
1961	49,961
1962	78,611
1963	42,929
1964	15,616
1965	16,447
1966	46,688
1967	51,147
1968	55,945
1969	52,625
1970	49,545
1971	50,001
1972	23,977
1973	12,579
1974	13,670
1975	8,488
1976	4,515
Year Ending September 30	
1977	4,548
1978	4,108
1979	2,644
1980	122,061
Total, January 1, 1959-September 30, 1980	793,856
Total April 1-December 31, 1980	125,118
1981-1984	38,600
1985	20,300
1990	9,400

Source: United States Department of Justice, Immigration and Naturalization Service (INS). (1980) "Cubans Arrived in the United States, by Class of Admission: January 1, 1959-September 30, 1980," October 1980, mimeo; and INS, Statistics Branch and *Statistical Abstract of the U.S., 1991.*

the early 1930s, then took on added importance again in the early 1960s. Between 1959 and 1980 approximately 800,000 people left Cuba for the United States (Table 4-4).

The most recent and largest wave of migrants came to the United States in the so-called Mariel sealift of April-September 1980. This emigration since 1960 from socialist Cuba is a reversal of the immigration pattern established in the nineteenth and early twentieth centuries. This most recent migration of Cubans to the United States has had a considerable impact on both Cuba and specific areas of the United States (Bach, 1980, 40), especially Florida. Continued migration to the United States is certainly possible, and will depend on conditions in both Cuba and the United States.

CHINA: THE WORLD'S FIRST DEMOGRAPHIC BILLIONAIRE

At the end of the thirteenth century, before the exploration of the New World had begun, the population of all of Europe was estimated at 75 million people. At the same time, the population of the Sung Dynasty numbered 100 million. In the next six centuries the population increased, reaching about 540 million in 1949.

Soon after this half-billion milestone was reached, noted China scholar John King Fairbank observed the following:

> . . . the Chinese people's basic problem of livelihood is readily visible from the air: the brown eroded hills, the flood plains of muddy rivers, the crowded green fields. . . all the overcrowding of too many people upon too little land, and the attendant exhaustion of the land resources and of human ingenuity and fortitude in the effort to maintain life. (cited in Jacobson, 1991, 265)

At about the same time that Fairbank observed this overcrowding, the revolutionary leader of China, Mao Zedong, had other thoughts. In the early years following the Revolution, official policy more-or-less followed Marxist ideology. Marx argued that in the socialist state there would be no unemployment and no overpopulation. He believed that overpopulation was a function of the type of social and economic organization, that overpopulation was a relative, rather than an absolute, situation.

However, following the 1953 census, which recorded a population of 583 million, there was an attempt to limit population growth. This early campaign was slow to get underway and was disrupted in 1958 by the Great Leap Forward. Following the Great Leap Forward, in 1962 the Chinese Government again began to promote family planning, but it lapsed into unpopularity again during the Cultural Revolution from 1966 to 1969. Since 1971 China has become what the demographer Tien (1983, 3) calls "a born-again advocate of population control."

In the early seventies, the Chinese government recognized the implications of population growth. The state family planning agency established a policy known as *wan-xi-shao*, which meant "later-longer-fewer." This policy encouraged Chinese couples to marry later, preferably later than the average age of 20, to increase the time between births, and to have fewer children overall. Another slogan associated with this family planning program was, "One is not too few, two is good, and three is too many." Targets were set by the government for both family size and age at marriage. Recognizing the economic and cultural differences between the rural and urban areas, rural targets were to have no more than three children after delaying marriage until age 25 for men and 23 for women. For urban dwellers only two children were encouraged after delaying marriage until age 28 for men and 25 for women.

In 1977 the government, after realizing that population growth would not fall as quickly as desired, began to promote the two-child family throughout the country. In 1979 a new slogan was adopted. It said, "One is best, at most two, never a third."

THE ONE-CHILD FAMILY

In 1980 China launched a crash program to stabilize population growth by limiting families to only one child. Then-Communist Party Chairman Hua Guofeng emphasized that coercion was not permissible, but that the country would rely on publicity and persuasion. Hua spoke out at the same time against overpopulation and bureaucratic obstacles to modernization.

Though some variation exists from place to place, a few specific family-planning regulations were well established early in the 1980s. Students and apprentices were not allowed to marry. Incentives were offered to couples that have only one child, whereas disincentives were used to discourage larger families. Incentives were also provided for sterilization after the birth of the first child. A monthly allowance was paid to couples that had only one child, and that child's medical and educational fees were waived. However, if a second child was born, a couple gave up all privileges accorded the one-child family and must repay all cash awards that were received. The third child is denied free education, subsidized food, and housing privileges. That child's parents were penalized by a 10 percent reduction in wages.

China's family planning program, sometimes called its Strategic Demographic Initiative (SDI), has a stated goal of keeping the total population "within 1.2 billion." This was established in the late 1970s when China had a total fertility rate of around 2.3. In order to meet this objective the one-child policy would require full implementation.

In 1981 China's Family Planning Commission was given cabinet status and a variety of scholars, including statisticians, system analysts, economists, and others, were called on to develop and set population policies (Song, et al, 1985). The impact of China's SDI has been substantial. According to Tien (1988, 7), "At the time of the program's start China's birth rate was nearly 34 and its total fertility rate exceeded 6 (the current total fertility rate is about 2.4)...the successes of China's initiative have been, and continue to be, impressive."

This one-child family initiative was an unprecedented attempt to change the reproductive behavior of an entire nation. The success of this program in lowering the rate, according to Jacobson (1991, 282), can be attributed to a variety of aspects of Chinese society:

> The population is nearly homogeneous with a 93 percent Han majority. The closed nature of the political system, the Confucian tradition and the strong sense of family it fostered, and a 3,000 year history of allegiance to authority made the one-child family program succeed where it might otherwise have failed.

■■■ TABLE 4-5. China Population Size: Birth, Death, and Natural Increase Rates; and Total Fertility Rates, 1970-1988

Year	Year-End Population (10,000)	Crude Birth Rate	Crude Death Rate	Natural-Increase Rate	Total Fertility Rate		
					Country	Urban	Rural
1970	82,992	33.43	7.60	25.83	5.81	3.27	6.38
1971	85,299	30.65	7.32	23.33	5.44	2.88	6.01
1972	87,177	29.77	7.61	22.16	4.98	2.64	5.50
1973	89,211	27.93	7.04	20.89	4.54	2.39	5.01
1974	90,859	24.82	7.34	17.48	4.17	1.98	4.64
1975	92,420	23.01	7.32	15.69	3.57	1.78	3.95
1976	93,717	19.91	7.25	12.66	3.24	1.61	3.58
1977	94,974	18.93	6.87	12.06	2.84	1.57	3.12
1978	96,259	18.25	6.25	12.00	2.72	1.55	2.97
1979	97,542	17.82	6.21	11.61	2.75	1.37	3.05
1980	98,705	18.21	6.34	11.87	2.24	1.15	2.48
1981	100,072	20.91	6.36	14.55	2.63	1.39	2.91
1982	101,590	21.09	6.60	14.49	2.65[†]	1.73*	n.a.[†]
1983	102,764	18.62	7.08	11.54	2.08[†]	1.61*	n.a.[†]
1984	103,876	17.50	6.69	10.81	2.03[†]	1.46*	n.a.[†]
1985	105,044	17.80	6.57	11.23	2.04[†]	1.23*	n.a.[†]
1986	106,529	20.77	6.69	14.08	2.44[†]	1.43*	n.a.[†]
1987	108,073	21.04	6.65	14.39	2.84[†]	1.38*	n.a.[†]
1988	-109,660	20.78	6.58	14.20	n.a.[†]	n.a.[†]	n.a.[†]

Source: From *Annals of the American Academy of Political and Social Science* by Greenhalgh, 510, July 1990, p. 75. Copyright © 1990 by Sage Publications, Inc. Reprinted by permission of Sage Publications, Inc.

Another important variable, according to Greenhalgh (1990, 80), was that the transition to socialism in the 1950s made a fundamental change in the relations between the state and society, thus state-dominated institutions could influence even the most private of decisions.

The one-child family policy has been criticized in a variety of ways. First, the compulsory nature of the program, at least in some areas, has been criticized and may, as some believe, produce a backlash like that which took place in India. Second, with fewer children, the next generation of elderly will have fewer laborers to support it, and third, most of the immediate economic benefits will not be realized in rural areas.

RURAL-URBAN DIFFERENCES

China's family planning policies were most successful in urban areas. While the total fertility rate (Table 4-5) declined precipitously in both urban and rural areas between 1970 and 1988, urban fertility fell a decade earlier and reached a lower level measuring 1.1 - 1.4 in the early 1980s, well below the level of 2.5-3.0 in the rural areas at the same time. In 1985 the "In-Depth Fertility Survey" showed that "Whereas the vast majority of both urban and rural couples wanted

Courtesy of Photo Disc

In an attempt to control population growth, China offers incentives to couples who only have one child.

two children, 80-90 percent of urban couples pledged not to have a second child, while fewer than 20 percent of rural couples signed the one-child pledge (Feng, 1989)."

There are several reasons for this disparity. Rural reforms in the early 1980s, which dismantled the collective farms, also reestablished the family as the principal unit of rural life and freed many farmers from the enforcement of the one-child policy by rural political cadres. In 1984 physical coercion was prohibited and could no longer be used by rural birth-planning cadres. With decollectivization, labor benefits of children also increased. Most rural dwellers are not included in a state-funded pension system and therefore have to rely on their children, especially their sons, to take care of them in old age.

This preference for sons among a large sector of the Chinese population is a major problem in the acceptance of one-child families in rural areas. Sons are perceived as more important because they can do heavy farm work and are providers of social security. Boys get preferential treatment in health care, food rations, and schooling. Some scholars have suggested that higher than normal sex ratios among reported live births in China is due to female infanticide or sex selective abortions, although a recent study by Johansson and Nygren (1991, 50) concluded that ". . . large numbers of adopted children are females and the statistical evidence cannot determine the extent of female infanticide." In the mid-1990s, however, numerous female infants in orphanages were found in starving condition, being left to wither and die.

RECENT BIRTH RATE CHANGES

Recent figures have indicated that the crude birth rate has risen in the past few years. After declining in the early 1980s and reaching a low in 1984, China's crude birth rate increased to 18 in the mid-1990s. According to the State Statistical Bureau (1988), China's crude birth rate increased by 20 percent between 1984 and 1987. Sixteen of the 30 provinces, municipalities, and autonomous regions experienced an increase exceeding 10 percent between 1981 and 1987, while 26 experienced an increase of over 10 percent between 1984 and 1987. A variety of explanations have been offered for this increase including relaxation of the one-child policy in rural areas, the weakening of local administrator's authority relative to family planning concerns and economic incentives to have more children (Feeney, et al., 1989 and Zeng, 1989).

The crude birth rate is a joint function of three variables—the population age distribution, marriage patterns, and age-specific marital fertility. In a recent study, Zeng, et al (1991) decomposed the recent changes in China's crude birth rate into these three components in order to assess and compare their contributions to China's recent increase in its crude birth rate. This decomposition of the crude birth rate into its three components was done for the country as a whole and for each province. Zeng, et al. (1991, 441) concluded that:

> . . . the effect of the much larger cohorts born in China's second baby boom of the 1960's reaching childbearing ages was to increase the crude birth rate by 1.59 births per thousand between 1981 and 1987. The effect of the declining age at marriage was to increase the crude birth rate by 2.59 births per thousand. Marital fertility in 1987 was significantly lower than in 1981 and its impact reduced the 1987 crude birth rate by 3.01 births per thousand.

The decrease in age at marriage and the changing age structure were the primary factors associated with the rise in the crude birth rate. China's population growth is not "out of control" as some have suggested and the recent increase in the crude birth rate does not imply a significant change in marital fertility.

THE 1990 CENSUS

On July 1, 1990 the Chinese government conducted a nationwide census. According to the State Statistical Bureau (1990), the total population for the People's Republic of China was 1,133,782,501 people; this figure did not include Hong Kong, Macao, or Taiwan. This was the fourth census taken since the coming to power in 1949 of the communist government and most experts believe it was their most accurate census. Future plans are to conduct a census every ten years. This total population figure was larger than had been estimated, probably because of previous underreporting, particularly in rural areas where increasing numbers of couples with one female child were allowed to have a second child (Tien, 1990).

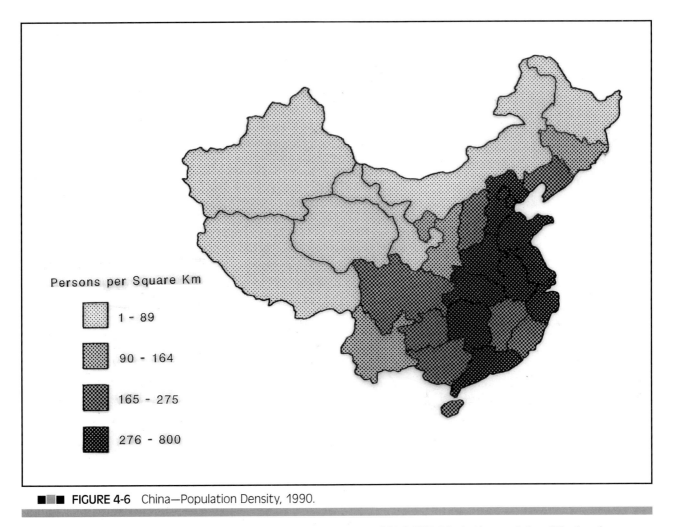

FIGURE 4-6 China—Population Density, 1990.

Source: From GEOGRAPHIC REVIEW, July 1991 by Pannel and Torguson. Copyright © 1991. Adapted by permission of the American Geographical Society.

A number of interesting spatial patterns and variations were revealed by the census. Most of the Chinese people live in the eastern half of the country, where more intensive agriculture is practiced due to more favorable climatic conditions and better land resources (Figure 4-6). Although there has been migration to the interior and far west regions, they remain relatively sparsely populated. A doubling of the population since the 1950s has significantly increased the density in the eastern part of the country.

Ethnic diversity is also found in China, although the vast majority of people (92 percent) are of the Han nationality. There are fifty-five recognized minority nationalities that accounted for approximately 8 percent of the population in 1990. This figure is up from 6.7 percent in 1982. A more liberal government policy on family size for these minority groups is probably responsible for this increase. Most of these minority peoples are found in the sparsely populated interior regions that border foreign countries. Conflicts with the Han majority have taken place, with tensions the highest in Tibet due to the ruthless repression and occupation of that country by the Chinese in the 1950s.

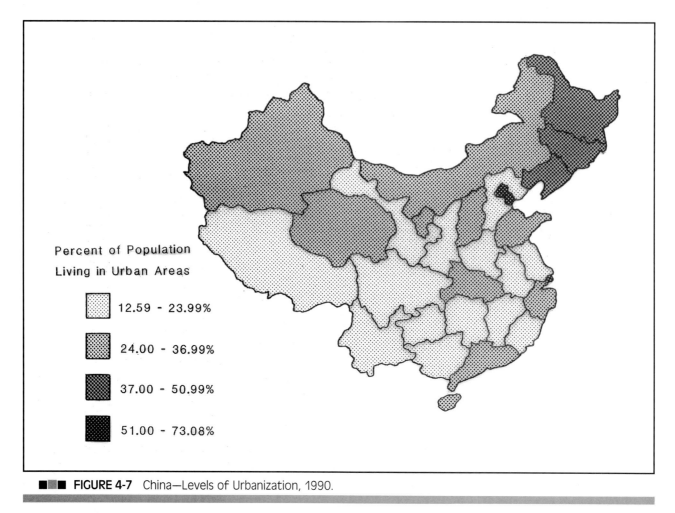

■■■ FIGURE 4-7 China—Levels of Urbanization, 1990.

Source: From GEOGRAPHIC REVIEW, July 1991 by Pannel and Torguson. Copyright © 1991. Adapted by permission of the American Geographical Society.

As part of the 1990 census, Chinese officials developed a realistic definition and index of the urban population. The urban population in 1990 was 26.2 percent of the total population, an increase of almost 6 percent from the 20.6 percent level in 1982; today it is 36 percent and rising.

Dynamic changes have occurred in Chinese cities in the past decade. China has three of the world's largest cities; Shanghai, with a population of over thirteen million, Beijing, with more than ten million, and Tianjin, with approximately nine million residents. Internal migration of redundant rural workers to neighboring urban centers has dramatically increased. This migrant population has increased the total urban population in several of China's largest cities by more than 10 percent (Goldstein and Goldstein, 1991). As a result of this migration, the cities of China are now beginning to have urban problems similar to that of other third-world countries.

The regional pattern of urbanization (Figure 4-7) indicates that the highest levels of urbanization are concentrated in the coastal and

Courtesy of Photo Disc

By glancing at this crowded street in Stockholm, Sweden, you would never guess that Sweden is among the few nations in the world that has almost stabilized its rate of population growth.

northeastern regions. According to Pannell and Torguson (1991, 315), "This pattern reflects the high degree of industrial activity and the comparatively well-established transportation networks in the regions in the aftermath of twentieth-century development."

THE 2000 CENSUS

Though most data from the 2000 census will not be released until at least 2002, China's National Bureau of Statistics reported some information in 2001. According to the 2000 census, China's population for that year was 1.266 billion—an increase of about 132 million over the 1990 census. The new census also showed an urban population of 446 million (about 36 percent of the population); an average family size of 3.44 (down from 3.96 in 1990); and a population of ethnic minorities that comprised about 8.4 percent of the population, up slightly from 1990.

Though Chinese censuses in the past have been considered quite accurate, demographers are concerned that the 2000 census may have been less accurate than previous ones, including 1990. Possible errors are thought to have occurred because of an undercount of children and because of the growing mobility of the Chinese population. More details about these errors and other problems with the census can be found in Kennedy (2001).

HEADING TOWARD ZPG: THE CASE OF SWEDEN

Sweden is among the few nations in the world that have almost stabilized population growth; in the near future it may face zero population growth (ZPG). Sweden's land area is approximately the same as California's and its low density of 52 persons per square mile is similar to that of the United States' 79 persons per square mile. Of course, Sweden's population is not evenly distributed across the country; it is primarily concentrated in the southern half. Most of the population, 83 percent, live in "densely populated areas" of at least 200 persons. The population of 8.9 million (2000) is quite homogeneous in its ethnic and social character; there are only a small number in ethnic minority populations (Gendell, 1980).

Sweden is the envy of many countries because of its high standard of living; few slums and little poverty are in evidence. It provides a wide range of social, educational, and health services and has one of the lowest infant mortality rates in the world. Sweden is one of a handful of countries, primarily located in Europe, where the number of births is almost equal to the number of deaths. It was a pioneer in the worldwide family planning movement and the first developed country to provide family planning aid to the developing nations of the world.

DEMOGRAPHIC CHANGES

Sweden and Finland were the first countries to collect national statistics on population, and they have been doing it continuously since 1749. In the middle of the eighteenth century Sweden's population was slightly under 1.8 million, and for the next two hundred years it grew at an average rate of 7 per thousand or 0.7 percent per year. In 1850 the population was 3.5 million, and it doubled again to 7.0 million by 1950. Since then the growth rate has varied only slightly, hovering around 0.1 percent in the 1990s.

Trends in birth, death, and migration rates are illustrated in Figure 4-8. Reliable figures are available for birth and death rates since the middle of the eighteenth century; information on migration rates, however, has only been available since the middle of the nineteenth century. During the second half of the eighteenth century, Sweden's crude birth rate was relatively high. It averaged 34 during that time, a rate similar to that found in many of today's developing countries. The death rate during that time was only 6 points lower.

The first half of the nineteenth century witnessed the beginning of Sweden's long-term demographic transition from high to low birth and death rates. The death rate declined steadily throughout the nineteenth century, with only periodic increases because of epidemics. The gap between births and deaths (demographic gap) began to widen during this period and accelerated to its peak in the latter half of the nineteenth century. Like many other European

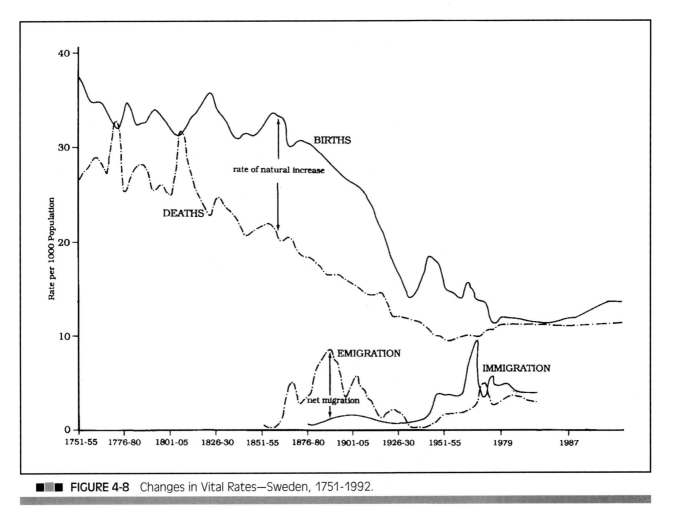

■■■ **FIGURE 4-8** Changes in Vital Rates—Sweden, 1751-1992.

Source: Compiled by authors from various statistical sources including selected United Nations *Demographic Yearbooks* and *World Population Data Sheets* published by the Population Reference Bureau. Reprinted by permission.

countries, Sweden lost large numbers of people to America, particularly during the 1880s and 1890s, thus relieving population pressures at home. The latter half of the nineteenth century also observed the inception of Sweden's sustained decline in fertility (Hofsten and Lundstrom, 1976).

Since the debut of the twentieth century, the principal features of Sweden's demographic character have been a continuing decline in both birth and death rates, as well as a marked shift since World War II from a nation of emigration to a nation of immigration. During the first three decades of the twentieth century, Sweden's birth rate dropped to a low of about 14 in 1933 and 1934, during the economic depression. The birth rate climbed during the 1940s and again in the early 1960s. But since the 1965 level of almost 16, the birth rate has fallen almost uninterruptedly to an all-time low of 10 in 2000.

The death rate declined throughout the twentieth century to a low of around 10 in the 1950s and 1960s. There has been a slight increase in the death rate beginning the 1970s because there has been an

■■■ TABLE 4-6. Life Expectancy at Birth 1851-2000 Sweden

| Year | Life Expectancy at Birth (Years) | | Sex Differential |
	Male	Female	(Female Minus Male)
1851-60	40.5	44.4	3.9
1901-10	54.5	57.0	2.5
1941-50	68.1	70.7	2.6
1951-60	70.9	74.1	3.2
1961-70	71.7	76.1	4.4
1971-75	72.1	77.7	5.6
1976	72.1	77.9	5.8
1977	72.4	78.5	6.1
1978	72.4	78.6	6.2
1984	73.8	79.9	6.1
2000	76.0	81.0	5.0

Source: Sweden: National Central Bureau of Statistics. (1980) *Population Changes 1978,* Part 3 (Stockholm), Table 4.17, p. 120; International Population Division, United Nations, United States: National Center for Health Statistics, *Vital Statistics of the United States,* Vol. II, Section 5, *Life Tables* (Hyattsville, MD: 1980); U.N. Demographic Yearbook 1985; and *World Population Data Sheet 2000.*

increasing proportion of elderly in the population—the rate was steady at 11 in 2000. Age-specific death rates, however, continued to decline, with concomitant increases in life expectancy (Table 4-6). Much of this gain in life expectancy was a result of very low infant and child mortality rates. Sweden has a long history of concern for infants and children, and the American demographer Tomasson noted that:

> . . . medical care, like education, has been regarded as both a communal responsibility and an individual right in Sweden [since] long before the advent of the modern welfare state. . . . Prenatal care, delivery, and postnatal care have been free and readily available throughout Sweden for many decades. Sustained professional care of pregnant women and later of their babies is universal (Tomasson, 1976, 970).

Sweden's negative rate of natural increase, around –0.1 percent, puts it very close to zero population growth. Immigration began to outweigh emigration during the 1930s, but substantial differences appeared after World War II, as Sweden's economy expanded and brought about a need for more workers. From 1943 through 1978 nearly 50 percent of Sweden's population growth resulted either directly or indirectly from net immigration (Gendell, 1980, 8).

COPING WITH AN AGING POPULATION AND ZPG

An analysis of a series of population pyramids (Figure 4-9) graphically shows the changes that have occurred in the Swedish population in the twentieth century. The pyramid for 1900 has the characteristic shape of a nation with a youthful population and a

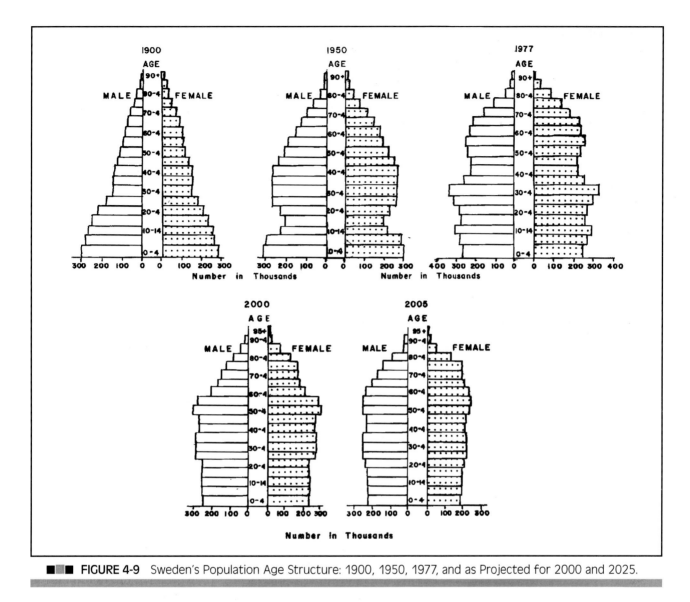

■ ■ ■ FIGURE 4-9 Sweden's Population Age Structure: 1900, 1950, 1977, and as Projected for 2000 and 2025.

moderately high birth rate. The slightly smaller number of people in the 30-40 age group is a reflection of the mass emigration of young people from Sweden during the 1880s and 1890s. The late 1950 pyramid shows the many changes that occurred in the first half of the twentieth century. The indentation in the 10-24 age groups is an indication of the drop in the birth rate during the depression years, and the bulge at the base reflects the rising birth rates in the post-World War II period. The relatively long period of low fertility can be seen in the rectangular shape of the 1977 pyramid.

The pyramids for the years 2000 and 2025 are based on population projections that assume a total fertility rate of 1.85 by 1985 and constant thereafter, with no net immigration. These projections could change slightly, but it is almost certain that there will be a continued equalization of age cohorts and an increase in the average age of the population.

The rising average age in Sweden means that there will be an increasing proportion of the population that will be elderly. The proportion of people 65 and over nearly doubled between 1900 and 1981, from 8.4 percent to 16.0 percent, compared with 11 percent for the United States. At the same time, the proportion of children under 15 declined by 12 percent.

These large numbers of elderly people are causing what the Swedish demographer Holmberg calls a "crisis" in finding enough medical staff and facilities to provide the long-term care needed by the very old (Holmberg, 1977). The share of Sweden's national budget that is devoted to the elderly is escalating rapidly and has much to do with its growing budget deficit (Downie, 1980). At the same time, the Swedish economy has run into problems, resulting in higher unemployment rates, which also burden the national budget. As Sweden approaches ZPG, discussions of population issues and their impact on Swedish society are becoming more numerous. Some people contend that population decline must be stopped, because the decline and associated aging of the population brings about economic stagnation and increasing inequalities in income distribution.

This argument is countered by those who believe that Sweden's economic difficulties, as well as the crisis in that society's ability to take care of the old and sick, has nothing to do with fertility decline. They feel that these problems can be solved by organizational changes, because ". . . a sharp increase in the birth rate would lead to difficulties for the educational system and for new entries to the labor market when the large number of children born grow up." (Holmberg, 1980, 4)

Sweden faces several problems as a result of demographic changes, but, fortunately, the importance of these problems has been recognized by much of the population and by many decision makers. How Sweden responds to the challenges ahead will make it a test case worth watching.

Mortality Patterns and Trends

Of the three basic population processes, **mortality** has in many ways been the easiest to begin with. Death still occurs only once (though modern medical science has been able to retrieve some from the precipice), and it is usually clearly defined, though recent medical technologies have complicated the picture in many ways. For example, physicians use terms such as "brain dead" to describe people who are still able to breath with the help of a respirator. Moral and legal issues surrounding the definition of death have drawn more attention in the last decade or so, as organ transplants have become commonplace (a central issue is whether or not to artificially keep someone alive in order to preserve a vital organ until it can be transplanted).

MEASURES OF MORTALITY

The crude death rate was introduced earlier and was defined as the annual number of deaths per 1,000 people in the midyear population. The rate is considered to be "crude" because the total population is included in the denominator, whereas the probability of dying in a particular time period is not equal for everyone. Thus, more refined mortality measures attempt to correct for the age structure of the population.

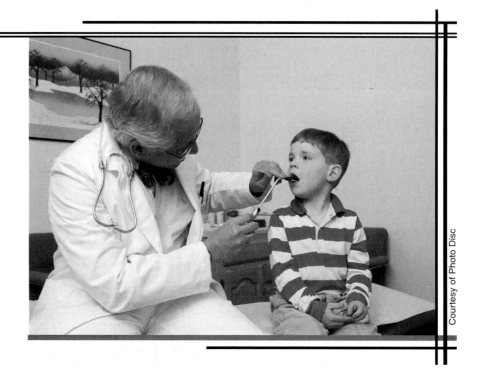

Having escaped the higher death rates of infancy, this youngster can look forward to a long life expectancy. Nonetheless, his chances will ultimately be affected by such things as socioeconomic class and access to further health care.

AGE-SPECIFIC DEATH RATE

The probability of dying within a given time interval is closely related to age. The **age-specific death** rate, then, takes age into consideration and may be defined as

$$ASDR = \frac{D_a}{P_a} \times 1000,$$

where D_a = number of deaths of people in the age group a, usually either a one- or five-year age group, and

P_a = midyear population in age group a.

Generally, death rates are lowest for adolescents and young adults. The typical **J-shaped mortality curve** is shown, separately for males and females, in Figure 5-1.

INFANT MORTALITY RATE

Infant mortality is unevenly distributed throughout the first year of life, with most infant deaths occurring in the first six months.

■■■

The **infant mortality** rate is defined as

$$IMR = \frac{D}{B} \times 1000.$$

where D = annual number of deaths of infants between birth and age one year, and
B = annual number of births.

Infant mortality is unevenly distributed throughout the first year of life, with most infant deaths occurring in the first six months. In

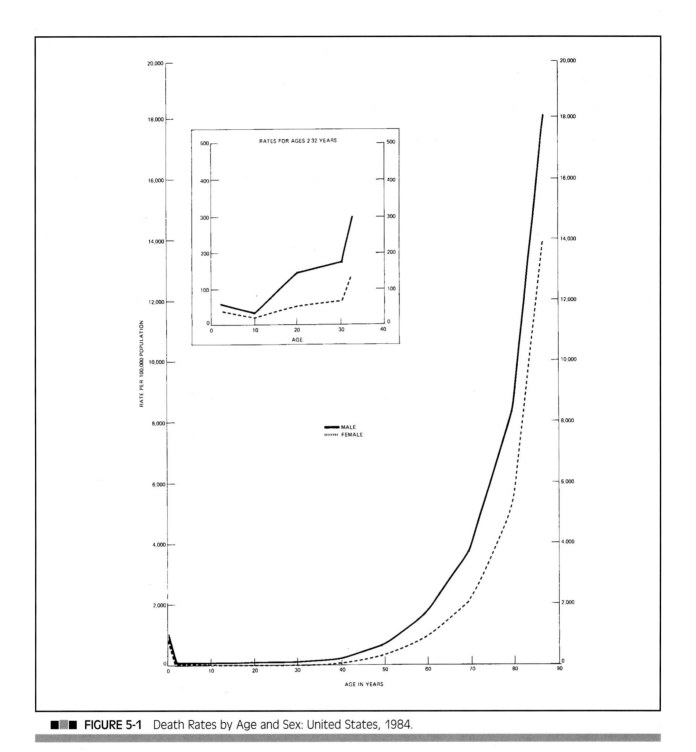

FIGURE 5-1 Death Rates by Age and Sex: United States, 1984.

Source: U.S. Department of Health and Human Services, Vital and Health Statistics, Series 20. No. 15, *Death Rates by 10-year Age Groups, Race, and Sex: United States, 1970 and 1972-1984.* (Washington, D.C.: U.S. Government Printing Office, 1987), p. 9.

areas where infant mortality rates are low, a high portion of infant deaths occur within the first 28 days of life. These early infant deaths often result from congenital defects or injuries at birth, deaths that modern medicine may be able to do little to stop.

NEONATAL AND POSTNEONATAL MORTALITY RATES

The **neonatal mortality** rate reflects the influence of congenital and birth-related problems on infant mortality and is calculated

$$NMR = \frac{D_j}{B} \times 1000,$$

where D_j = number of deaths to infants between birth and 28 days of age in a given year, and

B = number of births in that year.

As Thomlinson (1976, 157) noted, "Neonatal mortality . . . varies remarkably little from country to country, and has shown little susceptibility to reduction under pressure from modern medical science." In similar fashion the **postneonatal mortality rate** relates the number of deaths that occur to infants between ages 28 days and one year divided by the number of births in that year.

THE LIFE TABLE

Though various mortality rates are useful in describing mortality conditions, they tell us nothing directly about life expectancy at different ages nor about the numbers of people who will survive to different ages. Yet such information is extremely valuable. **Life tables,** however, are based on observed mortality, usually by both age and sex, and provide us with information on life expectancy and survivorship. A detailed discussion of life tables, as well as birth, death, and migration rates, may be found in Barclay (1958). The following discussion considers only an abridged form of the life table and its associated functions. It is abbreviated in the sense that not all functions of the life table are used.

COEFFICIENTS IN THE ABRIDGED LIFE TABLE

Four important coefficients are used in our abridged life table:

1. the life table mortality rate

2. the survival table

3. the number dying

4. the average remaining life

The Life Table Mortality Rate

The **life table mortality rate** is the primary coefficient in the life table. It gives the probability of dying within a given age interval and is represented by the symbol $^{n}q^{x}$, which gives the probability of dying between age x and age x + n. If n equals 1 and x = 0, then $^{1}q^{0}$ would be the probability of dying between birth and age 1. A graph of the life table mortality rate has the same J-shape discussed earlier for the age-specific death rate.

The Survival Table

The **survival table** gives us the numbers of people who survive to different ages. Usually the life table begins with 100,000 people at birth (regardless of the actual size of the population) and then shows the number who will survive to each age. The symbol 1^{x} is the number of people surviving to exactly age x.

Number Dying

This column shows the number of people dying in each successive interval and is designated $^{n}d^{x}$.

Average Remaining Life

The average remaining life, or expectation of life, designated by e^{x}, is the average number of years still to be lived by those who survived to age x. Table 5-1 shows an abridged life table for the United States, both for the total population and separately for males and females. Notice that a graph of either $^{n}q^{x}$ or $^{n}d^{x}$ would be a J-shaped curve.

Though it is of considerable usefulness in demography, the life table seldom enters into geographic studies of population. However, the life table is a valuable resource for students of population, regardless of disciplinary focus.

THE MAJOR DETERMINANTS OF MORTALITY

Mortality varies both from time to time and from place to place. In order to understand these variations it is necessary to consider the determinants of mortality. Ideally, vital registration data on mortality (in which deaths would be recorded by age, sex, and cause) would be most useful. In reality, however, most information from developing countries is far less complete. The United Nations has estimated that only 68 percent of deaths are recorded in Latin America, 15 percent in Asia, and less than 1 percent in Africa (Ruzicka and Lopez, 1990).

Progress has been made in the past decade regarding the use of sample surveys to ascertain both birth and death rates. Though there are still many gaps and limitations to mortality data, international agencies like the World Bank and the United Nations Population Division have been able to publish estimates of vital rates for most countries of the world (World Bank, 1991).

The **life table mortality rate** is the primary coefficient in the life table. It gives the probability of dying within a given age interval.

The **survival table** gives us the numbers of people who survive to different ages.

TABLE 5-1. Abridged Life Tables by Race and Sex: United States, 1988

Age Interval	Proportion dying			Average remaining lifetime
Period of life between two exact ages stated in years, race, and sex	Proportion of persons alive at beginning of age interval dying during interval	Number living at beginning of age interval	Number dying during age interval	Average number of years of life remaining at beginning of age interval
(1)	(2)	(3)	(4)	(7)
x to $x+n$	$_nq_x$	l_x	$_nd_x$	e_x
ALL RACES				
0-1	0.0100	100,000	999	74.9
1-5	.0020	99,001	198	74.7
5-10	.0012	98,803	120	70.8
10-15	.0014	98,683	134	65.9
15-20	.0044	98,549	431	61.0
20-25	.0058	98,118	565	58.3
25-30	.0061	97,553	596	51.6
30-35	.0074	96,957	717	46.9
35-40	.0096	96,240	924	42.2
40-45	.0126	95,316	1,204	37.6
45-50	.0189	94,112	1,77	33.0
50-55	.0300	92,335	2,766	28.6
55-60	.0473	85,569	4,238	24.4
60-65	.0728	85,331	6,208	20.5
65-70	.1055	79,123	8,344	16.9
70-75	.1568	70,779	11,096	13.6
75-80	.2288	59,683	13,654	10.7
80-85	.3445	46,029	15,858	8.1
85 and over	1.0000	30,171	30,171	6.0
MALE				
0-1	.0110	100,000	1,104	71.5
1-5	.0022	98,896	220	71.3
5-10	.0014	98,676	138	67.5
10-15	.0017	98,538	165	62.5
15-20	.0062	98,373	615	57.6
20-25	.0087	97,758	854	53.0
25-30	.0089	96,904	865	48.4
30-35	.0107	96,039	1,024	43.8

Age Interval	Proportion dying			Average remaining lifetime
Period of life between two exact ages stated in years, race, and sex	Proportion of persons alive at beginning of age interval dying during interval	Number living at beginning of age interval	Number dying during interval	Average number of years of life remaining at beginning of age interval
(1)	(2)	(3)	(4)	(7)
x to $x+n$	$_nq_x$	l_x	$_nd_x$	e_x
MALE				
35-40	.0134	95,015	1,276	39.3
40-45	.0170	93,739	1,592	34.8
45-50	.0245	92,147	2,261	30.3
50-55	.0385	98,886	3,457	26.0
55-60	.0610	86,429	5,273	22.0
60-65	.0941	81,156	7,639	18.2
65-70	.1360	73,517	9,996	14.9
70-75	.2022	63,521	12,842	11.8
75-80	.2931	50,679	14,852	9.1
80-85	.4239	35,827	15,189	6.9
85 and over	1.000	20,638	20,638	5.1
FEMALE				
0-1	.0089	100,000	890	78.3
1-5	.0018	99,110	175	78.0
5-10	.0010	98,935	101	74.2
10-15	.0010	98,834	100	69.2
15-20	.0024	98,734	240	64.3
20-25	.0028	98,494	272	59.4
25-30	.0033	98,222	321	54.6
30-35	.0041	97,901	405	49.8
35-40	.0058	97,496	567	45.0
40-45	.0084	96,929	817	40.2
45-50	.0135	96,112	1,293	35.5
50-55	.0219	94,819	2,077	31.0
55-60	.0347	92,742	3,217	26.6
60-65	.0537	89,525	4,810	22.5
65-70	.0793	84,715	6,716	18.6
70-75	.1210	77,999	9,435	15.0
75-80	.1843	68,564	12,640	11.7
80-85	.2981	55,924	16,671	8.7
85 and over	1.000	39,253	39,253	6.3

Source: United States Department of Health and Human Services. *Vital Statistics of the United States, 1988,* Vol. II, Section 6, *Life Tables* (Rockville, MD: National Center for Health Statistics, 1988), pp. 3-5.

MORTALITY, MORBIDITY, AND THE EPIDEMIOLOGIC TRANSITION

Morbidity, or sickness, may or may not result in death. The case fatality rate is a measure of the proportion of sick persons who die. Basically, diseases may be classified as communicable or degenerative. The role of these diseases in overall death rates has changed over time, leading to the development of an "epidemiologic transition" theory, first described by Abdel Omran (1977, 4) as follows:

> This theory focuses on the shifting web of health and disease patterns on population groups and their links with several demographic, social, economic, ecologic, and biological changes. Many countries have experienced a significant change or transition from high to low mortality accompanying either social development (as occurred in the West with the Industrial Revolution) or a combination of medical development and early social change (which was the story in many developing countries when antibiotics, insecticides, sanitation, and other medical technology were introduced after World War II). Common to all these countries is a clear shift in the kinds of diseases and causes of death that are prevalent at a fixed point in time: the Old World epidemics of infection are progressively (but not completely) replaced by degenerative diseases, diseases due to stress, and man-made diseases. Thus typhoid, tuberculosis, cholera, diphtheria, plague, and the like decline as the leading diseases and causes of death, to be replaced by heart diseases, cancer, stroke, diabetes, gastric ulcer, and the like, together with increased mental illness, accidents, disease due to industrial exposure, and, now, diseases which can be traced to a deteriorating environment.

Accompanying this **epidemiologic transition,** of course, has been a significant rise in the average life expectancy of those involved. For the world as a whole, average life expectancy has risen from about 30 years in 1900 to around 66 years in 2000. Worldwide, however, there are still great variations from nation to nation, as is apparent in Figure 5-2. Life expectancy at birth today ranges from a low of 37 years in the Zambia to a high of 81 in Japan. More than 300 million people live in nations where the average life expectancy is less than 50 years. In many of these countries one of every ten newborns dies before age one (United Nations Development Programme, 1990). Life expectancy tends to be related to the level of economic development (measured as per capita GNP), as is apparent on the map in Figure 5-2. AIDS has now reduced life expectancy in several African Nations.

The gains in life expectancy in the past four decades have not been shared equally among the world's nations. There is still a wide gap (about 11 years, 64 versus 75) between the life expectancies of the developing nations in Africa, Asia, and Latin America and the more developed nations.

Within the developing world itself there have also been divergent trends. Eastern Asia, dominated by China, had a remarkable gain of 26 years in life expectancy from 1950-1990, surpassing both the Commonwealth of Independent States and Latin America (Mosely

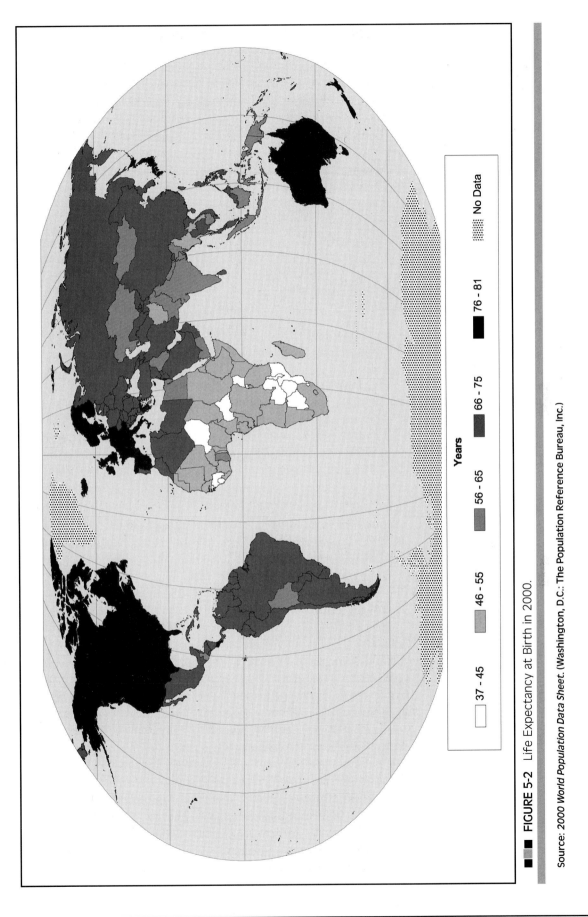

■ **FIGURE 5-2** Life Expectancy at Birth in 2000.

Source: *2000 World Population Data Sheet.* (Washington, D.C.: The Population Reference Bureau, Inc.)

and Cowley, 1991, 7). With a gain of only 14 years in life expectancy, Africa showed the least improvement over this forty year period. There are also great disparities among industrial countries, with Japan making the greatest gain among them. The gap between the countries with the lowest life expectancies in 1996 and those with the highest is approaching forty years (41 versus 79).

In order to better understand the importance of the epidemiological transition theory it is useful to return to Omran's study, in which he introduced four concepts. One is especially important at this point and it was stated by Omran (1977, 9) as follows:

> During the transition, a long-term shift occurs in mortality and disease patterns whereby pandemics of infection are gradually displaced by degenerative and man-made diseases as the chief forms of morbidity and primary causes of death. Typically, mortality patterns distinguish three major successive stages of the epidemiologic transition: 1. *The Age of Pestilence and Famine*, when mortality is high and fluctuating, thus precluding sustained population growth. In this stage the average life expectancy at birth is low and variable, vacillating between 20 and 40 years. 2. *The Age of Receding Pandemics*, when mortality declines progressively; the rate of decline accelerates as epidemic peaks become less frequent or disappear. The average life expectancy at birth increases steadily from about 30 to about 50 years. Population growth is sustained and begins to take off exponentially. 3. *The Age of Degenerative and Man-Made Diseases*, when mortality continues to decline and eventually approaches stability at a relatively low level. The average life expectancy at birth rises gradually until it exceeds 70 years. It is during this stage that fertility becomes the crucial factor in population growth.

Furthermore, Omran suggested that, because of variations in the timing of the transition, there are three basic models of the epidemiologic transition:

1. the classic or Western model

2. the accelerated model

3. the delayed model

The classic or Western model describes the transition as it occurred in Western societies over the past 200 years, a transition in which crude death rates decreased from around 30 to under 10. The mortality decline occurred gradually in response to a mosaic of economic, social, and environmental improvements. In the early stages of this mortality decline medical practices were relatively unimportant, while today downward shifts in mortality are strongly influenced by modern medicine (McKeown and Brown, 1969).

The accelerated model describes the mortality transition as it occurred in Japan, Eastern Europe, and the Soviet Union. The transition was socially determined at the outset, but it also benefited from the revolution taking place in medical science.

The delayed model may be applied to the observed conditions in most developing countries today, where rapid declines in mortality have occurred since the end of World War II. Mortality declines in the developing countries have been the result of modern medicine, the use of insecticides, and organized disease eradication programs; aid from the developed countries and the United Nations have had a considerable impact.

Subsequent to Omran's work a fourth stage—the Hybristic Stage—has been incorporated into the epidemiologic transition model (Rogers and Hackenberg, 1987, and Olshansky and Alt, 1986). The derivation of the term hybristic is the Greek word *hybris*, which means a feeling of invincibility or extreme self-confidence. This stage is one in which personal behavior and lifestyle strongly impact patterns of disease incidence and injuries. For example, in the United States homicide, suicide, cirrhosis of the liver, and AIDS remain major killers, and tuberculosis and other communicable diseases associated with poverty and poor lifestyle choices are on the rise (Rockett, 1999).

Arguments about the relationship between mortality and economic development have attracted considerable attention. Some people argue that advances in mortality conditions in developing countries today are not closely related to economic development, as they have been in the past; while others argue that economic development is still a major determinant of mortality conditions. In a study of health care in the Third World, Phillips (1990, 51) argued that, "the epidemiological transition is therefore apparently a concept that needs to be applied with caution in many Third World countries. Only . . . where health services are on the whole more uniformly available and data more reliable, can the concept be used with much confidence for present and future health care planning."

At first glance it seems that a nation's per capita income is a very important factor in determining mortality levels. The United States, and the nations of Northern and Western Europe, with per capita GNPs around $20,000 or more, have a life expectancy of 75 years, whereas many nations in Africa, with per capita incomes around $600 per year, have a life expectancy of 53 years. Although most wealthy countries have high life expectancies, there are a number of low-income countries that have life expectancies approaching that of the wealthier countries. Conversely, there are some countries with relatively high per capita incomes that rate relatively low on measures of health and survival (Mosley and Cowley, 1991, 31).

Table 5-2 lists five countries (Category A) that have relatively high life expectancies with low rates of GNP/per capita. Most noteworthy are China and Sri Lanka, which have incomes below $800 per capita yet have life expectancy rates approaching that of most developed countries. Most countries in that income range have life expectancies in the forties and low fifties. These five countries, however, have several features in common. They all have relatively high literacy rates for both men and women, an indication of an investment in education,

■■■ TABLE 5-2. Life Expectancies Relative to GNP Per Capita—1998

Category A—High Life Expectancy with Low GNP Per Capita		
Country	GNP/Capita U.S. dollars	Life Expectancy
China	$ 750	71 years
Sri Lanka	740	72
Jamaica	1,600	71
Costa Rica	2,640	76
Cuba	——	75
Category B—Lower Life Expectancy with High GNP Per Capita		
South Africa	$3,520	58
Fiji	2,470	63
Guatemala	1,470	65
Brazil	4,400	67
Algeria	1,520	67
Turkey	2,830	68

Source: *World Population Data Sheet 1998.* (Washington D.C.: The Population Reference Bureau, Inc.)

with particular attention to equity for women. These nations also have relatively low fertility rates.

The countries listed in Category B (Table 5-2) have per capita incomes over $1,400 but life expectancy rates under 70 years. In general, the literacy rates for these nations are lower than those in Category A and there is also a greater disparity between men and women. Total fertility rates for these nations are also high, indicating that women do not participate broadly in these societies. The role of women in many Islamic countries is limited to childbearing and childrearing.

According to Mosley and Cowley (1991, 32), "The per capita public expenditures for health services in Saudi Arabia ($248.) and Libya ($163.) are 20 to 50 times greater than those in China ($5.) or Sri Lanka ($7.), suggesting that good health is not necessarily produced by spending money on sophisticated hospital care." In an effort to find out how some of these poor nations have managed to achieve these remarkable gains in health, the U.S.-based Rockefeller Foundation sponsored an international conference in Bellagio, Italy. Conferees concluded that poor countries can achieve higher health standards and life expectancies if they have a political commitment to equality and if their programs focus on wide access to education, food, and basic health services (Halstead, et al, 1985).

However, there still is at least some tendency toward a relationship between income and mortality, though it is apparently weaker now than it was during the epidemiologic transition in the Western nations.

■■■ Table 5-3. Deaths, Death Rates, and Age-adjusted Death Rates for 1999 and Percent Changes in Age-adjusted Rates from Comparability-modified 1998 to 1999 for the 15 Leading Causes of Death: United States, Final 1998, Comparability-modified 1998, and Preliminary 1999

[Data are based on a continuous file of records received from the States. Rates per 100,000 population; age-adjusted rates per 100,000 U.S. standard population based on the year 2000 standard; see Technical notes. Figures for 1999 are based on weighted data rounded to the nearest individual, so categories may not add to totals]

Rank[1]	Cause of death (Based on the *Tenth Revision, International Classification of Diseases*, 1992)	Number	Death rate	Percent of total deaths	Age-adjusted death rate Comparability-modified[2] 1999	1998	1998	Percent change[3]
. . .	All causes	2,391,630	877.0	100.0	881.9	875.8	875.8	0.7
1	Diseases of heart	724,915	265.8	30.3	267.7	269.7	272.4	−0.7
2	Malignant neoplasms	549,787	201.6	23.0	202.6	204.4	202.4	−0.9
3	Cerebrovascular diseases	167,340	61.4	7.0	61.8	63.1	59.5	−2.1
4	Chronic lower respiratory diseases	124,153	45.5	5.2	45.8	44.1	42.0	3.9
5	Accidents (unintentional injuries)	97,298	35.7	4.1	35.7	36.1	35.0	−1.1
	Motor vehicle accidents	42,437	15.6	1.8	15.5	15.7	16.1	−3.1
	All other accidents	54,862	20.1	2.3	20.1	—	—	—
6	Diabetes mellitus	68,379	25.1	2.9	25.2	24.4	24.2	3.3
7	Influenza and pneumonia	63,686	23.4	2.7	23.5	24.2	34.6	−2.9
8	Alzheimer's disease	44,507	16.3	1.9	16.5	13.3	8.6	24.1
9	Nephritis, nephrotic syndrome, and nephrosis	35,524	13.0	1.5	13.1	12.1	9.8	8.3
10	Septicemia	30,670	11.2	1.3	11.3	10.6	8.9	6.6
11	Intentional self-harm (suicide)	29,041	10.6	1.2	10.6	11.3	11.3	−6.2
12	Chronic liver disease and cirrhosis	26,225	9.6	1.1	9.7	9.9	9.5	−2.0
13	Essential (primary) hypertension and hypertensive renal disease	16,964	6.2	0.7	6.3	6.0	5.4	5.0
14	Assault (homicide)	16,831	6.2	0.7	6.1	6.5	6.5	−6.2
15	Aortic aneurysm and dissection	15,806	5.8	0.7	5.8	6.1	6.1	−4.9
. . .	All other causes	380,503	139.5	15.9	—	—	—	—

. . . Category not applicable.
— Data not available.
[1]Rank based on number of deaths; see Technical notes.
[2]Age-adjusted rate modified with the comparability ratios shown in Table II of the Technical notes.
[3]Percent change between the comparability-modified 1998 age-adjusted death rate and 1999 age-adjusted death rate.
Source: National Center for Health Statistics, *National Vital Statistics Report,* Vol. 49, No. 3, June 26, 2001, p. 5.

CAUSES OF DEATH

The epidemiologic transition helps explain minor shifts in causes of death from various diseases, though other causes of death must also be considered. Table 5-3 shows leading causes of death in the United States for one recent year. Diseases of the heart were the major single cause of death, followed by malignant neoplasms. In 1900 the leading causes of death in the United States were influenza and pneumonia, but today almost 60 percent of all deaths in the United States result from two causes, heart disease and malignant neoplasms.

■ ■ ■ TABLE 5-4. Life Expectancy at Specified Ages, Massachusetts, 1850 to 1900-1902

Year	At Birth Male	At Birth Female	Age 20 Male	Age 20 Female	Age 40 Male	Age 40 Female	Age 60 Male	Age 60 Female
1850	38.3	40.5	40.1	40.2	27.9	29.8	15.6	17.0
1855	38.7	40.9	39.8	39.9	27.0	28.8	14.4	15.6
1878-82	41.7	43.5	42.2	42.8	28.9	30.3	15.6	16.9
1890	42.5	44.5	40.7	42.0	27.4	28.8	14.7	15.7
1900-02	46.1	49.4	41.8	43.7	27.2	28.8	13.9	15.1

Source: United States Bureau of the Census (1975) *Historical Statistics of the United States, Colonial Times to 1979.* Part 1 (Washington, D.C.: U.S. Government Printing Office), p. 56.

■ ■ ■ TABLE 5-5. Life Expectancy at Birth by Race and Sex, U.S., 1900-1995

Year	White Male	White Female	Nonwhite Male	Nonwhite Female
1900	46.6	48.7	32.5	33.5
1910	48.6	52.0	33.8	37.5
1920	54.4	55.6	45.5	45.2
1930	59.7	63.5	47.3	49.2
1940	62.1	66.6	51.5	54.9
1950	66.5	72.2	59.1	62.9
1960	67.4	74.1	61.1	66.3
1970	68.0	75.6	61.3	69.4
1980	70.7	78.1	65.3	78.6
1985	71.8	78.7	67.2	75.2
1990	72.7	79.6	67.5	75.7
1995	73.4	79.6	67.5	75.8

Source: National Center for Health Statistics. *Vital Statistics of the United States, 1970.* Vol., 2, Mortality, Part A, Table 5-5 and *Statistical Abstract of the United States, 1997.* (Washington, D.C.: U.S. Government Printing Office), Table 117.

Mortality conditions in the United States prior to 1900 are not well documented, though it is safe to assume that mortality was well above what it has been in most of the twentieth century. Most likely, mortality conditions during the eighteenth and nineteenth centuries in the United States were similar to those experienced in Western Europe during the same period. Wars, famines, and epidemics caused fluctuations in year-to-year death rates. Early data for Massachusetts suggest that some improvement in mortality conditions probably occurred in the nineteenth century, as is apparent in Table 5-4. These data suggest that improvements were affecting mainly infant and childhood mortality, since major gains in life expectancy appear only for life expectancy at birth. Undoubtedly, during this period considerable regional variations in death rates could be observed, as well as differences for whites and nonwhites. Life expectancy during the twentieth century also changed dramatically in the United States, as is apparent in Table 5-5.

diseases. Only 5 percent of deaths in these nations are among the elderly older than age 75. Some African villages, for example, have reported deaths among infants and young children at ten times the rate of the aged.

In the developed nations, on the other hand, only 2 percent of deaths are among children under age 5, whereas almost two-thirds occur to those over 75 years of age.

INFANT MORTALITY IN THE UNITED STATES

Infant mortality in the United States has consistently decreased, from a level of 100 or more at the turn of the past century (as is apparent in Figure 5-3) to 7.0 in 2000. Figures for infant mortality rates during the eighteenth and nineteenth centuries are unavailable, but most likely they were always in excess of 100 and may at times have been twice that.

Significant disparities in infant mortality rates exist within the United States today. A study reported that the infant mortality rate among blacks was 17.6 per 1,000, which was more than double the white rate of 8.5 per 1,000 (National Center for Health Statistics, 1990). The nation's capital, Washington, D.C., had an infant mortality rate of 21.2 per 1,000, a rate greater than less developed countries such as Cuba, Chile, and Costa Rica.

The reason for the discrepancy in infant mortality along racial lines is primarily because blacks have a variety of social disadvantages. Black mothers are less likely to have received prenatal care and black infants are twice as likely to be of low birth weight. Significant cost-effective programs to overcome these problems are available, but funding has not kept pace with the growing number of eligible women and children.

INFANT MORTALITY IN OTHER PARTS OF THE WORLD

Despite the affluence and advanced technology in the United States, infant mortality conditions here are not as good as they are in a number of other countries (see Table 5-6), though they are improving.

Worldwide the pattern of infant mortality is as shown in Figure 5–4. Infant mortality rates range from a low of 3.5 in Japan to a high of 150 in Afghanistan (where it has actually improved during the 1990s). Among major world regions infant mortality is highest in Africa, with an overall rate of 88. Asia is next highest with an overall rate of 56. Western Europe is the lowest with a rate of only 5, though North America with 7 is very close behind.

Infant and child mortality is primarily a problem associated with the developing countries. In the mid-1990s approximately 98 percent of the deaths among children under five years of age occurred in the developing countries. The tragedy, according to UNICEF estimates, is that 95 percent of the estimated 14.5 million infant and child deaths in developing countries in 1990 were preventable (Grant, 1990). Even in the developed countries, where only 300,000 children under age five

Diet and Longevity (Continued) and some forms of cancer. Interestingly, though Japan is far from the Mediterranean, the Japanese diet is similar in many ways to the above—cereals, fruits, and vegetables comprise the bulk of the diet, supplemented with smaller quantities of fish, and even smaller quantities of red meat.

Is this the perfect diet? Further research is needed before anyone can answer that with certainty, but the Mediterranean Diet or something similar would seem to be a prudent guide for us all. Other dietary items may prove to have particular health benefits; for example, soy is being studied right now—its prevalence in the Japanese diet makes it a potential candidate.

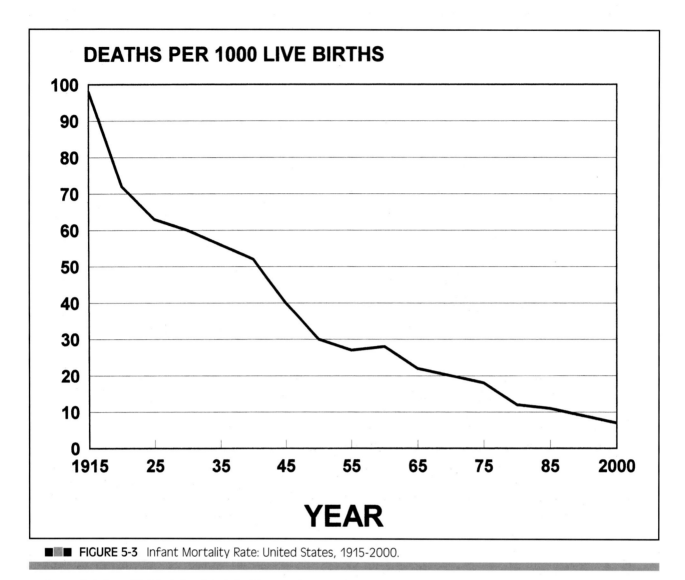

DEATHS PER 1000 LIVE BIRTHS

■□■■ **FIGURE 5-3** Infant Mortality Rate: United States, 1915-2000.

Source: Data from *Vital Statistics of the United States,* Vol. II, Mortality Part A. Section 2, Table 2.1 and *2000 World Population Data Sheet.*

died (mostly in Eastern Europe and the former Soviet Union), more than sixty percent of those deaths were potentially preventable as well.

It has been estimated that 24 percent of the world's newborns died before age five in 1950; by 1990, however, the infant and child mortality rate (ICMR) had declined more than 55 percent to 10.5 percent. Perhaps the most impressive decline occurred in China, where the rate went from 26.6 percent to 4.1 percent, an 85 percent reduction. Japan also had an impressive drop from 7.5 percent to 0.8 percent, a drop of 89 percent.

Researchers have identified several factors that are thought to be related to infant mortality risks. High risk babies include those born to mothers who are adolescents, over age 40, or have had more than seven births, and when the interval between births is less than seven years (Mosley and Cowley, 1991, 10). In the United States the number of babies born to drug-addicted mothers has become increasingly

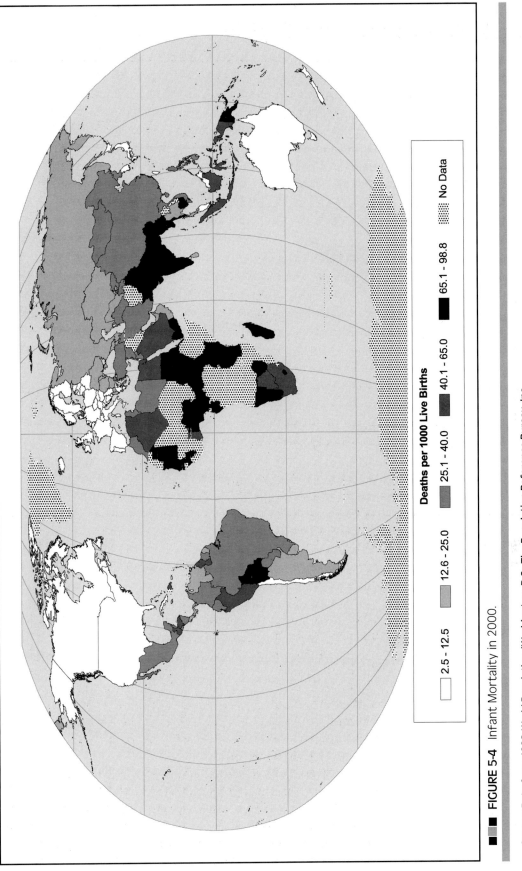

Deaths per 1000 Live Births

2.5 - 12.5	25.1 - 40.0	65.1 - 98.8
12.6 - 25.0	40.1 - 65.0	No Data

FIGURE 5-4 Infant Mortality in 2000.

Source: Data from 1998 *World Population*. (Washington, D.C: The Population Reference Bureau, Inc.,

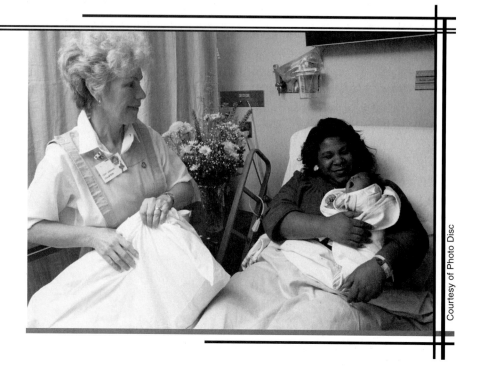

The more educated a mother is, the more likely she is to use maternal and child health services. In turn, this helps lower infant and child mortality rates.

Courtesy of Photo Disc

problematic. National policies aimed at discouraging early marriage and childbearing, along with aggressive family planning programs, can prevent births to high-risk mothers and substantially reduce infant and child mortality (Rutstein, 1991).

In almost every country studied, the ICMR fell as the education of mothers improved. Figure 5-5 depicts changes in under-five mortality rates as it relates to mother's education for selected countries. According to Mosley and Cowley (1991, 11), "Maternal education appears to affect child health in a number of ways. . . . The more educated a mother is, the more likely she is to use maternal and child health services—specifically prenatal care, delivery care, childhood immunizations, and oral rehydration therapy for diarrhea."

A few major diseases are responsible for a large proportion of deaths among children under five years of age (Table 5-7). These diseases, for the most part, are preventable at a relatively low cost. Oral Rehydration Therapy (ORT) has been effectively used to combat childhood diarrhea and UNICEF estimates that in 1990 more than 1 million diarrhea deaths were prevented by ORI, although another 2.5 million preventable deaths still occurred from diarrhea. Immunization coverage has greatly increased in the past decade, although it varies widely in the developing world from 95 percent in China to only 53 percent in sub-Saharan Africa (Grant, 1991). Complacency and other complications have led to decreasing immunization rates in parts of the United States; even wealthy nations still have problems.

■■■ TABLE 5-6. Countries with Lowest Infant Mortality Rates, 1998	
Country	**Deaths per 1000 Live Births**
Finland	3.5
Japan	3.8
Singapore	3.8
Sweden	3.9
Hong Kong	4.0
Norway	4.0
Spain	4.7
Switzerland	4.7
Austria	4.8
Germany	4.9
Luxembourg	4.9
France	5.1
Australia	5.3
Ireland	5.5
Iceland	5.5
Netherlands	5.7
Denmark	5.8
Belgium	5.8
Italy	5.8
United Kingdom	6.1
Canada	6.3
Taiwan	6.7
United States	7.0

Source: World *Population Data Sheet 1998*. (Washington, D.C.: The Population Reference Bureau, Inc.).

Another more intractable problem is that of childhood malnutrition. Although malnutrition is not listed in many cases as a cause of death, estimates are that it is a contributing factor in up to 40 percent of child deaths. An estimated 177 million children in the developing world are malnourished (Carlson and Wardlaw, 1990)—others put the number at closer to 200 million. Even in the United States as many as 12 million children are considered to be malnourished (Brown and Pollitt, 1996).

There is little doubt that malnourished children are at a disadvantage; their chances of success in life are already diminished. Physically, they are hindered by stunted growth, a reduced resistance to infections, and a variety of other health problems; ultimately, they are more likely to die before reaching adulthood. Mentally, malnourished children suffer both direct consequences—slower brain development, for example—and indirect consequences, which include persistent tiredness and lack of energy—making learning more difficult. Improved nutrition for children needs to be a high priority

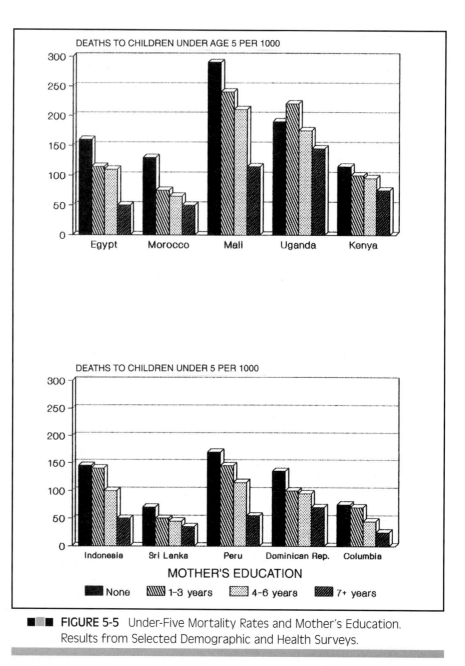

DEATHS TO CHILDREN UNDER AGE 5 PER 1000

DEATHS TO CHILDREN UNDER 5 PER 1000

MOTHER'S EDUCATION

■ None   ▨ 1-3 years   ▨ 4-6 years   ▨ 7+ years

■■■ **FIGURE 5-5** Under-Five Mortality Rates and Mother's Education. Results from Selected Demographic and Health Surveys.

Source: W. Henry Mosley and Peter Cowley. (1991) "The Challenge of World Health," *Population Bulletin* 46:10. Courtesy of The Population Reference Bureau, Inc.

everywhere, not just in the developing countries. As Brown and Pollitt (1996, 43) concluded:

> Enriched programs for children in economically impoverished communities can often ameliorate some of the problems associated with previous malnutrition...it seems clear that prevention of malnutrition among young children remains the best policy—not just on moral grounds but on economic ones as well....Steps taken today to combat malnutrition and its intellectual effects can go a long way toward improving the quality of life—and productivity—of large segments of a population and thus society as a whole.

■■■ TABLE 5-7. Causes of Death to Children under Age 5 in Developing Countries

Causes of Death	Number (thousands)	Percentage of Deaths
Diarrheal diseases	4,000	27.4
Immunizeable diseases	3,700	25.3
Measles	2,000	13.7
Whooping cough	600	4.1
Tuberculosis	300	2.1
Neonatal tetanus	775	5.3
Polio	25*	0.2
Acute respiratory infections	2,375	16.3
Malaria	750	5.1
Other infections/parasitic diseases	450	3.1
Perinatal causes	2,425	16.6
Injuries	200	1.4
Other causes	700	4.8
Total	14,600	100.0

*Polio causes about 250,000 paralytic cases annually, of which about 10 percent die.
Source: W. Henry Mosley and Peter Cowley. (1991) "The Challenge of World Health," *Population Bulletin* 46:11.

Not only do life expectancies continue to increase but also the quality of life for the elderly has increased as well, especially in the developed countries.

SEX DIFFERENTIALS

Sex differences in mortality have been of considerable interest to demographers. The risk of death appears to be greater for males at all ages. Even among fetal deaths males are more likely to die. No simple explanation of these sex mortality differentials has been accepted.

Some scholars have argued that the female is biologically superior to the male (an argument that most females, of course, have no trouble accepting!). Support for this argument comes from reported sex differentials favoring females among other species, including rats, and from the higher fetal mortality for males in humans. Madigan's study of mortality among religious teaching orders of brothers and sisters also supported the idea of female biological superiority (Madigan, 1957). His sample of residents in religious communities was an attempt to control for sociocultural differences that influence mortality and to view mortality under conditions that were as similar as possible for both sexes. Both sexes were found to have greater life expectancies than the general population (partly as a result of lower stress levels), but the differences in life expectancy between the sexes was as apparent for this sample as it was for the entire population. That demographers are not generally convinced of this argument is apparent in the following comment from demographer Thomlinson (1976, 148):

> Although some persons believe that women are biologically superior to men, this conclusion cannot be properly deduced from the factually supportable premises. One piece of contrary evidence is the higher mortality rates among underdeveloped areas and in some nonliterature societies. Although the evidence is insubstantial, what we can learn from excavations of ancient burial sites implies a higher mortality among women.

However, much of the higher mortality experienced by women in less developed countries stems from high fertility rates; repeated pregnancies and childbirths increase the probability of dying from complications related to them.

A different explanation of mortality sex differences is to attribute the differences to differentials in occupation, status, and role. Males are employed in more hazardous occupations than are females, for example, and military deaths are primarily male. Males are believed by some to be under greater stress than females, though this is changing in most developed countries as more females move into higher-stress occupations. Males generally smoke more, drink more, and even drive more than females; they are also more often murdered. All of these are likely to raise male mortality rates. At the same time, as societies move through the demographic transition, women may benefit differentially because fewer childbirths undoubtedly increase the probability of living longer. However, both these social differences and biological differences probably contribute to observed sex differentials.

Indications of mortality sex differentials, as well as changes in these differentials, may be seen in Tables 5-8 and 5-9 and are certainly worth some attention at this point.

COLOR AND CLASS DIFFERENTIALS

Many, if not most, racial and ethnic mortality differentials reflect differences in socioeconomic status. No convincing evidence of significant biological differences in the resistance to diseases can be identified

TABLE 5-8. Selected Life Table Values: 1979-1994

Age and Sex	Total 1979-1981	Total 1990	Total 1994	White 1979-1981	White 1990	White 1994	Black 1979-1981	Black 1990	Black 1994
Average expectation of life in years									
At birth: Male	70.1	71.8	72.4	70.8	72.7	73.3	64.1	64.5	64.9
Female	77.6	78.8	79.0	78.2	79.4	79.6	72.9	73.6	73.9
Age 20: Male	51.9	53.3	53.6	52.5	54.0	54.4	46.4	46.7	47.1
Female	59.0	59.8	59.9	59.4	60.3	60.4	54.9	55.3	55.5
Age 40: Male	33.6	35.1	35.5	34.0	35.6	36.0	29.5	30.1	30.5
Female	39.8	40.6	40.7	40.2	41.0	41.1	36.3	36.8	37.0
Age 50: Male	25.0	26.4	26.9	25.3	26.7	27.2	22.0	22.5	23.1
Female	30.7	31.3	31.5	31.0	31.6	31.7	27.8	28.2	28.5
Age 65: Male	14.2	15.1	15.5	14.3	15.2	15.6	13.3	13.2	13.6
Female	18.4	18.9	19.0	18.6	19.1	19.1	17.1	17.2	17.2

Source: *Statistical Abstract of the United States, 1997.* (Washington, D.C.: U.S. Government Printing Office), Table 118.

TABLE 5-9. Men per 100 Women by Age and Race, 1900-1990

Age and Race	1900	1930	1960	1970	Projection 1990
All Races:					
Over 65	102.0	100.4	82.6	72.1	67.5
Over 75	96.3	91.8	75.1	63.7	57.8
White:					
Over 65	101.9	100.1	82.1	71.6	67.7
Over 75	97.1	92.0	74.3	63.2	58.4
Black:					
Over 65	102.9	105.7	30.1	79.8	68.2
Over 75	89.6	89.6	87.6	74.7	62.6

Source: Leon Bouvier, Elinore Atlee, and Frank McVeigh. (1975) "The Elderly in America," *Population Bulletin,* 30(3):9. Courtesy of the Population Reference Bureau, Inc., Washington, D.C.

among different racial and ethnic groups, though a few diseases are largely restricted to certain groups. Probably the best example of such a disease is sickle-cell anemia, which is mainly restricted to blacks.

Crude death rates for the United States in 1995 for whites and nonwhites appear in Table 5-10. They suggest heavier mortality rates among whites, and especially among white females. However, keep in mind that the crude death rate can be deceptive. Age-specific death rates reveal a different picture entirely, as is apparent in Table 5-11.

■■■ **TABLE 5-10.** Crude Death Rates by Color and Sex, United States, 1995

| | Crude Death Rate | |
Population	White	Black
Total	9.1	8.6
Males	9.3	9.7
Females	8.9	7.5

Source: *Statistical Abstract of the United States, 1997,* Table 120. (Washington, D.C.: U.S. Government Printing Office).

■■■ **TABLE 5-11.** Death Rates by Age, Color, and Sex, United States, 1988

Color and Sex	Under 1 Year	1-4 Years	5-14 Years	15-24 Years	25-34 Years	35-44 Years	45-54 Years	55-64 Years	65-74 Years	75-84 Years	85 Years and Over
White											
Total	8.3	0.5	0.2	1.0	1.2	1.9	4.4	11.7	26.7	62.8	158.8
Male	9.3	0.5	0.3	1.4	1.7	2.6	5.6	15.3	35.0	82.0	188.1
Female	7.3	0.4	0.2	0.5	0.6	1.2	3.2	8.5	19.1	51.3	147.6
Nonwhite											
Total	17.5	0.7	0.3	1.3	2.4	4.1	7.8	16.8	32.5	66.9	127.9
Male	18.1	0.8	0.4	2.0	3.5	5.9	10.7	21.9	41.8	84.0	145.5
Female	16.0	0.6	0.3	0.6	1.3	2.6	5.5	12.7	25.5	56.3	119.6

Source: United States Department of Health and Human Services. *Vital Statistics of the United States, 1988,* Vol. II, Mortality—Part A (Washington, D.C.: U.S. Government Printing Office), Table 1-3, pp. 3-4.

■■■ **TABLE 5-12.** Death Rates in 1975 by Size of Landholdings, Companiganj, Bangladesh

Size of Land Holding (acres)	Death Rate
None	35.8
.01-.49	28.4
.50-2.99	21.5
3.00+	12.2

Source: Lester R. Brown. (1976) *World Population Trends: Signs of Hope, Signs of Stress.* Worldwatch Paper No. 5 (Washington, D.C.: Worldwatch Institute) October, p. 20. Used by permission.

For most groups nonwhite mortality is considerably above white mortality. Table 5-11 also illustrates sex differentials of mortality.

In most countries social status and mortality are related, though the degree of the relationship varies in different types of political and economic systems. One example of this relationship is shown in Table 5-12.

However, such class differentials have probably increased as countries passed through the epidemiological transition. As Pressat (1970, 37) commented:

It is certainly true that, until recent times, social inequality did not too seriously aggravate the risks of mortality among the lower

Life expectancy in the United States is affected by many different variables, including race and socioeconomic status. Persistent gaps remain between whites and blacks and between rich and poor.

classes. Because of the ineffectiveness of medical care, the poor lost very little by not being able to take advantage of it. The great epidemics spared no one and the standard of hygiene at the court of Versailles was probably no higher than in country districts. The crisis of subsistence was probably the only plague which did not affect the rich, whose dietary habits could be disastrous anyway.

Today many differences in mortality may be attributed to differences in socioeconomic factors, but many other differences are also important and they cannot be overlooked.

MEDICAL GEOGRAPHY

Medical geography, the study of the spatial aspects of health and illness, is an area of geographic inquiry that has seen a considerable amount of recent research activity. Its focus has been twofold:

1. on disease ecology, the "manner and consequences of interaction between the environment and the causes of morbidity and mortality" (Birdsall, 1991, 392)

2. on the health care delivery system and its spatial aspects

More generally, geographer Robin Kearns (1993) has argued the necessity of placing geographic concerns with public health issues more directly within the context of place. Such an orientation toward place in medical geography, she argued, would aid both our understanding of illness and health service provision systems and the collective experience of place in particular communities—health services

Medical geography is the study of the spatial aspects of health and illness.

are clearly related to quality-of-life issues. Kearns emphasizes health, rather than illness, and argues that (1993, 145):

> A central requirement for a geography of health will be models of society as well as of health that recognize the contingent relations that pertain for individuals and groups at particular locations.... Only a further integration of issues of place, identity, and health into medical geography will firmly (re)locate the field within social geography.

As Kearns and others have noted, geographers would do well (and are beginning to work toward) to incorporate more consideration of gender, race, ethnicity, and place into the practice of medical geography.

The mapping of diseases has received much attention and studies such as those on the techniques of mapping cancer by Boyle, et al. (1989) and Clayton and Kaldor (1989) have provided useful geographic insights into that disease, as has a detailed historical study of the diffusion of smallpox in Finland by Wilson (1993). A more traditional atlas and compendium of analytical techniques was developed by Cliff and Haggett (1988).

The analysis of geographical variations in health care provision and consumption is another area of concern to geographers. Analytical models in this area have focused on both the individual level (Senior and Williamson, 1990) and the national level (Mohan, 1988). According to Jones and Moon (1991, 443), writing in a recent review about the progress made in medical geography, "Sophisticated modelling of this nature, together with informed qualitative work, provides medical geography with a means to complement its traditional regional-scale analyses. In disease mapping it allows the assessment of environmental effects taking into account individual characteristics. In health-care geography it provides a basis for in-depth analyses of the relationship between need and provision/consumption."

ACQUIRED IMMUNE DEFICIENCY SYNDROME (AIDS)

The past decade has seen an alarming rise in mortality attributable to **AIDS.** In many ways, AIDS is like the bubonic plague of medieval times—it has become the first modern pandemic. Both diseases were initially incurable, and virtually everyone who contracts AIDS will eventually die from it until medical science develops better treatments, and ultimately, with luck, a real cure.

Currently, researchers have identified two serotypes of HIV; HIV-1, which is found worldwide, and HIV-2, which is found almost exclusively in West Africa (largely within former Portuguese colonies).

The past decade has seen an alarming rise in mortality attributable to **AIDS.** In many ways, AIDS is like the bubonic plague of medieval times—it has become the first modern pandemic.

Though transmission routes and risk factors are similar for both forms of HIV, the latency period for HIV-2 appears to be longer.

Data from the World Health Organization pointed out that over 500,000 AIDS cases were reported worldwide by mid-1987, and between five and ten million people worldwide had been infected by the virus (Marx, 1987, 1523). By the end of 1986, 32,560 cases had been reported in the Western Hemisphere, with 90 percent of them in the United States. In 1983 a little over two thousand AIDS cases were reported by the United States Centers for Disease Control (CDC)—by 1988 this number had surpassed thirty thousand, and by the end of 1995 AIDS cases in the United States had reached 500,000 (about the number reported worldwide only eight years earlier). By 1995 the World Health Organization officially reported a total of about 1.2 million cases, but estimated that a more realistic figure (because of severe under-reporting) was around 4.5 million; the number infected by HIV is thought to be more than 36 million, with the greatest numbers in Africa (perhaps as many as 70 percent of the world's AIDS cases). Because of its prevalence among both sexes, researchers in Africa have been searching for correlates of the disease in an effort to check its further diffusion. Caldwell and Caldwell (1996, 68) found a high level of correlation between lack of male circumcision and the distribution of AIDS, concluding that ". . . in the AIDS belt, lack of male circumcision in combination with risky sexual behavior, such as having multiple sex partners, engaging in sex with prostitutes and leaving chancroid untreated, has led to rampant HIV transmission."

Although the exact geographic origin of the AIDS virus has not been determined, most researchers believe that the retrovirus responsible for AIDS originated in Africa. Six years before its identification in the United States, the disease had been isolated in central Africa. Since its initial recognition in Africa around 1972, the disease has diffused to most other parts of the world. In a study of AIDS in Asia, for example, geographer Bimal Paul (1994, 377-378) found that:

> The rate of HIV infection and the prevalence of AIDS vary widely among Asian countries. No country can consider itself free of the dread infestation. . . . If the AIDS epidemic continues unchecked in Asian countries, the disease is expected to have severe economic, social, and demographic effects on future population increments, age structure, and family composition. . . . Heterosexual transmission of HIV between prostitutes and customers seems to be a main cause of the spread of AIDS in many Asian countries.

Many researchers are already revising population projections for many countries, raising death rates somewhat, and even decreasing birth rates sometimes as well, because of the spread of AIDS. As Way and Stanecki (1994, 24) noted:

> AIDS has already begun to substantially revise our thinking about patterns and trends of mortality in countries around the world. . . . The cumulative effect of national AIDS epidemics will

Hemorrhagic Fever

Late in 1995 American travelers were warned that dengue fever was spreading through Mexico, Central America, and the Caribbean. By October of that year more than 35,000 cases of dengue fever had been reported, including some 500 cases of its most severe form, hemorrhagic dengue. This dengue epidemic began in 1994; few doubt that there has been a resurgence, and some argue that the pace of the current epidemic is accelerating.

In May, 1993, a young couple in New Mexico died mysteriously just a few days apart—high fevers, headaches, violent coughs, and muscular cramps marked their final days. Similar cases were recorded in Colorado and Nevada. The Centers for Disease Control (CDC) in Atlanta analyzed samples from several comparable cases, finally determining that they had all died of an unknown type of hantavirus.

Other viral outbreaks have been identified as well in various parts of the world: Ebola in Congo (in Yambuku in 1976 and most recently in Kikwit in 1995), Marburg in Germany, Rift Valley fever in Egypt, Lassa fever in Nigeria, Machupo in Bolivia, and Junin in Argentina are examples. What all of these viruses have in common is membership in a broad class of hemmorrhagic fever viruses. Most are not new; rather, they are often mutations or recombinations of known viruses termed "emerging pathogens." They can be deadly, and we are likely to hear more about them in the decades ahead.

Hemorrhagic fever includes several "families" of viruses, including the following: arenavirus, filovirus, animal filovirus, flavivirus, hantavirus, and bunyavirus (Rift Valley fever).

Hemorrhagic Fever (Continued)

According to virologist Bernard Le Guenno (1995, 58):

The primary cause of most outbreaks of hemorrhagic fever viruses is ecological disruption resulting from human activities. The expansion of the world population perturbs ecosystems that were stable a few decades ago and facilitates contacts with animals carrying viruses pathogenic to humans.

In his best-selling book, *The Hot Zone*, journalist Richard Preston provided a graphic, compelling description of the Ebola virus at work in the human body:

The liver bulges up and turns yellow, begins to liquefy, and then it cracks apart. . . . The kidneys become jammed with blood clots and dead cells...the blood becomes toxic with urine. The spleen turns into a single huge, hard blood clot. . . . The intestines may fill up completely with blood. The lining of the gut dies and sloughs off into the bowels and is defecated along with large amounts of blood.

Much remains unknown about the various hemorrhagic fever viruses; scattered outbreaks have not lead to the identification of hosts in most cases, yet the viruses must be surviving somewhere. As with HIV, much remains to be learned about how to prevent or minimize such viral outbreaks in a world in which population growth and ecosystem disturbances are certain to continue.

be staggering. Government health programs and facilities, with meager budgets already stretched, will be unable to cope with the numbers of people with AIDS and other HIV-related illnesses. . . . In the countries most affected by HIV and AIDS, a quarter or more of the adult urban populations are infected with HIV. As these infections progress to AIDS and death, the impact of AIDS will be felt by the majority of the urban population as friends and relatives begin to die of AIDS.

AIDS kills! That message has been with us for more than a decade now, yet large numbers of people, even in the developed countries, remain either ignorant of it or hope to dodge it (it's never easy to convince young people that they are not immortal). Though abstinence (not very popular) or the use of condoms (still too often shunned, especially by males) would help reduce the spread of AIDS/HIV by sexual contact, their use shows little if any sign of increasing. This is especially apparent when you look at the incidence of other sexually transmitted diseases (STDs) in the United States. In addition to AIDS, but suggestive of the low use of condoms in general, in the mid-1990s Americans were annually experiencing about 4 million cases of chlamydia, 3 million cases of trichomoniasis, 1 million cases of gonorrhea, and over 100,000 cases of syphilis. On top of that about 40 million Americans have genital warts (human papillomavirus) and 35 million have genital herpes—whatever your moral or political leanings, sex is "in" and "protection" is lagging far behind. Geographers would do well to look more closely at spatial aspects of some of these STDs, but little has been done so far.

AIDS IN THE UNITED STATES

In the United States the initial appearance of AIDS had three prominent foci—the metropolitan areas of San Francisco, Los Angeles, and New York City. According to a study by Dutt, et al, (1987, 457-458) "Prior to 1983, 67 percent of the AIDS victims in the country were confined to these areas." The rate map for 1989 (Figure 5-6) shows the emergence of another focal point in Florida, probably the result of its close connections with both the Caribbean countries and New York City where the disease was prevalent.

The number of people diagnosed as having AIDS has risen sharply since its first identification (Figure 5-7). In 1981 only 318 persons were diagnosed as having AIDS; by the mid-1990s that number had risen to more than half a million, with about 40,000 new cases per year (and, sadly, about as many AIDS-related deaths each year).

An analysis of AIDS victims by age and gender points out some interesting relationships (Table 5-13). Most AIDS victims are between the ages of 20 and 49, but almost half of them are in the 30-39 year age cohort. Most of these younger victims contracted the disease through blood transfusions. AIDS victims are still predominantly male, but the disease is affecting an increasing number of females as well

Courtesy of Photo Disc

Though American society has acted in ambivalent ways with respect to AIDS, the good news right now is that AIDS deaths have been decreasing in the United States.

(though they are about four times less likely to be infected than males). In the mid-1980s only 7.4 percent of American AIDS victims are female. These age and gender characteristics were similar throughout the nation and changed little between 1981 and 1987 (Dutt, et al., 1987, 487). In the early 1990s some evidence suggested an increasing incidence of AIDS among women, mainly prostitutes and those in relationships with heterosexual or bisexual IV drug users.

AIDS is a disease that is fatal to almost everyone who contracts it. An analysis of the racial nature of mortality shows that mortality has been higher among Hispanics and African-Americans than among whites. These differences are probably due to more and better health-care facilities being available to whites than to African-Americans and Hispanics.

Although AIDS first appeared in the metropolitan areas of San Francisco, Los Angeles, and New York, it has now spread to other large metropolitan areas as well as to small cities and towns. According to Dutt, et al., (1987, 471) "Once confined to the peripheries of the country, the disease has now spread to its heartland. AIDS is also becoming more prevalent among heterosexuals, who constitute the overwhelming bulk of the sexually active Americans. If this trend continues, the disease will intensify and no part of the country, indeed no group in it, will be AIDS-free in the near future. AIDS has no racial, political, social, or physical barriers. Without the development of a curative drug or preventive vaccine, the potential for destruction from the spread of AIDS cannot be measured or predicted. It is not an overstatement to assert that humanity is in danger."

Hemorrhagic Fever (Continued)

In her book, *The Coming Plague: Newly Emerging Diseases in a World Out of Balance*, journalist Laurie Garrett warned us that these viruses are our predators and that we had better improve our understanding of where they come from, where they hide, and how they can be dealt with—we have been too comfortable with the notion that "plagues" are a thing of the past. Beyond medical science and forays into microbial ecology, Garrett argues that we must consider changes in social behavior as well.

Geographers could focus more attention on the human-environmental relationships that exist in places where viral outbreaks have occurred. We could also focus more attention on the severe consequences of another doubling of the world's population between now and perhaps the year 2050. Most of that growth will occur in the countries of the Third World, where rapid urbanization is already creating massive centers of urban deprivation, high population densities, and growing anarchy.

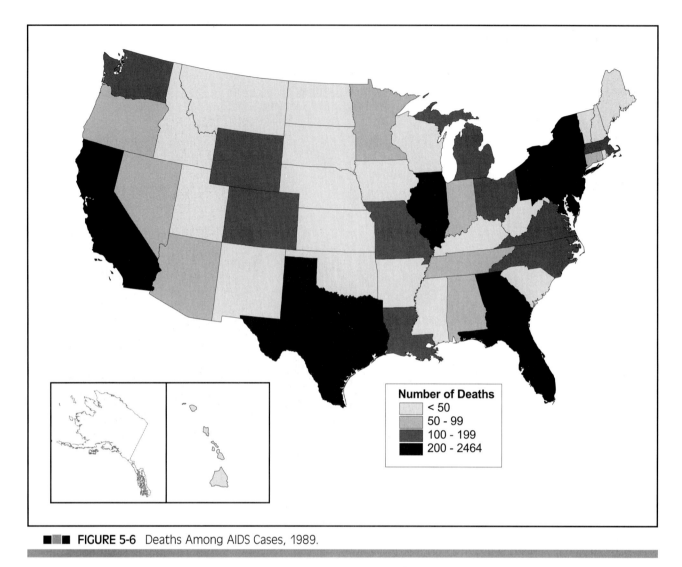

FIGURE 5-6 Deaths Among AIDS Cases, 1989.

Number of Deaths
- < 50
- 50 - 99
- 100 - 199
- 200 - 2464

Source: Centers for Disease Control, Center for Infectious Diseases, AIDS Program.

> By 1993 **AIDS** had become the number one cause of death for people in the 25-44 year age group—a measure of seriousness that is compelling.

By 1993 **AIDS** had become the number one cause of death for people in the 25-44 year age group—a measure of seriousness that is compelling. By the mid-1990s it was estimated that as many as one in every 92 American men could be infected with HIV. Minorities have been especially vulnerable, with as many as one in 33 African-American males carrying the virus. The following actual rates of AIDS in 1993 showed how differently the disease was affecting different groups: African-Americans, 101 per 100,000; Latinos, 51 per 100,000; Anglos, 17 per 100,000, Native Americans, 12 per 100,000; and Asians, only 6 per 100,000. These figures alone suggest the importance of more carefully considering racial and ethnic factors in medical geography studies. New AIDS cases are most common among African-American and Hispanic males.

Considerable resources have been marshalled in the struggle to understand and treat AIDS, both in the United States and elsewhere.

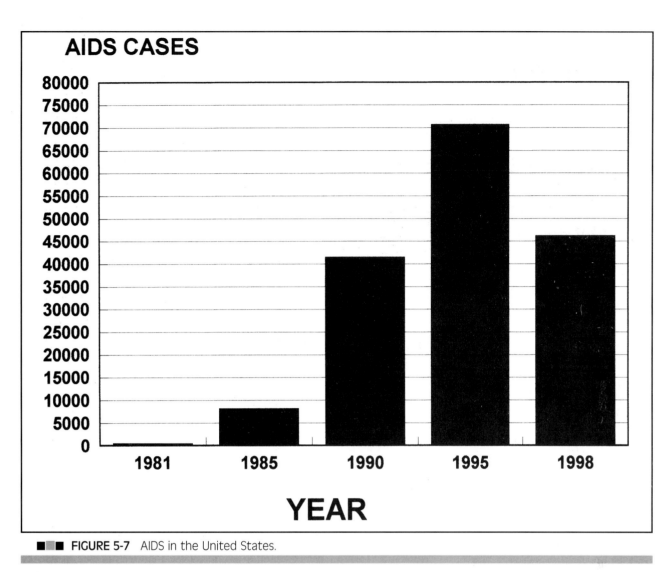

AIDS CASES

	80000
	75000
	70000
	65000
	60000
	55000
	50000
	45000
	40000
	35000
	30000
	25000
	20000
	15000
	10000
	5000
	0

1981   1985   1990   1995   1998

YEAR

■■■ **FIGURE 5-7** AIDS in the United States.

Source: Data from National Center for Health Statistics, *Health U.S. 2000,* Table 53.

By the mid-1990s one encouraging development was that of protease inhibitors—powerful antiviral AIDS drugs. These drugs (the first approved in 1995 under the names inverase or saquinavir) are different than earlier drugs such as AZT, DDI, DDC, D4T, and 3TC. Success of the earlier drugs was only temporary; HIV gradually mutated, escaping their effects. Protease inhibitors have been developed to interfere with a different part of the HIV replication cycle; theoretical results suggest that replication attempts will result not in new HIV viruses, but rather in new and harmless viruses instead. Thanks mainly to new drug combinations, AIDS deaths in 1997 in the United States dropped by 45% from a year earlier, and 1996 was the first year in which AIDS deaths declined at all.

In an effort to understand the spatial dimensions of the diffusion of AIDS Gould, et al. (1991, 80) have developed modeling techniques to portray the geographic dimensions of AIDS diffusion and have attempted to "predict the next maps, not just numbers down the time

■□■ TABLE 5-13. Number and Percentage of Persons with AIDS, by Selected Characteristics and Period of Report—United States, 1981–2000

Characteristic	1981–1987 No.	(%)	1988–1992 No.	(%)	1993–1995 No.	(%)	1996–2000 No.	(%)
Sex								
Male	46,251	(92.0)	177,132	(87.5)	211,909	(82.4)	204,730	(77.4)
Female	4,029	(8.0)	25,387	(12.5)	45,353	(17.6)	59,672	(22.6)
Age group (yrs)								
0–4	649	(1.3)	2,763	(1.4)	2,105	(0.8)	1,355	(0.5)
5–12	101	(0.2)	667	(0.3)	650	(0.3)	618	(0.5)
13–19	199	(0.4)	759	(0.4)	1,381	(0.5)	1,722	(0.7)
20–29	10,523	(20.9)	38,507	(19.0)	43,445	(16.9)	36,252	(13.7)
30–39	23,239	(46.2)	92,178	(45.5)	116,335	(45.2)	114,072	(43.1)
40–49	10,472	(20.8)	46,922	(23.2)	67,475	(26.2)	78,032	(29.5)
50–59	3,684	(7.3)	14,494	(7.2)	19,153	(7.4)	23,980	(9.1)
≥60	1,413	(2.8)	6,230	(3.1)	6,718	(2.6)	8,373	(3.2)
Race/Ethnicity								
White, non-Hispanic	30,033	(59.7)	102,130	(50.4)	109,101	(42.4)	88,896	(34.0)
Black, non-Hispanic	12,796	(25.5)	63,319	(31.2)	97,742	(38.0)	118,665	(44.9)
Hispanic*	7,044	(14.0)	35,116	(17.3)	47,442	(18.4)	52,092	(19.7)
Asian/Pacific Islander	312	(0.6)	1,342	(0.7)	1,927	(0.8)	2,147	(0.8)
American Indian/ Alaska Native	68	(0.1)	437	(0.2)	870	(0.3)	962	(0.4)
Region†								
Northeast	19,541	(38.9)	62,102	(30.7)	78,000	(30.3)	81,466	(30.8)
North Central	3,772	(7.5)	20,416	(10.1)	25,778	(10.0)	25,532	(9.7)
South	12,933	(25.7)	65,754	(32.5)	89,559	(34.8)	102,576	(38.8)
West	13,502	(26.9)	46,303	(22.9)	55,586	(21.6)	45,574	(17.2)
U.S. territories	524	(1.0)	7,883	(3.9)	8,812	(3.2)	8,829	(3.3)
Vital status								
Living	2,103	(4.2)	20,572	(10.2)	96,998	(37.7)	203,192	(76.9)
Deceased	47,993	(95.5)	181,212	(89.5)	159,048	(61.8)	59,807	(22.6)
Total§	**50,280**	**(6.5)**	**202,520**	**(26.2)**	**257,262**	**(33.2)**	**264,405**	**(34.1)**

*Persons of Hispanic origin may be of any race.

†Northeast = Connecticut, Maine, Massachusetts, New Hampshire, New Jersey, New York, Pennsylvania, Rhode Island, and Vermont; North Central = Illinois, Iowa, Kansas, Michigan, Minnesota, Missouri, Nebraska, North Dakota, Ohio, South Dakota, and Wisconsin; South = Alabama, Arkansas, Delaware, District of Columbia, Florida, Georgia, Kentucky, Louisiana, Maryland, Mississippi, North Carolina, Oklahoma, South Carolina, Tennessee, Texas, Virginia, and West Virginia; West = Alaska, Arizona, California, Colorado, Hawaii, Idaho, Montana, Nevada, New Mexico, Oregon, Utah, Washington, and Wyoming.

§Includes persons for whom sex, age, race/ethnicity, region, or vital statistics are missing.

Source: National Centers for Disease Control.

line." A time-series of maps was produced for the State of Pennsylvania (Figure 5-8) and, according to Gould, et al. (1991, 86), "If you know your human geography of Pennsylvania, you can see the hierarchical diffusion, the jumping from city to city; you can see the spatially contagious diffusion, the wine stain on the tablecloth; and you can also see the way in which the structure of the human landscape, the major turnpikes and roads, channels the epidemic."

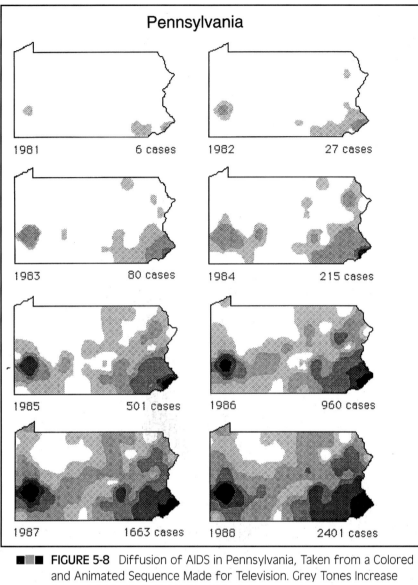

Pennsylvania

1981   6 cases   1982   27 cases

1983   80 cases   1984   215 cases

1985   501 cases   1986   960 cases

1987   1663 cases   1988   2401 cases

■■■ **FIGURE 5-8** Diffusion of AIDS in Pennsylvania, Taken from a Colored and Animated Sequence Made for Television. Grey Tones Increase Geometrically, 1, 3, 9, 27, 81, 243, etc.

Source: Reprinted by permission of Peter Gould, "AIDS: Predicting the Next Map,"
INTERFACES, Volume 21, Number 3, May-June 1991. Copyright © 1991 by the
Operations Research Society of America and the Institute of Management Sciences,
290 Westminster Street, Providence, Rhode Island 02903 USA.

Similar maps have also been developed for the spread of AIDS in the western United States from 1981 to 1988 (Gould, 1991).

AIDS IN THE DEVELOPING COUNTRIES

Today it is nearly impossible for most of us to comprehend the terrible extent to which HIV/AIDS is affecting life in the developing world, especially in Africa, which has more than 25 million people with HIV. As Carol Ezzell (2000, 96) tells us, "AIDS is destined to alter history in Africa—and, in fact, the world—to a degree not seen in humanity's past since the Black Death." She may well be correct.

Approximately 95 percent of all HIV/AIDS cases today occur within the developing nations, where poverty, poor health care, and limited public health systems are unable to cope (Figure 5-9). Though HIV/AIDS is far more prevalent in Africa than anywhere else in the world today, the rate of infection there has actually slowed in Sub-Saharan Africa because so many people already have the disease. On the other hand, areas of rapid growth of new HIV infections include North Africa, Eastern Europe, and Central and East Asia. In places such as China HIV/AIDS data are scarce, but mounting evidence suggests that parts of China, for example Henan Province, may in fact have far more cases than has generally been acknowledged.

Needless to say, HIV/AIDS has wreaked havoc on the demography and economy of many Third World nations. For example, the United Nations has projected that life expectancies in southern Africa, which rose from 44 in about 1950 to 59 in the early 1990s, is likely to drop to around 45 between 2005 and 2010. Unlike degenerative diseases such as heart attacks and various cancers, AIDS kills working-age adults and even children. In its wake we can expect to see millions of orphans—one estimate suggests as many as 40 million by 2010—as young parents succumb. The United Nations found that by 2000 more than 11 million infants had lost their mothers.

By 2001 the starkness of HIV/AIDS was receiving considerable media attention, much of which focused on two themes: drugs for AIDS patients in poor countries and prevention of AIDS transmission. The cost of AIDS drugs is extremely high—a drug regimen for an AIDS patient in the United States can run $1,000 or more per month. Though some drug companies have offered cheaper drugs to poor countries, the prices are seldom affordable in places where people subsist on as little as $1.00 a day. Humanitarian arguments suggest, and even a few economists now support, a program in which the rich nations pool money (and such a pool was started in 2001) to buy drugs and distribute them in poor nations. Aside from prohibitive costs, many professionals feel that poor nations lack the necessary medical infrastructure to utilize AIDS drugs effectively. Ezzell (2000, 103) notes, for example, that "Another hurdle for widespread anti-retroviral drug use in Africa is the sheer difficulty of most of the AIDS drug regimens."

Though both must be considered, prevention seems more likely to pay sizable dividends, and its cost is primarily for education. As Nadine Gordimer (2000, B5) tells us, ". . . the future of the world's children is no future if they are born HIV-positive or contract the virus while they are schoolboys and schoolgirls. We have to break irrevocably the silence that stigmatizes victims and inhibits frankness in education for prevention."

There are signs of hope. In Uganda, for example, the rate of new HIV infections declined during the 1990s, at least in part because of a successful campaign to halt spread of the disease. Support for that

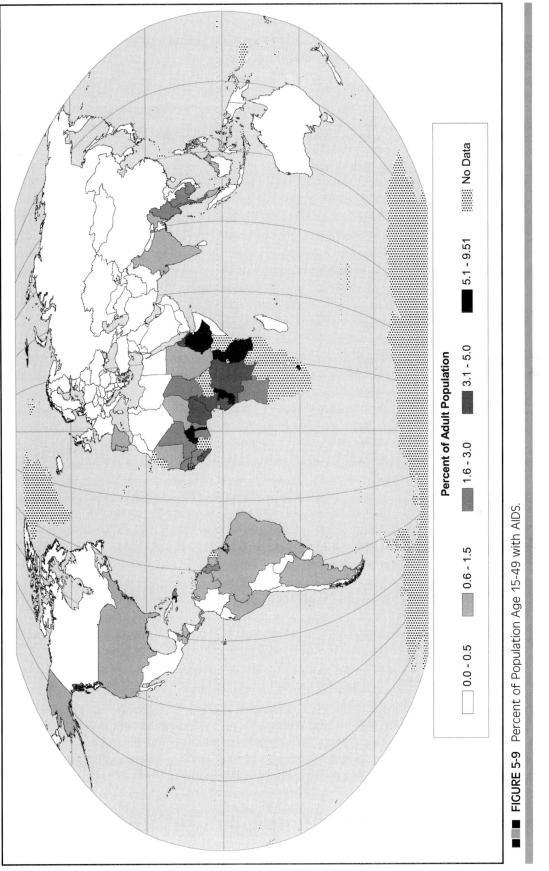

FIGURE 5-9 Percent of Population Age 15-49 with AIDS.

Percent of Adult Population

☐ 0.0 - 0.5	▨ 0.6 - 1.5	
▨ 1.6 - 3.0	▨ 3.1 - 5.0	
■ 5.1 - 9.51	▨ No Data	

Source: *2000 World Population Data Sheet.* (Washington, D.C.: The Population Reference Bureau, Inc.).

campaign came from President Yoweri Museveni, and has been widespread. Similarly, concerted government action in Thailand has helped decrease new HIV/AIDS infections by promoting safe behaviors, including condom usage.

THE PATTERN OF WORLD MORTALITY

The current world pattern of mortality is shown in Figure 5-10. Any interpretation of this pattern must be approached cautiously because the crude death rate served as the measure of mortality, and the crude death rate is greatly affected by the age structure of a population. For example, the crude death rate in the United States is higher than that of Mexico; yet few people would expect mortality conditions to be better in Mexico. Two other measures, life expectancy at birth and the infant mortality rate, are better indicators of actual mortality conditions. For Mexico the infant mortality rate is more than four times higher than that of the United States; and the life expectancy at birth in Mexico is less than that of the United States. Maps of both world infant mortality rates and world life expectancy at birth have been discussed already. The main reason for now considering a map of world crude death rates is that it provides us with a look at one of the two rates that are used to calculate the rate of natural increase. Areas of low crude death rates may, if they also have high crude birth rates, be areas of rapid population growth. Areas with high crude death and birth rates are potential areas for fast growth in the future, since typically crude death rates have fallen earlier and faster than crude birth rates.

In 2000 the estimated crude death rate for the world was 9. Death rates of 9 or less are found in many developed countries, from the United States to Australia, as well as in some of the developing countries of Latin America. Most of Western Europe and Asia, along with parts of North Africa, have death rates between 10 and 20. The highest death rates, those above 20, are found in Africa, as well as in Afghanistan. Barring major catastrophes, such as widespread famines, epidemics, and nuclear holocaust, it is likely that the epidemiological transition will continue to reduce world mortality rates. However, some signs of checks on the downward trend in death rates are appearing. In discussing the decreasing world growth rate Brown (1976, 6) commented that:

> Tragically, the slowdown in population growth is not due entirely to falling birth rates. In some poor countries population growth is being periodically checked by hunger-induced rises in death rates. These recent upturns in the national death rates represent a reversal of postwar trends, one which political leaders in the affected countries are not eager to discuss.

Food scarcity tends mainly to increase the death rates for infants and the elderly, though others are also affected. The world food situation is discussed in more detail in Chapter 10. The impact of rising

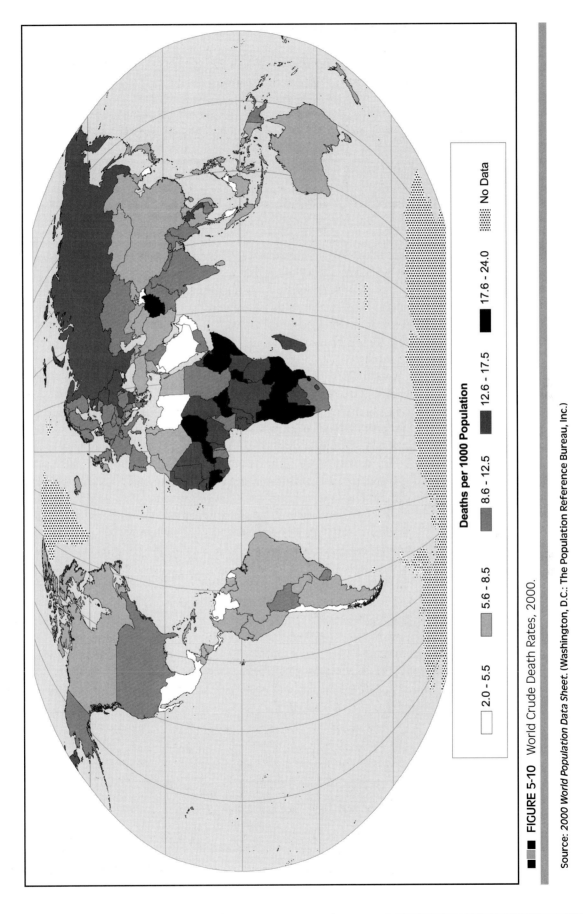

FIGURE 5-10 World Crude Death Rates, 2000.

Deaths per 1000 Population

2.0 - 5.5	5.6 - 8.5	8.6 - 12.5	12.6 - 17.5	17.6 - 24.0	No Data

Source: *2000 World Population Data Sheet.* (Washington, D.C.: The Population Reference Bureau, Inc.)

death rates from starvation and malnutrition need to be mentioned here, however. It is especially worthwhile to consider Brown (1976, 25), who stated that although death rates measure the principal demographic effect of prolonged hunger, they do not measure the social impact. For every person who dies, scores or even hundreds may suffer and lie close to death. The true social cost of food shortages must take into account these hapless individuals, as well as the millions of infants and children who somehow survive prolonged periods of semi-starvation only to suffer irreparable brain damage in the process.

No one wishes to see death rates increase anywhere, but clearly the planet is becoming more strained in its ability to deal with population growth, most of which occurs in the developing countries of the Third World (those least able to deal with rapidly increasing numbers). Given that some balance between birth rates and death rates must be achieved in order to curtail world population growth, we continue to believe that lowering fertility rates is much preferable to raising death rates.

As we move on to a discussion of fertility, however, it will become obvious that we are moving neither far enough nor fast enough to prevent the world's population from at least doubling again during the next century. At the same time, despite its important connections to economic, social, and environmental problems around the globe, discussions of population growth were virtually absent from the agenda at the Earth Summit that was held in Rio de Janeiro in June of 1992; fertility reduction was not pushed very hard at the 1994 International Conference on Population and Development in Cairo, either. The latter conference gave human rights and well-being center stage; gender equity, the empowerment of women, and the integration of family planning services shared time with discussions related to decreasing population growth rates.

Fertility: Patterns and Trends

The causes and consequences of fertility patterns are of major concern in today's world. At one time relatively high reproductive rates were characteristic of most societies, mainly as a response to prevailing high death rates—survival was at stake. However, with death rates either already low or falling in most regions of the world today, fertility patterns differ considerably from place to place. Explanations of spatial variations in fertility are drawn from many sources and differ from place to place as well as from time to time. As Robertson (1991, 24) noted, ". . . the processes and values of reproduction are inextricably tied up with the processes and values of economic production and political control." Leete (1999) provides more on values and fertility change.

Fecundity refers to the biological capacity for reproduction.

MEASURES OF FERTILITY

Before proceeding, it is useful to differentiate between two terms that are sometimes confused, namely fertility and fecundity. **Fecundity** refers to the biological capacity for reproduction, whereas **fertility** refers to actual reproductive behavior. One measure of fertility has already been introduced, the crude birth rate; others now need to be considered.

Fertility refers to actual reproductive behavior.

CHILD-WOMAN RATIO

The **child-woman ratio** is an indirect measure of fertility that is used to estimate fertility in situations where birth records are deficient or nonexistent, mainly in the underdeveloped countries. It is "indirect" in the sense that it does not employ births in the measure at all. It is calculated as

$$CWR = \frac{P_{0-4}}{F_{15-44}} \times 1000,$$

Where CWR = child-woman ratio,
P_{0-4} = the total number of children under five years of age, and
F_{15-44} = the total number of females between 15 and 45 years of age. (This age group is considered to be the child-bearing age group sometimes the 15–49 age group is used instead.)

The child-woman ratio, then, indicates the number of children under five years of age per 1,000 females that are in their child-bearing years. One disadvantage of using this ratio as a measure of fertility is that the number of surviving children is affected by prevailing mortality rates. Seldom, however, will you encounter this particular measure in the literature today.

GENERAL FERTILITY RATE

The **general fertility rate** is a more refined measure of fertility than either the crude birth rate or the child-woman ratio. As with the crude birth rate, actual numbers of births are used in the numerator. In the denominator, however, the total population is replaced by the total number of females in the child-bearing age group, usually considered to be 15–44 years of age. The **general fertility** rate is calculated as

$$GFR = \frac{B}{F_{15-44}} \times 1000,$$

where B_a = total number of births in a one-year period, and

F_{15-44} = total number of females between the ages of 15 and 45 at midyear.

This measure gives us the number of live births per 1,000 women in the child-bearing age group.

AGE-SPECIFIC BIRTH RATE

Age-specific birth rates are useful because child-bearing varies considerably with age. The age-specific birth rate is analogous to the general fertility rate, but instead of having the total number of females in the child-bearing age group as the denominator, it has the total number of women in a smaller age group, such as a one-year or five-year age group. The numerator, then, is the total number of children born in any given year to mothers in that specified age group. Normally, five-year age groups are used. There can be either six or seven age groups, depending upon the upper age limit that is used. In generalized form the calculation of age-specific birth rates is

$$\text{ASBR} = \frac{B_a}{F_a} \times 1000,$$

where B_a = the number of births to females in the age group designated by "a" (e.g., a = 15–19 year olds), and

F_a = the total number of females in the age group "a" at midyear.

If the child-bearing ages of 15–44 are used in five-year age groups, then there are six age-specific birth rates, one for each of the age groups 15–19, 20–24, . . . , 40–44. Often a seventh is added for the age group 45–49. For many countries sufficient data for calculating the age-specific birth rates are not available. Births occurring to women who are outside of the normal child-bearing age range are arbitrarily assigned to either the 15–19 year age group, if they occurred to females below age 15, or to the 45–49 year age group if they occurred to females above fifty years of age. For most populations there are relatively few births outside of the 15–49 age range.

TOTAL FERTILITY RATE

The **total fertility rate** is the most useful way to summarize the age-specific birth rates for a population. It is the average number of children a woman would have *if* she were to have children at the prevailing age specific rates as she passed through her reproductive years. The total fertility rate is calculated as

$$\text{TFR} = 5 \sum_{a=1}^{7} \frac{B_a}{F_a}$$

where B_a = the number of births to females in age group "a" in a one-year period,

F_a = the midyear number of females in age group "a," and

a = five-year age groups as follows: 15–19, 20–24, 25–29, 30–34, 35–39, 40–44 and 45–49, for a total of 7 different age groups.

The five preceding the summation sign is there because we are using five-year age groups. If one-year age groups were used, the five would disappear from the equation.

The **age-specific birth rate** is analogous to the general fertility rate, but instead of having the total number of females in the child-bearing age group as the denominator, it has the total number of women in a smaller age group.

■■■

Total fertility rate is the average number of children a woman would have *if* she were to have children at the prevailing age specific rates as she passed through her reproductive years.

■■■

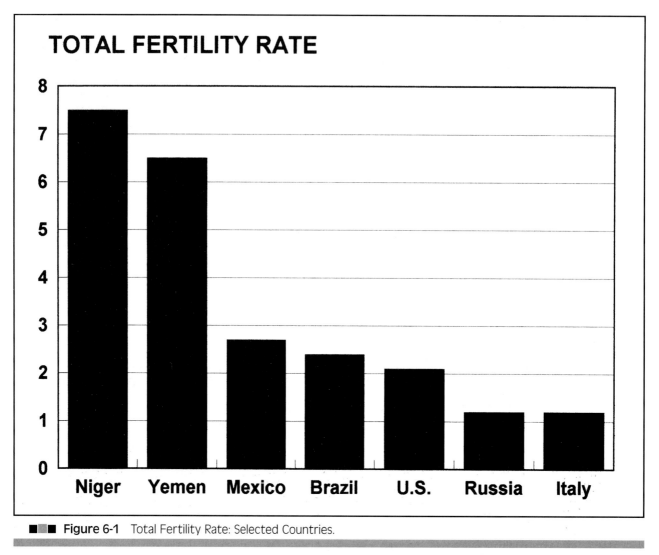

TOTAL FERTILITY RATE

■■■ **Figure 6-1** Total Fertility Rate: Selected Countries.

Source: Data from *2000 World Population Data Sheet*. (Washington, D.C.: Population Reference Bureau, Inc.)

In the United States in 2000 the total fertility rate was 2.1, considered to be "replacement" level, but above the 1.8 that it had been only a few years earlier. In 2000 the world's total fertility rate was estimated to be 2.9, continuing a gradual decline. Figure 6-1 provides some idea of the range of values that countries show for total fertility rates. Countries that maintain a total fertility rate of less than 2.1 for a sustained period of time will experience negative population growth, as we saw earlier in our discussion of different population projections for the United States. Both the causes and consequences of negative population growth have received considerable attention (examples include Teitelbaum and Winter, 1985 and Davis, Bernstam, and Ricardo-Campbell, 1987). Bongaarts and Feeney (1998) suggested a way to modify the total fertility rate to include changes in the timing of childbearing.

GROSS REPRODUCTION RATE

The **gross reproduction rate** is the same as the total fertility rate except that only female births are counted. Thus, it gives the average number of daughters that a woman would have *if* she passed through her entire reproductive life at the prevailing age-specific birth rates. The gross reproduction rate can be found by multiplying the total fertility rate by the proportion of births that is female. The gross reproduction rate is calculated as

$$GRR = 5 \sum_{a=1}^{7} \frac{FB_a}{F_a}$$

where FB_a = the number of female births to females in age group "a" in a one-year period,
F_a = the midyear number of females in age group "a," and
a = five-year age groups.

NET REPRODUCTION RATE

The **net reproduction rate** is the same as the gross reproduction rate except that it is reduced somewhat to allow for the fact that not all women will live through their entire reproductive period. A net reproduction rate of 1.0 would indicate that on the average females are exactly replacing themselves.

THE MAJOR DETERMINANTS OF FERTILITY

The reproductive behavior of a population results from a complex of biological, social, and economic factors that operate differently from place to place as well as from time to time. In order to explain spatial and temporal variations in fertility, it is necessary to consider its determinants.

BIOLOGICAL DETERMINANTS OF FERTILITY

As noted earlier, it is necessary to distinguish between fertility, actual reproduction performance, and fecundity, the biological capacity for reproduction. However, they are obviously related. Fecundity may be affected by a number of physical factors including age, health and nutritional status, and even the physical environment.

Age and Fecundity

In most societies reproduction is accomplished mainly by young adults. In calculating fertility rates, the female reproductive years are generally assumed to be ages 15–44 or 15–49. Puberty marks the onset of reproductive capacity. For females, menarche, or the beginning of menstruation, denotes puberty. The change from infecund to fecund is not a sudden change from one state to another, but rather a more gradual increase through a period of adolescent subfecundity to a mature fecundity, reached at roughly ages 26–30 (Doring, 1969). The female's fecund period ends with menopause, when menstruation ceases. For males puberty also marks the beginning of reproductive capacity; however, the end of the male reproductive period is not so clearly demarcated.

Modern science is helping older women extend their childbearing years. For example, multiple births have increased significantly in the past two decades, mainly among women in their forties. These women find it difficult to become pregnant, so they turn to fertility drugs for help, which in turn increases the likelihood of multiple births. More twins were born to women in their late forties in 1997, for example, than in the entire decade of the 1980s.

Health, Nutritional Status, and Fecundity

A person's health may affect fecundity for varying periods of time. In general, good health and fecundity go together. A variety of diseases may impair fecundity either temporarily or permanently. The most important diseases in this respect are venereal diseases, which may, if left untreated, cause permanent sterility. Nutritional status affects fecundity as well, especially in cases of severe malnutrition, which may even lead to temporary infecundity. In turn, a woman's health may be adversely affected by reproduction, especially if she was already experiencing health or nutritional problems. As the National Research Council (1989, 15) noted, "By avoiding pregnancy, women with health problems may substantially improve their own chances for survival and good health."

Environment and Fecundity

Most of the literature on environment and fertility is concerned with the effect of various environments on sexual activity and related behavior rather than on fecundity. One environmental variable thought by some to be related to fecundity is altitude, though the evidence is so far inconclusive (James, 1966).

SOCIAL DETERMINANTS OF FERTILITY

Given a normally fecund population, fertility will be determined to a considerable degree by variations in social variables, including at least the following: marriage patterns, contraceptive practices, and abortions. Demographers Kingsley Davis and Judith Blake (1956) provided a framework that identified "intermediate variables" through which various social factors operate to affect fertility. They identified three sets of such variables under the following headings:

1. factors that affect exposure to intercourse

2. factors that affect exposure to conception

3. factors that affect gestation and successful parturition

Although marriage is not a necessary prerequisite for reproduction, most societies recognize marriage as the union in which procreation generally occurs. The age at which couples marry and the percent of the population that marries directly affect the fertility of the population.

Fertility variations among nations, then, can be explained by understanding how social differences are being translated into fertility behavior through these "intermediate variables." Major sources of such variations include those discussed subsequently: the following three subheads correspond closely with the three suggested originally by Blake and Davis. John Bongaarts has noted that the four most important factors in fertility declines have been the use of effective contraceptives, the age of marriage, the length of time after child-birth that a woman cannot conceive (because of either sustained breast-feeding or abstinence), and abortion (Robey, Rutstein, and Morris, 1993).

Marriage and Fertility

In most societies today marriage is the socially recognized union within which procreation occurs. However, marriage is obviously not a necessary prerequisite for reproduction, thus illegitimacy and variations in cultural norms from place to place must also be considered when possible. The formation of any sexual union, however temporary or permanent, can potentially result in reproduction, though data for unions other than marriage are often difficult to obtain. As Robertson (1991, 6) commented, "The way we reproduce seems to make the family a logical necessity, but the enormous variation in family patterns around the world is a reminder that these elementary relations of reproduction may be organized in many different ways." Nonetheless, whatever the nature of sexual unions may be, entering into them is a prerequisite for reproduction.

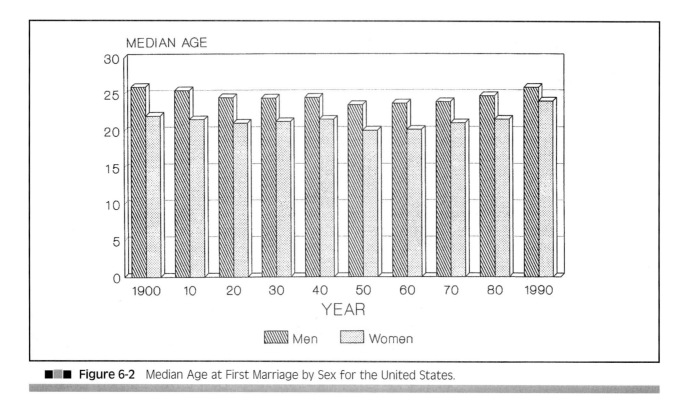

■■■ Figure 6-2   Median Age at First Marriage by Sex for the United States.

Source: *Statistical Abstract of the United States,* 1997, Table 148.

In non-contraceptive populations, especially, the age at which females marry or enter into sexual unions of any kind is an important determinant of fertility. Assuming that the ages 15–44 are the child-bearing years for most women, it is logical that those who marry later will have fewer children on the average, providing that there are few illegitimate births. Typically, in the developing countries today the average age at marriage is fairly low, especially in comparison with the pattern in most of the developed countries. In some nations, such as Panama, Honduras, and Bolivia, the legal marriage age is as low as twelve for females and fourteen for males. Where women marry young and practice little or no contraception, fertility is likely to be high unless other factors operate to lower it. In preindustrial European countries the average age at marriage was also relatively low, though there was a rise in the age at marriage beginning as early as the eighteenth century (Hajnal, 1965).

In the United States the median age at first marriage has not changed considerably during the last one hundred years, though it has moved gradually upward since World War II, as is apparent in Figure 6-2. This rising age at first marriage has also been associated with an increase in the percentage of those never married by given ages, as is shown in Figure 6-3.

One of the steps that the People's Republic of China has taken to decrease fertility has been to encourage people to postpone marriage, ideally to around age 25 for females and 28 for males. Coupled with

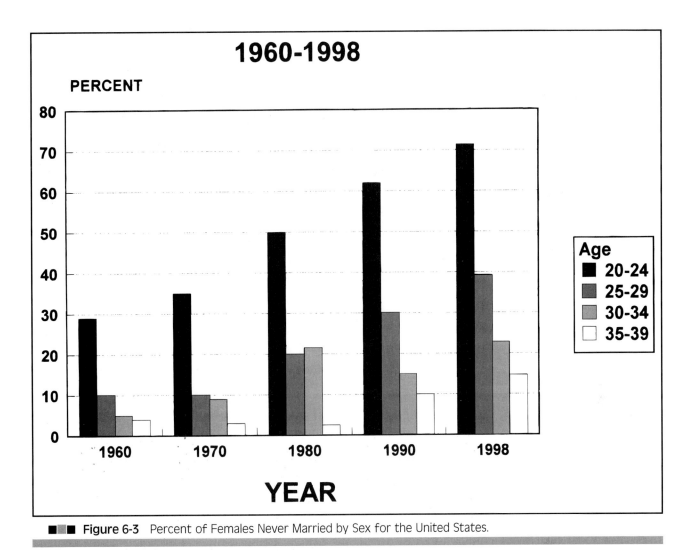

■■■ Figure 6-3 Percent of Females Never Married by Sex for the United States.

Source: Data from United States Bureau of the Census, *Marital Status and Living Arrangements*: March 1990, Current Population Reports, Population Characteristics, Series P-20, No. 450, and Fertility of American Women, Series P-20-526, Table D.

strong sanctions against premarital sex, there is little doubt that the higher age at marriage has been a factor in China's recent fertility decline. However, it is far from being the only factor.

Another determinant of the fertility of a population is the percent of the population that marries. A high rate of celibacy is unusual in most populations, though there are a few exceptions, such as Ireland.

Related to marriage patterns, and also influencing fertility to varying degrees in different societies, is the amount of time spent between marriages after a divorce, separation, or death of a spouse. In general, the rate of dissolution of marriages is lower in preindustrial societies than in advanced ones, with such exceptions as the Islamic peoples. Societies have varying attitudes toward the remarriage of widows.

Given exposure to sexual intercourse within marriages or other types of sexual unions, fertility is then influenced primarily by sexual behavior and contraceptive practice. Both the frequency of intercourse and abstinence for varying periods of time may have effects on fertility. Periodic abstinence may result from sickness or other reasons. Furthermore, some societies forbid intercourse during certain times, for example, while a woman is breast-feeding a child (Saxton and Serwada, 1969). Such practices would tend to increase the spacing between children and reduce fertility. Geographers have seldom studied variations in sexual behavior, though a pioneering study by Ford and Bowie (1989) provided evidence that such studies would be a rewarding addition to the literature in population geography.

Contraception and Fertility

A major determinant of fertility today is the degree to which contraception is practiced. Contraceptive practice is most common in the industrialized nations; however, with the increasing influence of family planning programs in many countries, more and more people are beginning to control their fertility, as noted some time ago by Nortman (1977, 3):

> Among the most far-reaching and visible developments going on in the world today is the change in contraceptive knowledge, attitudes, and practices—popularly known as the "KAP" of contraception. The past decade has seen a rapid acceleration in the historical trend to ever-increasing adoption and use of contraception. Contraception is now being practiced where it never was before, and people everywhere are turning to new and much more efficient methods.

Nortman estimated that around one-third of the married couples in the world now practice some form of contraception, and that percentage is likely to rise during the 1990s and beyond.

Contraceptives may be classified into the following groups:

1. those that prevent the entry of sperm

2. those that avoid or suppress ovulation

3. those that prevent implantation (Segal and Nordberg, 1977, 24)

Included in the first group are coitus interruptus (withdrawal), condoms, spermicides, douches, and diaphragms, along with sterilization; the second group includes both "rhythm" methods and oral contraceptives; and the third category is that of the intrauterine device, or IUD. Abortions are in a separate category because they are used only after conception has occurred, so they can serve as a back-up to contraceptive failure.

Though contraceptive technology has lagged considerably since the innovations of the 1960s, there are some new contraceptive methods currently under investigation. Physiologist Nancy Alexander

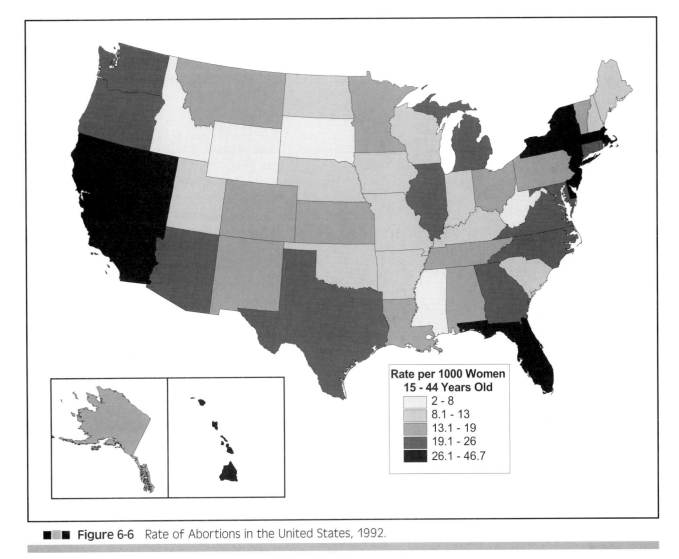

Figure 6-6 Rate of Abortions in the United States, 1992.

Rate per 1000 Women
15 - 44 Years Old
- 2 - 8
- 8.1 - 13
- 13.1 - 19
- 19.1 - 26
- 26.1 - 46.7

Source: *Statistical Abstracts of the United States, 1999.* Table 125.

a general relationship between crude birth rates and both GNP per capita and the level of urbanization, variables that serve at least to some degree as measures of the level of development.

One notable area of fertility research is concerned with developing an economic theory of fertility. Espenshade (1977, 3) pointed out that:

> Children are valuable. Most fundamentally, they provide for the continuation of the human species. If births were to cease, mankind would become extinct within the span of our lifetime. But this reason for wanting children is usually not uppermost in the minds of parents. To them, children are sources of joy and happiness, companionship, and pride. In some circumstances, children may also be prized because they are a potential means of support and security once parents are no longer able to provide for themselves. At the same time, children are costly. They put added pressure on family resources, and they can in other ways curtail the activities and opportunities of parents.

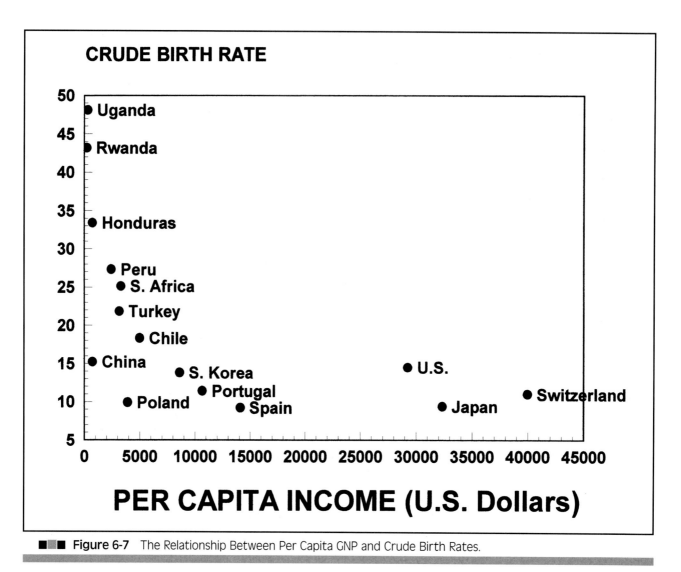

CRUDE BIRTH RATE

PER CAPITA INCOME (U.S. Dollars)

■■■ **Figure 6-7** The Relationship Between Per Capita GNP and Crude Birth Rates.

Source: Data from *2000 World Population Data Sheet.* (Washington, D.C.: Population Reference Bureau, Inc.)

The essence of an economic theory of fertility is that a couple's decision to produce, or not to produce, a child is based on the costs and benefits of the child, as perceived by the couple. Such a theory, then, begins with a consideration of the costs and benefits of children. Costs and benefits have both economic and noneconomic components.

The Value of Children

Hoffman and Hoffman (1973, 20) suggested that the value of children may be thought of as "the functions they serve or the needs they fulfill for parents." A variety of terms for this concept have been used in the literature, including satisfaction, benefits, utilities, rewards, gains, gratifications, advantages, and positive general values. As Espenshade (1977, 4) commented, "The value of children is used to mean that collection of good things parents receive from having children."

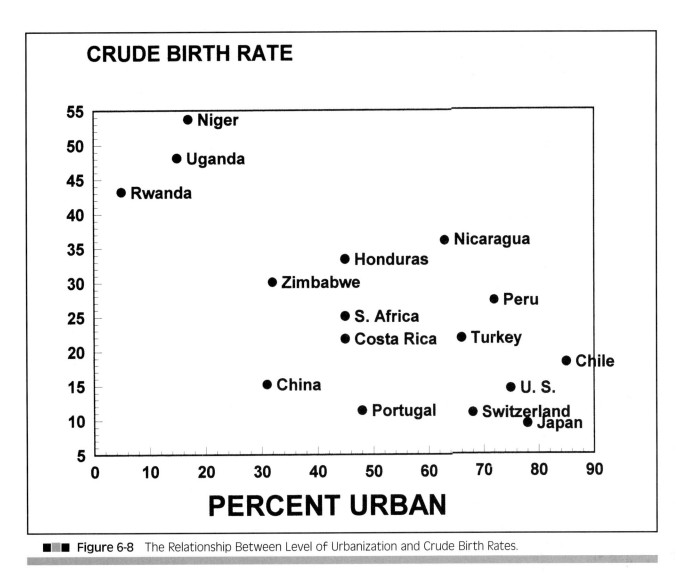

CRUDE BIRTH RATE

(scatter plot of Crude Birth Rate versus Percent Urban)

Niger
Uganda
Rwanda
Nicaragua
Honduras
Zimbabwe
Peru
S. Africa
Costa Rica
Turkey
Chile
China
U. S.
Portugal
Switzerland
Japan

PERCENT URBAN

■■■ **Figure 6-8** The Relationship Between Level of Urbanization and Crude Birth Rates.

Source: Data from *2000 World Population Data Sheet*. (Washington, D.C.: Population Reference Bureau, Inc.)

Children may have both economic and noneconomic values. The noneconomic values are mainly psychological in nature and include the psychic satisfactions that parents derive from having and rearing children. Among these noneconomic values are the following: adult status and social identity; expansion of the self; morality; primary group ties; stimulation, novelty, and fun; creativity and accomplishment; power and influence; and social comparison and competition.

Two types of economic benefits of children are the following:

1. children as a source of financial security in old age and in emergencies (Rendell and Bachieva, 1998)

2. the value of children as productive agents (Leibenstein, 1963)

Courtesy of Photo Disc

Espenshade (1977, 5) stated that:

> There is a tendency to assume that the economic values of children are most salient in the developing countries, especially in rural areas. In fact, it is widely believed that this is the major reason parents in such regions want large families. As a society modernizes and achieves higher levels of economic and social development, the economic value of children declines in importance. The extension of compulsory schooling and the enactment of child labor laws reduce the economic contribution from children. Similarly, to the extent that social security becomes institutionalized in such programs as public health and welfare measures, pension plans, and private annuity and life insurance programs, parents can relax their dependence on children as a source of old-age support.

More than two decades ago Mahmood Mamdani (1972, 113) made the importance of children in Third World countries clear in the following discussion of the failure of birth control in Manupur, a small village in India's Punjab:

> To Hakika Singh, the solution to his financial troubles is not to reduce the size of his family he has to support, but to *increase* it. It is the family that will support him, will even be his salvation. Admittably, he must take one chance: the next baby may be a girl instead of a boy. But since he is faced with utter financial disaster if he does anything else, Hakika Singh is willing to take that chance. Even with a number of sons, he may fail; yet they are his only hope. Even if the chances for success are low, his sons are his

only route to success. As we concluded our conversation, his parting words were: "A rich man invests in his machines. We must invest in our children. It's that simple."

The Cost of Children

Costs may be considered as the disadvantages of children. Among the terms used for the costs of children are dissatisfactions, disadvantages, disvalues, penalties, and negative general values. These costs may be both economic and non-economic.

Among the non-economic costs are included the emotional and psychological problems that children impose on parents. Any parent is aware that raising children causes anxieties about such matters as the child's health and behavior. Among the economic costs are

1. direct maintenance costs

2. opportunity costs

Direct maintenance costs are actual monetary outlays required for the support of children. These costs include food, clothing, housing, educational expenses, and medical expenses. **Opportunity costs** measure opportunities that parents must sacrifice to have and raise children.

According to Espenshade (1977) three types of opportunity costs may be recognized. First, a lower standard of living may result as certain consumption expenditures are foregone. Second, children may reduce a family's ability to save and invest. Third, and of increasing importance in the United States as well as elsewhere, the wife may sacrifice her earnings. Espenshade (1977, 6) noted that:

> The relative importance of the three kinds of economic opportunity cost is likely to vary according to the level of economic development. In the less developed countries, for example, there is a tendency to think that children scarcely affect consumption standards or the ability to save and invest. But whether or not this is actually the case depends on the level of aspirations among the population.

The costs and benefits of having children vary from time to time and place to place. In very general terms, in preindustrial, primarily agrarian societies, the perceived costs of having and rearing children are low, whereas the perceived benefits of having and rearing children are high. Thus, in such societies couples will tend, all else being equal, to have large families.

With economic development, industrialization, and urbanization underway in a society, the perceived costs and benefits of having and rearing children may be altered. As people increasingly become urban dwellers, and more frequently pursue non-agricultural jobs, they may begin to see that the costs of children appear to be rising and the benefits of children seem to be diminishing. Families in urban areas may find housing, education, and medical costs all more expensive. At the same time they may find more goods available to them.

Direct maintenance costs are actual monetary outlays required for the support of children. These costs include food, clothing, housing, educational expenses, and medical expenses. **Opportunity costs** measure opportunities that parents must sacrifice to have and raise children.

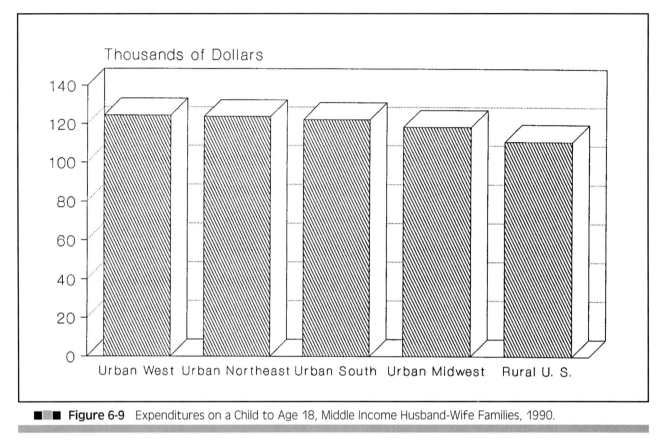

■■■ **Figure 6-9**   Expenditures on a Child to Age 18, Middle Income Husband-Wife Families, 1990.

Source: Data from *Family Economics Review 1991,* 4(1):33.

Thus, they may find that children are not as attractive to them as they are to their rural counterparts.

In the developed countries a large percentage of the population is urban, and most of the labor force is employed in the secondary, tertiary, and quaternary sectors of the economy. Females usually find more job opportunities outside the home in these developed countries, and their participation in the labor force has increased greatly during the last two decades. The costs of having and rearing children are likely to be quite high. With the costs of children high and their perceived benefits low, the modern couple in an urban area of an industrialized country is likely to desire a small family. Figure 6-9 shows the direct costs of raising a child to age 18 in the United States. Add on top of that a substantial figure for sending a child to college and it's no wonder that most couples today in the United States are opting for small families, as they are in Europe and Japan (Foster, 2000).

According to Peggy Orenstein (2001) young women in Japan are increasingly opting to remain single, live at home with their parents, and spend their entire income on themselves. These young women have become what sociologist Masahiro Yamada has termed "Parasite Singles." Orenstein (2001, L1) noted that "More than half of Japanese women are still single by 30. . . . Although they earn, on average, just $27,000 a year, they are Japan's leading consumers, since their entire income is disposable."

Though these young women help drive the Japanese economy today, their very low levels of reproduction are apt to hurt it tomorrow. Low fertility is leading to labor shortages, leaving the Japanese economy weaker than it might be otherwise, especially when Japan has little interest in promoting labor immigration. Parasite singles are thought to be both a remnant of the bubble economy that burst in the 1980s and a projection of the nation's pessimistic view of the Japanese economy.

One final consideration of an economic view of fertility is what we might learn about population growth itself and the likelihood of controlling it. McKenzie and Tullock (1989, 97) argued that:

> Economists are in general agreement that the optimum quantity produced of anything is that quantity at which the marginal cost of the last unit is equal to the marginal benefits of it. This will be the case if all costs and benefits are actually considered by the individual making the decision on the output. And this rule of thumb holds for the production of children.

They go on to note, however, that seldom do couples have to assume the total costs of their fertility decisions—some of those costs are born by others, either directly in such forms as increased congestion or indirectly through taxes for such things as schools. At the same time, they note that population growth will be controlled in response to changes that are likely to occur in the child production process, including the following:

1. resource scarcity will force up the costs of children

2. less favorable tax structures and the availability of contraceptives and abortion

3. greater incentives for improved contraceptive technology with the rising costs of children.

KENYA: A LOOK AT HIGH FERTILITY

With a total fertility rate of 8.0 in 1988, Kenya began the new decade with the dubious distinction of having the highest fertility level of all countries in the world. According to a study, Mott and Mott (1980, 7) noted that "Tribal loyalties and the high value placed on children by Kenya's vast population majority of rural families, and especially its women, combine to dampen national efforts to reduce population growth." Kenyans perceive that the relative place of a tribe in the political arena is primarily a function of tribal size, a situation that frustrates efforts to lower fertility.

Courtesy of Photo Disc

Kenya has a high fertility level. Low actual birth costs and low opportunity costs, combined with high perceived benefits of children, lead to high fertility.

"Tribal pressures affect Kenyan's behavior more than pronouncements arriving from the national seat of government," according to Mott and Mott (1980, 7), "but what ultimately counts is what an individual perceives in his or her own best interest." In a country in which the vast majority of the people work on the land, children are viewed as essential. They are the key to survival and are closely tied to status. They are a ready supply of labor, especially during peak times in the agrarian life cycle. Because much of the farm work is done by females, they perceive children as the major way in which the burden of farm chores can be relieved. Thus, to them the benefits of children are clear and tangible, whereas the costs are minimal. Of course, children also represent a major source of security for both males and females. Again according to Mott and Mott (1980, 9), "the vast majority of rural parents must still look to children for financial—and physical—support in old age."

To rural Kenyans children have a considerable prestige value as well. More children increase the chances that at least one of them will turn out to be a success or to marry well. This reinforces the tendency for women to have lots of children.

Because opportunities outside of the home are extremely limited for rural Kenyan women, the opportunity cost of having children remains low. Low actual costs and low opportunity costs, combined with high perceived benefits of children, lead to a perpetual cycle of high fertility. Nonetheless, by 2000 Kenya's total fertility rate had dropped to 4.7 as couples were responding to changing perceptions about the costs and benefits of children and to an increasing availability of information about family planning and contraceptives.

200   Population Geography

Robey, Rutstein, and Morris (1995, 65) found that, "Analyses of the surveys in Ghana and Kenya show that mass-media campaigns have shaped women's family-planning decisions."

FERTILITY DIFFERENTIALS

The fertility of a particular population results from a complex of different factors, and different groups within a population may respond to similar factors in different ways. Thus, different groups within a population are likely to display different fertility levels. Such differential fertility within a nation may in turn help explain regional fertility patterns. Important observed **fertility differentials** include rural-urban, income, education, and ethnic differentials.

RURAL-URBAN FERTILITY DIFFERENTIALS

As noted already, fertility tends to be related to the level of urbanization when viewed at the world scale. Differential fertility between rural and urban areas within a nation is also common. In the United States in 1986, for example, the general fertility rate was 70.3. However, in metropolitan areas the general fertility rate was only 67.4, compared with a rate of 81.5 in nonmetropolitan areas. Furthermore, within metropolitan areas the general fertility rate was 68.2 in the central cities and 66.8 in the areas outside of the central cities.

Similarly, in a study of fertility differentials for Japanese women by place of residence, Matsumoto, Park, and Bell (1971, 18) found that "as expected, within each sample area, fertility decreases with increasing urbanization."

INCOME DIFFERENTIALS

Previously we observed that fertility tended to be related to levels of economic development, as measured by GNP per capita for a sample of countries. Within a nation fertility differences also tend to exist for various income groups. In general, fertility tends to be highest for the lowest income groups and to decrease with increasing income levels, though occasionally it will rise again at very high income levels. Part of the economic explanation for this rise is that those at high income levels can afford more of everything, including children. Data for the United States appear in Figure 6-10. These data show a somewhat more complex pattern of relationships between fertility and income, though fertility is still highest for the lowest income group.

EDUCATIONAL DIFFERENTIALS

Like income, education tends to be negatively related to fertility. As Hawthorn (1970, 92) commented, "It is true of all societies that awareness of methods of birth control varies directly with urban background

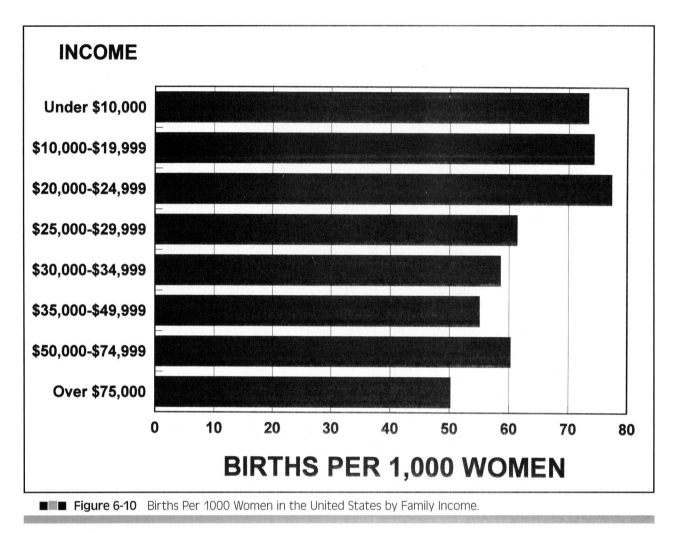

INCOME

Under $10,000	
$10,000-$19,999	
$20,000-$24,999	
$25,000-$29,999	
$30,000-$34,999	
$35,000-$49,999	
$50,000-$74,999	
Over $75,000	

BIRTHS PER 1,000 WOMEN

■■■ **Figure 6-10** Births Per 1000 Women in the United States by Family Income.

Source: **United States Bureau of the Census,** *Fertility of American Women: June 1998,* **Current Population Reports, Population Characteristics, Series P-20-526.**

or residence, a higher than average education, and a higher than average income. . . ." However, beyond the knowledge of contraceptives, education may motivate couples to limit family size because of their greater awareness of the costs and benefits of children. Also, for more educated females the opportunity costs of having and rearing children are likely to be higher than for their less-educated counterparts and they are likely to marry at somewhat later ages. Figure 6-11 shows the relationship between educational attainment and fertility for 1990 in the United States.

RACIAL AND ETHNIC DIFFERENTIALS

Different racial and ethnic groups often have different fertility levels than the national populations of which they are a part. For example, in the United States African-Americans, American Indians, and Mexican-Americans tend to have higher fertility rates than Anglos. Such differentials tend to exist in most nations where

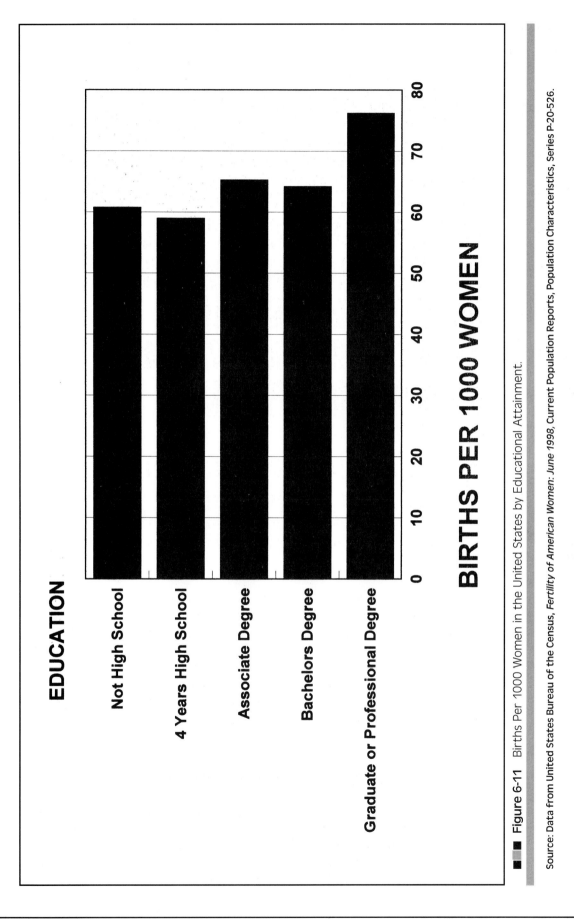

Figure 6-11 Births Per 1000 Women in the United States by Educational Attainment.

Source: Data from United States Bureau of the Census, *Fertility of American Women: June 1998*, Current Population Reports, Population Characteristics, Series P-20-526.

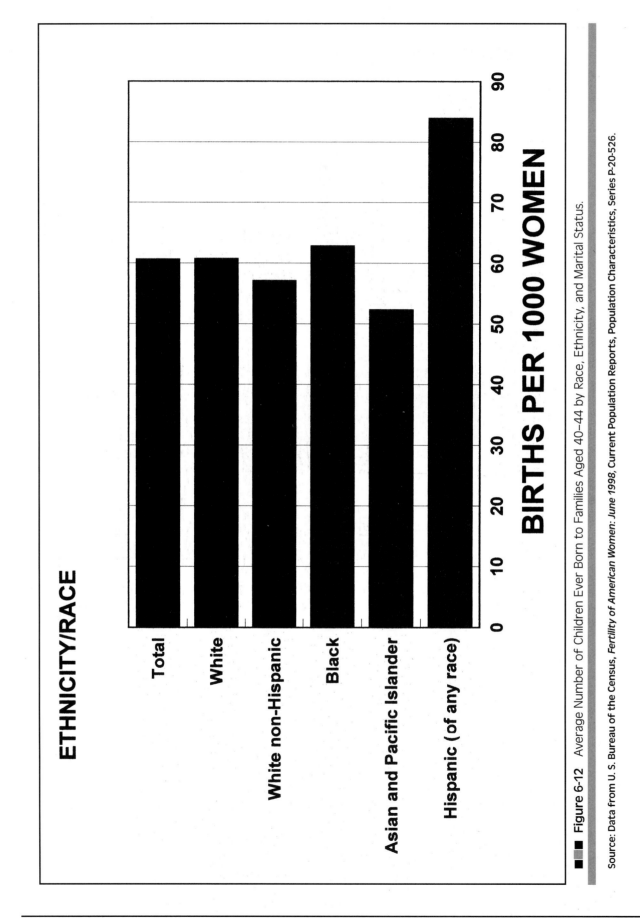

Figure 6-12 Average Number of Children Ever Born to Families Aged 40–44 by Race, Ethnicity, and Marital Status.

Source: Data from U. S. Bureau of the Census, *Fertility of American Women: June 1998, Current Population Reports, Population Characteristics,* Series P-20-526.

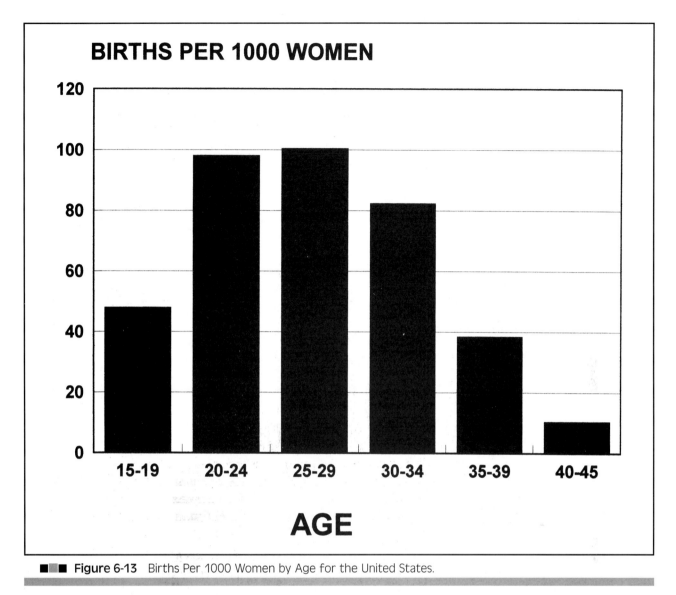

BIRTHS PER 1000 WOMEN

■■■ **Figure 6-13** Births Per 1000 Women by Age for the United States.

Source: Data from United States Bureau of the Census, *Fertility of American Women: June 1998,* Current Population Reports, Population Characteristics, Series P-20-526.

there are significant numbers of the national population who belong to different racial or ethnic groups. Figure 6-12 provides information for such differentials in the United States.

AGE DIFFERENTIALS

Women of child-bearing age in a society are not equally likely to have children at any given age. In most societies fertility is more concentrated in certain age groups, as is apparent for the United States in Figure 6-13.

Fertility differentials are useful and important for explaining spatial and temporal differences in fertility. We must keep in mind, however, that fertility differentials of different types are probably not independent. Educational differences in fertility, for example, may also be reflecting income differences as well. Ethnic differentials also reflect income and educational differences. Furthermore, some patterns, once established, may be self-perpetuating. For example, Kahn and Anderson (1992, 54) found that ". . . it appears that both teen marriage and childbearing behaviors tend to be reproduced across generations." Finally, statements about the causes of fertility differentials should be made with care. Furthermore, many other fertility differentials can be noted. Examples include religious and occupational differentials.

SPATIAL FERTILITY PATTERNS AND TRENDS

Fertility varies in both space and time. Such variations result from similar variations in the determinants of fertility, as well as variations in the response to particular determinants. The map in Figure 6-14 shows the recent spatial pattern of crude birth rates in the world. The rates vary from a high of 52 in Uganda to a low of 9 in Greece, Italy, and Estonia. In general, the world fertility pattern reflects the pattern of economic development, as shown in Figure 6-15. Lutz (1989) has carefully examined fertility differences among nations.

Africa has a crude birth rate of 38, and many African nations have crude birth rates of at least 40. Among the major exceptions are Tunisia and South Africa, with crude birth rates of 22 and 25, respectively. Bledsoe (1990) provided a detailed look at how marriage patterns and fertility in much of Africa are related and how education and other variables are changing traditional marital arrangements.

For Asia, with close to 60 percent of the world's population, the fertility pattern is more varied than it is in Africa. Asia has a crude birth rate of 22, and the highest Asian crude birth rates are found in southern and western Asia. The United Arab Emirates is one of many examples where our generalization about an inverse relationship between income and fertility fails to hold. This is true for several oil-rich nations, where the sudden wealth has not yet been paralleled by general modernization and a sharp contrast exists between extreme wealth (often concentrated in a few hands) and the traditional socio-economic fabric of the majority of residents.

In broad regional terms, East Asia, with a crude birth rate of 15, stands in marked contrast to Southwest Asia. Even here fertility varies from a crude birth rate of 20 in Mongolia to one of only 9 in Japan. Of major importance in this highly populated region is the recent decline of fertility in the People's Republic of China, which in 2000 had a crude birth rate of 15, a figure that represents a slight increase from the beginning of the 1980s. However, estimates of China's population and vital rates differ considerably. There is little doubt among researchers that Chinese fertility has fallen since the beginning of the 1970s, but with insufficient data the actual rate is unknown, probably even to the Chinese. Extremely low crude birth rates have been reported in many urban areas. When discussing

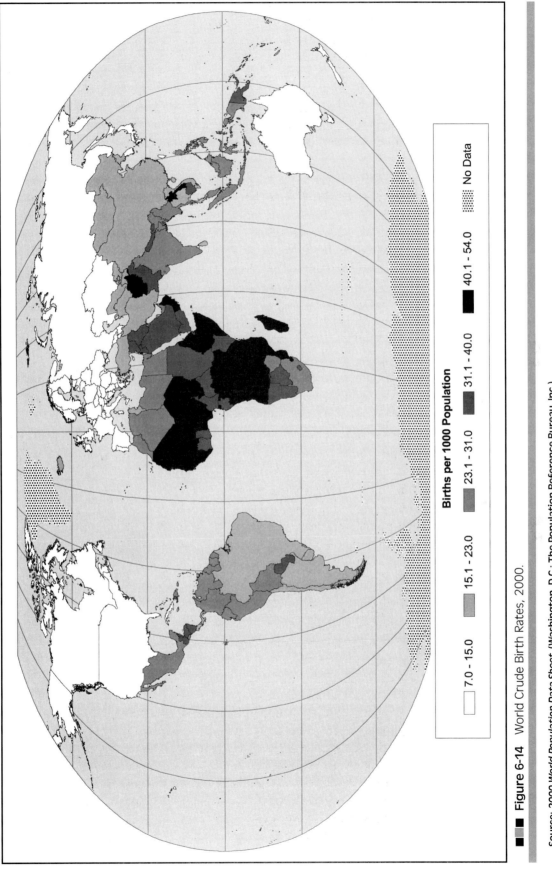

Figure 6-14 World Crude Birth Rates, 2000.

Births per 1000 Population

7.0 – 15.0	23.1 – 31.0	40.1 – 54.0	
15.1 – 23.0	31.1 – 40.0	No Data	

Source: *2000 World Population Data Sheet.* (Washington, D.C.: The Population Reference Bureau, Inc.)

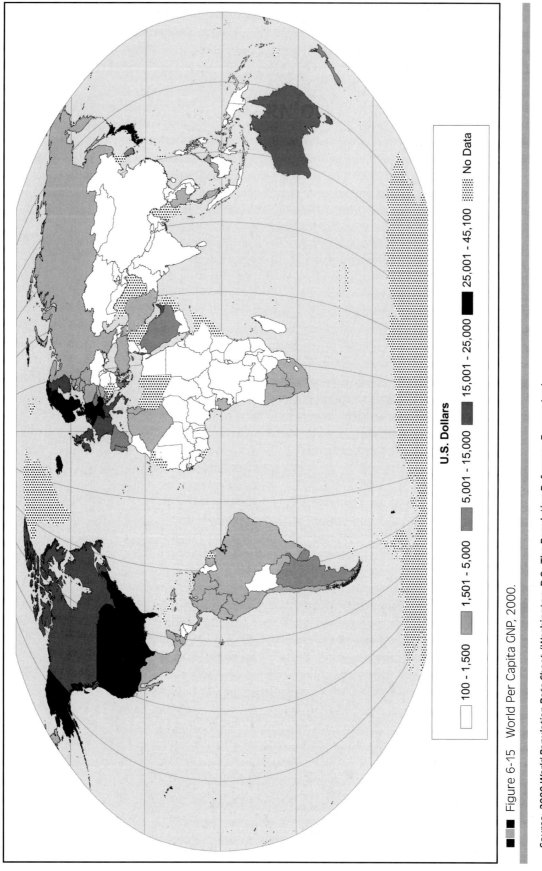

Figure 6-15 World Per Capita GNP, 2000.

U.S. Dollars

100 - 1,500   1,501 - 5,000   5,001 - 15,000   15,001 - 25,000   25,001 - 45,100   No Data

Source: *2000 World Population Data Sheet.* (Washington, D.C.: The Population Reference Bureau, Inc.)

Chinese population, despite its major demographic significance, we are still left at a point described by Orleans (1976, 56) as follows:

> It is easy to sympathize with anyone who has to choose a particular population figure or series of figures for China, but unfortunately one can offer little more than consolation. It is possible to agree or disagree with some of the assumptions and conclusions and to support either a particular estimate or a divergent point of view, but there is no way to prove that one set of figures is more accurate or more reliable than another.

In Southern Asia the crude birth rate is 33. India, Asia's other population giant, has a crude birth rate of 27. According to Mauldin (1976, 245), ". . . the evidence is unmistakable that fertility has fallen significantly in India during the past 15 years even though the level of fertility is not known with precision." With an annual natural increase of 1.8 percent in 2000 India continued to grow steadily, despite three decades of attempts to control fertility. Krishnan (1989) found that there were widespread differences in fertility within India, with crude birth rates ranging from only 20.4 in Goa to 39.7 in Rajasthan. India recently became the second nation to reach a population of one billion.

In 2000 Latin America had a crude birth rate of 24 and, like Asia, showed considerable regional variations in fertility. According to Palloni (1990, 143), "Most Latin American countries are currently undergoing what could be considered the third revolution in their demographic patterns . . . the downward shift in fertility." In Central America the crude birth rate is 26, with a high of 37 in Guatemala. Costa Rica has the lowest crude birth rate in Central America, 22, a figure that is down somewhat from recent years. Only in temperate South America and parts of the Caribbean do crude birth rates get much lower. Growth rates are high throughout most of Latin America because of the prevalence of low crude death rates, though declining birth rates should reduce growth rates and have already done so in many countries.

In North America, Canada has a crude birth rate of 11 and the United States has one of 15. The annual rate of natural increase for the two together is around 0.60 percent. Though these growth rates are slow compared to those in the developing countries, growth in the United States and Canada contributes substantially to the growth in demand for energy and other resources.

Similarly, Europe has an overall crude birth rate of 10. Crude birth rates within Europe vary from a low of 7 in Iceland to a high of 18 in Albania, which also has the lowest per capita GNP in Europe. In general, crude birth rates are lower in Western Europe and higher in Eastern Europe, though such differences are not large.

In Oceania the crude birth rate is 18. Within the region the rate varies from a low of 13 in Australia to a high of 37 in the Solomon Islands. Again, the rates tend to reflect differences in economic development.

Clearly, despite several exceptions, fertility patterns at the world scale reflect the spatial distribution of economic development. Within the realm of the developed countries crude birth rates have fallen to quite low levels, sometimes even below crude death rates, as in Germany and Austria. Total fertility rates are currently below replacement level in many developed countries, including the United States, Denmark, Sweden, Norway, Hungary, and Switzerland.

Fertility has begun to decline in a number of developing countries also, including the People's Republic of China, Hong Kong, Taiwan, South Korea, Panama, Costa Rica, and Mexico. Using considerable information from area studies, Thornton and Lin (1994) provide an engaging account of fertility changes in Taiwan, noting the way that the Chinese population was able to maintain considerable continuity in a society that was undergoing rapid modernization; the family and fertility changes are considered within the broader context of social change. Without a doubt, during the 1990s more developing countries were added to the list of those with declining levels of fertility.

Everyone interested in population keeps an eye on China because of its size, if for no other reason. In 1979 the Chinese government decided to promote the one-child family in a dramatic effort to bring population growth to a halt, even though Chinese fertility had dropped during the 1970s. By 2000 China's crude birth rate was 15, a very low figure considering that country's low income level, but up somewhat from what it had been in 1980. However, the rate of annual increase was still almost 1.0 percent, which is considered problematic because of the base population of more than one billion people. Though the one-child policy seems acceptable in urban areas, a greater concern in the early 1980s was the considerable resistance to it that was apparent in rural areas. Goodstadt (1982, 53) noted, "In the countryside, the government's measures to solve chronic agricultural bottlenecks have led to more income for the larger family; and the family itself has considerable survival power because even urban China has not developed economic and social substitutes that offer better safeguards for the individual against old age and the other hazards of life."

In 2000 the Population Reference Bureau reported a world population of 6,067,000,000, a figure that was well beyond the 1990 high variant projection of the United Nations—5,327,000,000. As the Population Reference Bureau (1988, p. 3) earlier noted, "The differences in these absolute numbers mean far less than the actual progress of rates, in particular the total fertility rate (TFR), or average number of children per woman. The increase in fertility in China in 1986 and 1987, along with a slower than projected decrease in fertility in India, has led to a slightly higher than projected total fertility rate."

Fertility: Family Planning Programs

Family planning programs and national policies regarding population issues were a new phenomenon in the early 1960s. India was the only country in 1960 that had such a program. In the last forty years we have seen a significant increase in support for family planning activities in both the developed and developing countries. Today, the vast majority of the planet's people live in nations that have formulated and adopted national policies to reduce population growth. In a notable study of family planning programs, Nortmann (1988, 32) determined that 76 percent of the population in the less developed countries lived in countries with official policies and national family planning programs to reduce population growth—the percentage of couples using contraceptives in developing countries continues to grow steadily, reaching levels of 50 percent or more in many of them.

The global diffusion of family planning practices has been the result of efforts by both public and private organizations. In the early 1960s only about 18 percent of the women of childbearing ages in the developing nations were practicing family planning. The total fertility rate was slightly over 6 children per woman in the developing world compared with 2.7 children for the industrialized nations. Today, the total fertility rate for the less developed countries has dropped to 3.2 while that for the developed nations is 1.5 children per woman. Half of all couples of childbearing age in the developing world and 70 percent in the developed world use some means of birth control (United Nations, 1990).

Nonetheless, researchers estimate that every day about 400,000 conceptions occur worldwide, as many as half of them not intended,

> In the last forty years we have seen a significant increase in support for family planning activities in both the developed and developing countries.

Courtesy of Photo Disc

and a smaller portion of those truly unwanted. As Potts (2000, 90) succinctly pointed out, "The quality of life of a large proportion of humanity during the coming century—and the future size of the global population—will depend critically on how quickly the world can satisfy the currently unmet demand for family planning."

In 1951 India became the first country to adopt an official policy to support family planning and slow population growth. In the early 1960s four more countries (Table 7-1), primarily in Asia, announced population growth control policies. Since the mid-1960s, many more developing countries have adopted policies, although 50 developing countries (39 percent of the total) do not have specific policies to lower fertility and another 18 (14 percent) have policies to maintain or raise fertility. The countries are mostly small nations and represent only about 18 percent of the developing world's population (Figure 7-1).

Two terms that are often (but not accurately) used interchangeably need to be introduced and defined—namely, the terms *policy* and *program*. A national **population policy** is an official government policy that is specifically designed to affect the size and growth rate of a population, the distribution of a population, or its composition. The important feature of such a policy is that the government *intentionally* plans to control one or more demographic variables. As noted in a United Nations (1969, 178) report:

A national **population policy** is an official government policy that is specifically designed to affect the size and growth rate of a population, the distribution of a population, or its composition.

> All governments design policies, adopt administrative programs, and enact laws which intentionally or unintentionally directly influence the components of population growth—fertility, mortal-ity, and international migration—as well as the internal redistribution

TABLE 7-1. First Governments to Support Family Planning and Slower Population Growth		Annual Growth Rate	
Country	Year Policy Adopted	1960–1965	1985–1990
India	1951	2.26	2.07
Pakistan	1960	2.69	3.44
Republic of Korea	1961	2.64	0.95
China	1962	2.07	1.45
Fiji	1962	3.27	1.78
Egypt	1965	2.51	2.39
Mauritius	1965	2.64	1.17
Singapore	1965	2.81	1.25
Sri Lanka	1965	2.43	1.33
Turkey	1965	2.49	2.08

Source: Peter J. Donaldson and Amy Ong Tsui. (1990) "The International Family Planning Movement," *Population Bulletin* 45:5.

of the nation's inhabitants. However, such measures represent national population policy only when implemented for the purpose of altering the natural course of population movements.

Once a government policy is adopted and a set of objectives has been specified, a population *program* may be initiated. The program includes the various means and measures that must be utilized in order to achieve the objectives of the population policy. For example, a government establishes a policy for the expressed purpose of lowering fertility. A family planning program may then be established to provide information and methods for controlling conception. Such programs may be official government programs or they may be private programs aimed at aiding the government in operationalizing its policy. According to Ross and Frankenberg (1993, 13):

> The principal functions of the programs are to provide services, information, persuasion, and legitimation; countries vary greatly in the relative emphasis they give to these aspects. They also differ in coverage of the population for each function. Some programs operate only in clinics in the larger cities; at the other extreme, some deliver a variety of services to the doorstep in the villages.

Explicit population policies are aimed at changing selected demographic variables. However, these demographic variables may also be influenced indirectly by other economic and social policies, including tax laws (income tax deductions for children, for example), public education, welfare (currently a hot-button issue in the U.S.), health, and various development programs. Furthermore, not all governments adopt national population policies, though they may, as in the case of the United States, support, or at least permit, population-related programs.

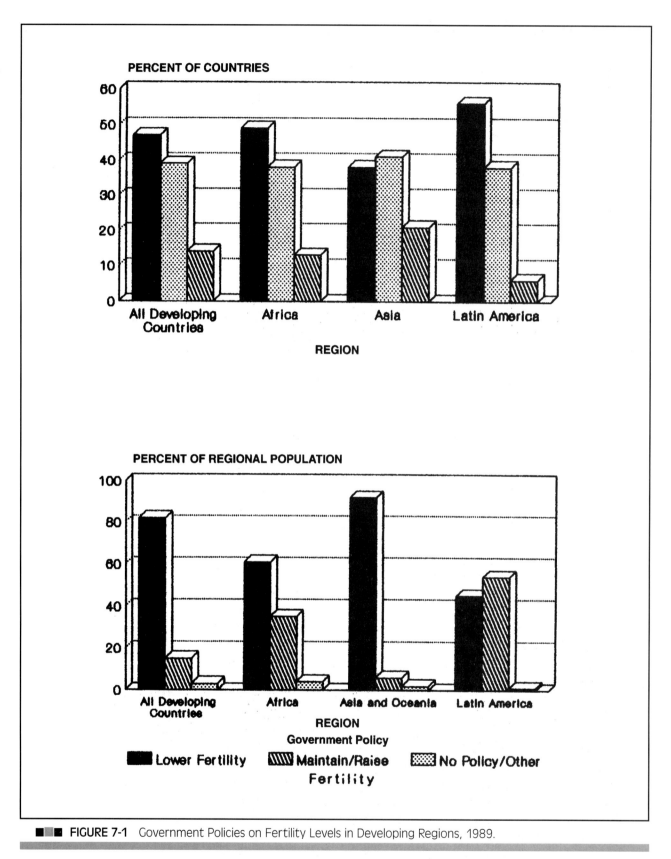

FIGURE 7-1 Government Policies on Fertility Levels in Developing Regions, 1989.

Source: Peter J. Donaldson and Amy Ong Tsui. (1990) "The International Family Planning Movement," *Population Bulletin* 45(3):6. Courtesy of the Population Reference Bureau, Inc.

OBJECTIVES OF POPULATION POLICY

Most national population policies, especially in the developing countries, have been conceived to decrease the prevailing high rates of population growth. The central focus of such policies is upon decreasing fertility. However, some earlier population policies in the developed countries were designed to encourage higher birth rates. As Schroeder (1976, 6-7) noted:

> Declining fertility led governments to proscribe abortion and contraception, and to adopt measures aimed at promoting large families, early marriages, and increased immigration. In most countries, the decision to promote more rapid population growth was reinforced by other factors—religious, military, or economic.

Examples of attempts to encourage higher birth rates include egalitarian and welfare measures in Sweden, the pronatalist policy of Hitler's Germany, and family subsidies in France. Similarly, for the old Soviet Union, Thomlinson (1976, 604) commented that " . . . worries over the low birth rate and consequent slow rate of population increase were among the reasons that led Soviet leaders to revoke legalization of abortions in 1936. . . ." Though changes in some of these earlier policies have occurred in recent years, many developed countries still have policies, direct or indirect, that could be described as pronatalist. Between 1976 and 1989 the number of governments that limited access to modern methods of birth control fell from 15 to 6 (Table 7-2). The only nations that currently restrict access to modern birth control methods are Iraq, Kampuchea, Laos, Mongolia, Saudi Arabia, and the Vatican. The number of nations that do not directly support modern birth control fell from 28 to 20 countries. However, most recent policy trends are increasingly antinatalist, especially in the developing countries.

Government policies designed to reduce mortality are virtually universal. The bases of such programs are mainly economic, but they are also humanitarian in nature. Programs for carrying out death control policies are many and varied. Unlike programs to decrease fertility, those designed to increase health and welfare meet with universal acceptance. Examples include disease control programs, work-safety programs (though these are regularly denounced by conservative politicians in the United States), and medical care programs.

Government immigration policies are also common in most countries, though they are designed for different purposes in different places; most have become quite restrictive. Their impact is mainly on the size and composition of the national population, though arguments that favor immigration are more generally made in the realm of labor market considerations. Current high rates of immigration in the United States have stirred contentious debates in recent years; we'll explore some of the relevant issues in the next chapter.

The development of population policy is usually slow and is often fraught with difficulties; fertility decisions are personal and private, though the social consequences of such decisions are easy to see. The

■■■ TABLE 7-2. Trends in Government Policies Regarding Modern Methods of Birth Control, 1976–1989

Date of Assessment	Total	Access Limited	No Support Provided	Access Not Limited	
				Indirect Support Provided	Direct Support Provided
Number of Governments					
1976	158	15	28	18	97
1978	158	13	27	20	98
1980	165	11	27	22	105
1983	168	7	32	28	101
1986	170	6	18	24	122
1989	170	6	20	20	124
Percentage of Governments					
1976	100	9	18	11	61
1978	100	8	17	13	62
1980	100	7	16	13	64
1983	100	4	19	17	60
1986	100	4	11	14	72
1989	100	4	12	12	72

Source: United Nations. (1989) *Trends in Population Policy,* Population Studies No. 114 (New York: United Nations), table 16.

Population Council has suggested the following series of eight steps that might ordinarily characterize the sequential development of a policy to limit fertility (Nortman, 1973, 11):

1. Government interest in population as it impinges on development planning;

2. A pronouncement on population matters by some responsible public official;

3. The establishment of a commission to study the demographic situation;

4. Surveys to determine the extent of knowledge, attitude, and practice of contraception (KAP surveys); the establishment of pilot population projects; and course work in demography at the university level;

5. The allocation of a new or increased budget for family planning, usually through official health programs;

6. The elaboration of specific demographic targets and time frames;

7. Establishment of a family planning apparatus somewhere in the government to implement programs;

8. Integration of all previously existing population related programs and development of subsidiary activities such as public information and evaluation.

It should be apparent that the path to government adoption of a fertility control policy, as well as the development of programs designed to meet the goals of that policy, is a lengthy and controversial one. Still, the increasing awareness of such policy needs is helping to create the necessary atmosphere in which to adopt and implement growth control policies.

Out of the adoption of fertility control policies has come the need for family planning programs; they are normally government supported and administered. Their stated primary purpose is to provide birth control information and services on a voluntary basis to people who desire such information. The difficulties of the task set out for family planning programs are many, but the perceived need seems to justify their existence. Some time ago Berelson (1972, 201) made the following comment on the state of family planning: "In my view, the present state is simultaneously impressive, frustrating, uneven, inadequate, and doubtful or unknown. The prospects are promising and dubious." Berelson suggested that the difficulties facing family planning programs could be classified as political, bureaucratic, organizational, economic, cultural, religious, and personal, along with the problem of sheer size in many countries.

FAMILY PLANNING PROGRAMS

With a continual lowering of the world's death rate, the only humane path to slowing population growth (for those who believe that slow growth, or even no growth, would be beneficial to both the earth and its residents) is to lower fertility. Those countries that have not yet completed the demographic transition need to do so. As we have already seen, modernization in the developing countries, as important as it is, may not be sufficient to bring about needed fertility declines. Though family planning programs seem relatively new, their history is not; on the bright side, however, we are observing lower total fertility rates almost everywhere.

BRIEF HISTORICAL SKETCH

Perhaps the earliest attempt to advocate the use of contraceptives to control family size occurred early in the Industrial Revolution in England with the publication of *Illustrations and Proofs of the Principle of Population* by Francis Place in 1822. Place's argument in favor of controlling family size was purely economic. He argued that workers could receive higher wages and better working conditions if only they could limit the supply of laborers available. Following Place, similar books were published in England and in the United States. However, during the 1870s legal attempts to suppress such literature were made in both countries.

In the United States the **"Comstock Law,"** which forbade the dissemination by mail of information about birth control, was passed by Congress in 1873. Similar laws against the distribution of birth control literature were also passed by many states and the importation of such literature was outlawed in 1890.

In America the chief voice in the early campaign for family planning was that of Margaret Sanger. Following the efforts of Sanger and others, the Comstock Laws were finally repealed; however, the last of the Comstock Laws was repealed by the Massachusetts legislature only three decades ago, in 1966. By then the first oral contraceptives were being distributed.

In 1916, the same year in which the Comstock Law was repealed, Margaret Sanger opened America's first birth control clinic in Brooklyn, New York. A similar clinic, aimed at serving poor women, was opened five years later by Marie Stopes in England. By the mid-1930s the medical profession voiced its acceptance of family planning through the qualified endorsement of birth control by the conservative American Medical Association (AMA). The history of landmark events in the international family planning movement is outlined in Table 7-3.

After World War II there was a significant increase in both the stature and the number of private groups that promoted family planning activities. The **International Planned Parenthood Federation (IPPF)** was established, the United Nations established the Population Commission, and John D. Rockefeller, 3rd, established the Population Council (which remains one of the most important agencies dealing with population issues, especially population control and family planning). According to Donaldson and Tsui (1990, 10):

> The **IPPF** is the most prominent and widely recognized private sector effort to support family planning internationally. Its objectives were (and are) to promote family planning and population education, to support family planning services, and to stimulate and disseminate research on fertility regulation. In 1990, IPPF had 107 family planning member affiliates representing 150 countries.

RECENT DEVELOPMENTS IN FAMILY PLANNING PROGRAMS

A surge of activity and rapid expansion of the family planning movement occurred during the 1960s and 1970s. This support was evident in both the developing and the industrialized nations. In 1968 the United States Congress specifically allocated funds to support family planning activities. Family planning became a central component of the United States Agency for International Development (AID) program. In the next two decades AID contributions to population and family planning assistance throughout the Third World totaled $3.9 billion, making it the largest single donor of population aid (Gillespie and Seltzer, 1990, 562).

Another important event during this period was the establishment of the United Nations Fund for Population Activities (UNFPA)

TABLE 7-3. Landmark Events in International Family Planning

Early Advocates

1860 Malthusian League founded in England to spread information on birth control.

1873 Comstock law in the United States enacted to prohibit advertising or prescription of contraception (later repealed in 1916).

1916 First free birth control clinic opened in Brooklyn, New York, by Margaret Sanger.

1921 First birth control clinic opened in England by Marie Stopes.

1937 American Medical Association gives qualified endorsement to birth control.

Post-World War II Developments

1946 United Nations Economic and Social Council establishes a population commission representing member governments and a population division within the secretariat.

1951 India adopts family planning as part of its economic program.

1952 John D. Rockefeller, 3rd establishes the Population Council

Constitution for the International Planned Parenthood Federation (IPPF) drafted at an international conference in Cheltenham, England (ratified in 1953).

Rapid Program Expansion

1960 Oral contraceptives are introduced.

1961 Plastic IUDs become available.

1967 The trust fund for the United Nations Fund for Population Activities (UNFPA) is created.

1968 United Nations International Conference on Human Rights issues the Teheran proclamation of which article 16 states: "Parents have a basic human right to determine freely and responsibly the number and spacing of their children."

Pope Paul VI issues **Humanae vitae** banning the use of artificial conception.

Paul Ehrlich publishes **The Population Bomb.**

U.S. Congress first allots foreign aid funds for family planning.

1969 First non-eugenic, non-restrictive sterilization laws are passed in Singapore and the state of Virginia, U.S.A.

1973 U.S. Supreme Court upholds **Roe v Wade** decision, limiting the right of states to interfere in the private decision of a woman and her doctor to terminate a pregnancy.

1974 United Nations holds first world conference on population in Bucharest. The United States urges countries to adopt policies aimed at slowing growth, primarily through family planning.

1979 The People's Republic of China begins campaign for "one child for one couple."

Political Realignments and New Technologies

1984 The Second UN World Conference in Population is held in Mexico City. Most third world countries favor slower population growth and family planning. The United States shifts its position,stating that population growth is a neutral phenomenon.

1985 A law is enacted that prohibits the U.S. government from supporting any organization that supports or participates in the management of a program of coercive abortion or involuntary Sterilization. Support for the UNFPA and IPPF is suspended.

1988 RU-486 is approved for terminating early pregnancy in France.

Source: Peter J. Donaldson and Amy Ong Tsui. (1990) "The International Family Planning Movement," *Population Bulletin* 45:8. Courtesy of the Population Reference Bureau, Inc.

in 1967. The goals of UNFPA are ". . . to provide an awareness of population problems and their relationship to social and economic development, the implementation of population policies, and the spread of family planning to better the health and well-being of the individual, family, and community" (Donaldson and Tsui, 1990, 11).

Americans became more aware of population issues through the publication of Paul Ehrlich's book, *The Population Bomb*, and a number of organizations were founded to publicize the problems associated with rapid population growth (examples include Zero Population Growth, the Population Crisis Committee, and the Population Institute). Population studies also increased at the nation's academic institutions. Population studies programs were started or expanded at a number of universities, including Brown, Chicago, Michigan, North Carolina, Pennsylvania, and Princeton.

The 1980s saw many new developments regarding population and family planning. According to Donaldson and Tsui (1990, 12):

> This period has been characterized by the increasing commitment of the public and policymakers in the Third World to family planning, both to slow population growth and to improve the health of women and children. The period has also witnessed the waning of U.S. Government concern about the effects of rapid population and of its support for family planning.

During the 1980s a rising tide of political conservatism in the United States, led by then-President Ronald Reagan, brought about a drastic change in United States policy relative to population issues. That change led the United States to withdraw its financial support for UNFPA and the IPPF. The central problem, which has by no means disappeared, is the association that is made between population policy and abortion. Interestingly, though abortion is legal in the United States, it has been primarily arguments that abortions were being performed as a part of some family planning efforts that successfully curbed support for them.

Fortunately for UNFPA and the IPPF (and probably for the earth itself), the cutting of funds by the United States was more than matched by additional funding support from the nations of Scandinavia, Western Europe, and Japan. A comparison of support and changes from 1984–1989 are shown in Table 7-4. Under the Clinton administration funds for population activities have expanded, though not without considerable debate in Congress. Support has again diminished under the new Bush administration.

IMPLEMENTING FAMILY PLANNING PROGRAMS

The success of family planning programs depends on continuing financial support. In addition there must also be public support, strong political support, and good management and organization. These family planning programs have been described as ". . . organized efforts to assure that couples who want to limit family size and space their children have access to contraceptive information and services and are encouraged to use them as needed" (Simmons, 1986, 175).

Ten Largest Donor Governments	**1984 Millions of US$**	**Ten Largest Donor Governments**	**1989 Millions of US $**
United States	$38.0	Japan	$40.2
Japan	19.8	Netherlands	24.5
West Germany	12.5	West Germany	21.1
Norway	10.9	Norway	19.6
Netherlands	9.5	Sweden	17.6
Canada	8.8	Finland	14.9
Sweden	5.7	Denmark	11.6
Denmark	4.5	Canada	9.8
United Kingdom	3.9	United Kingdom	8.7
Italy	1.8	Switzerland	4.4
Other governments	7.0		7.1
Total contributions	122.4		179.5
Percentage from 10 largest donors	94		96

■■■ **TABLE 7-4.** Government Support for UNFPA, 1984-1989

Source: UNFPA, 1984 and 1989 *Annual Reports* (New York: United Nations).

Five types of activities have been particularly successful in the development of family planning activities, particularly in the developing nations. These are:

1. Demonstration projects that were able to establish that there was a demand for family planning services and that these services could be delivered in a manner that was both medically safe and culturally acceptable;

2. The provision of contraceptives such as the IUD, condoms, spermicides, or the pill;

3. The support of training programs in population and family planning;

4. Assistance with surveys and national censuses; and

5. Other technical assistance, including that which would help the staff of family planning programs improve counseling on contraceptive side effects, expand the types of contraceptive methods offered, or conduct clinical trials to evaluate new contraceptive methods (Donaldson and Tsui, 1990, 18–20).

Though some combination of these and other activities may help meet the needs of growing numbers of potential parents, numerous obstacles remain. As Potts (2000, 90) noted, "An estimated 120 million couples in developing countries do not want another child soon but have no access to family-planning methods or have insufficient information on the topic. Consequently, pregnancy too often brings despair instead of joy."

FAMILY PLANNING
IN DEVELOPING COUNTRIES

Family planning programs in developing countries, for the most part, have sprung up in the last thirty years. Family planning services were originally introduced in developing countries in the 1940s and 1950s by private physicians and private groups to women who were self-motivated. By the late 1950s and early 1960s, however, governments became aware of rapid population growth and its socioeconomic consequences; this led to an enthusiasm for family planning.

Government support of family planning programs is now widespread among the developing nations. The number of men and women using contraception has grown more than six-fold since the 1960s, with over 400 million of the estimated 850–880 million married couples now using contraception. Nearly 70 percent of these couples live in the developing world.

Recent estimates (Figure 7-2) suggest that roughly 50 percent of married women of childbearing age in the developing nations use some form of family planning. There is considerable regional variation, however, ranging from a low of 17 percent in Africa to a high of 75 percent in East Asia. Even in Africa, more women are learning about family planning methods and searching for alternatives to repeated pregnancies and births. Fertility is declining in many African countries, even though economic development and wealth transfers would not predict such a result (Caldwell, 1994; Dow, et al, 1994).

This rapid increase in contraceptive use is primarily the result of increases in sterilization and the use of modern contraceptives such as the pill and the IUD; newer contraceptives, especially Norplant, may become important as well. The most widely used form of contraception in the world today is sterilization. In almost 45 percent of Third World couples who use birth control, one partner has been sterilized. Female sterilization outnumbers male sterilization in Third World countries by almost three to one, even though the procedure is still more complicated for women than it is for men. These figures are likely to change in the years ahead, however, as feminist scholars and activists become both more numerous and more powerful. One group of such women, DAWN (Development Alternatives with Women for a New Era), promotes broad concerns with gender and development, many of which are chronicled in Corrêa and Reichmann (1994). The second most prevalent type of contraceptive in the Third World is the IUD, used by over seventy million women, followed by the pill, used by 38 million women. There can be no argument that males have not shared equally in contraceptive responsibilities.

Methods of contraceptive use vary regionally throughout the world. Essentially, half of the birth control practiced in Asia and the Pacific is through sterilization, whereas the comparable figure in Latin America is a little over a third and in Africa less than 8 percent. Latin America also relies heavily on oral contraceptives, which are practi-

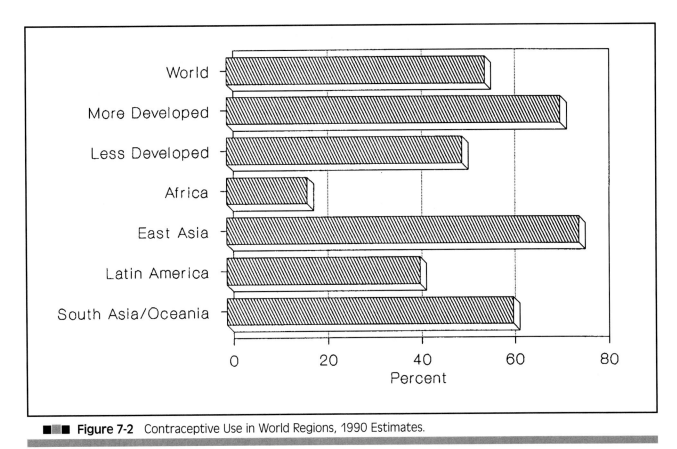

Figure 7-2 Contraceptive Use in World Regions, 1990 Estimates.

Source: Peter J. Donaldson and Amy Ong Tsui. (1990) "The International Family Planning Movement," *Population Bulletin 45* (3):20. Courtesy of the Population Reference Bureau, Inc.

cally as popular as sterilization. In Africa, 38 percent of contraceptive users rely on the pill, with a significant proportion, 23 percent, relying on traditional methods. Given the widespread incidence of AIDS in Africa (where it is about equally prevalent among males and females), it makes sense to promote condoms more than other methods.

In an effort to look at regional variations in family planning programs within the Third World, Freedman (1990) divided these countries into three categories:

1. less developed countries with rapid development

2. countries with moderate development

3. countries with relatively little development

In countries with rapid development—such as Korea, Taiwan, Mexico, and Singapore—fertility rates have fallen rapidly. For example, in Taiwan in 1960, just before both family planning and development programs started, the total fertility rate was about 6 and contraceptive use was at low levels. By 1986, according to Freedman (1990, 37), ". . . only 26 years later, the fertility rate had fallen by 70 percent to 1.7—below replacement levels—and virtually all women were using contraception before the end of their childbearing years." Although Taiwan's family planning program coincided with rapid social and economic development, it is ". . . unlikely that the disadvantaged masses—the poor, rural, and illiterate—would have adopted family planning so rapidly without the family planning program." (Freedman, 1990, 37) In 2000 Taiwan's total fertility rate was 1.5, still well below replacement level.

Freedman's second category, those countries with moderate development, included China, Indonesia, and Thailand. In these countries rapid fertility decline has taken place under initially rather unfavorable social and economic conditions. Freedman attributes this decline of fertility and increased use of contraception to a variety of factors, including the mobilization of the bureaucratic infrastructure and the ability of governments to incorporate the village masses into family planning programs.

The third category includes those countries with little socioeconomic development and weak infrastructure. Most of the countries of sub-Saharan Africa, as well as India, Pakistan, Bangladesh, and the smaller countries of Southwest Asia, are in this category. Although some of these countries have family planning programs, they have not been very successful in increasing contraceptive use enough to make a major decrease in fertility. Freedman believes that the reason for this lack of success is because ". . . the family planning system is weak, with poor-quality services provided in ways that are inappropriate for the local culture." (Freedman, 1990, 39)

Though there are continuing debates about the efficacy of family planning programs, increasing evidence suggests that they are having a significant impact on fertility in many places. Most of the debate focuses on whether economic development and its concomitant social and economic transformations are a necessary precondition for fertility decline. Family planning advocates argue that an increase in the supply of birth control information and services can bring about significant fertility reductions without substantial economic development.

A variety of studies have attempted to evaluate the role of family planning in fertility decline. For example, Lapham and Mauldin (1987) measured the quality of national family planning programs in 91 less developed countries. They concluded that high-quality family planning programs play a significant role in contraceptive use and fertility decline, independent of social and economic change. Another significant conclusion from their study was that family planning goes hand-in-hand with steady economic development and that higher levels of contraceptive use and greater and more rapid birth rate declines were a joint product of the two forces.

Family planning programs cannot take all the credit for fertility declines, but in many places they have had an impact. Other factors in recent fertility declines include improvements in literacy, improvements in the status of women (though they remain considerably disadvantaged relative to males in most developing countries), rural development programs, income redistribution, and population education programs. Further improvements in the lives of women are necessary, however, not just for fertility control but also for the betterment of the societies in which they live. Schultz (1995) provides a stimulating series of essays focused on the topic of investing more in human capital for females, not just in labor markets but also in health and education.

Improved communications are important in determining the effectiveness of family planning programs. Various barriers to communication, from cultural sensitivities to language differences, have impeded the family planner's ability to get the message across. Various forms of communication have been tried, from traditional means to mass media approaches. Posters showing small, happy, healthy families have been common in several countries, including India, Taiwan, and Jamaica. In one study Worral (1977) suggested that family planning communication tended to proceed at stages that were dependent on the government's attitude and policy.

After observing that fertility rates are coming down in a number of developing countries, and after discussing improvements in communicating family planning activities, it is useful to look at the implications of these fertility trends for future contraceptive use in the developing world. Estimates of the number of women using contraception in selected future years were made by Bongaarts (1986, 131) and appear in Table 7-5. In 1984 approximately 256 million women were practicing contraception; of that number, more than half lived in East Asia (China). Estimates are that by the year 2025 there will be a threefold increase to 736 million. The largest percentage increase is projected to be in South Asia (from 81 to 307 million) and Africa (from 11 to 168 million). According to Bongaarts (1986, 131): "These projections make clear that a rapid expansion of family planning services provided by both the public and private sectors will be required for a least the next several decades. This need for additional services is greatest in South Asia and Africa."

In a more recent study Bongaarts (1991) analyzed the unmet need for contraceptive services; he found that the total unmet need was strongly correlated with the fertility level (Figure 7-3). The average total demand was nearly 60 percent, of which 42 percent represented unmet use and the remainder, 17 percent, was unmet need. According to Bongaarts (1991, 311):

> Differences by region and level of fertility in these variables are as expected: the total demand in sub-Saharan Africa and in countries with high fertility is lower than elsewhere. In these countries the unmet need actually exceeds the current level of contraceptive use by a substantial margin. The unmet need represents a smaller but still important proportion of total demand in the Asian and Latin American populations and in countries with low fertility.

A variety of factors are responsible for this unmet need for contraceptives, including poor quality (and often poorly-financed) family planning services, limited access to these services, lack of contraceptive knowledge, weak motivation, and opposition from

The Happening in Cairo

Cairo seemed a perfect place to get delegates to a United Nations conference more concerned about the effects of rapid population growth. Greater Cairo, for example, contains some 13 million residents; a half million Cairenes have taken up residence in a group of cemeteries, creating "The City of the Dead," and another half million or so live on rooftops around the city. Teeming crowds are everywhere, yet conference results, though encouraging a variety of admirable goals, seem unlikely to reduce population growth very rapidly.

Under the auspices of the United Nations, the International Conference on Population and Development was held in Cairo between September 5 and 13, 1994. Before Cairo, similar conferences had been held in Mexico City, 1984, and Bucharest, 1974. Though population is the central theme in each, they have differed considerably in many ways.

More than 10,000 people came to Cairo, representing about 180 countries. For three years or so prior to the conference, deliberations took place among government representatives, community leaders, experts on demography and development, and a host of non-governmental organizations (NGOs)—their goal was to develop a Program of Action to guide conference proceedings, which they did.

Out of the Program of Action, modified by deliberations about it before and during the actual conference, came an outline of plans for the next two decades—a strategy for population and development. Despite criticisms from every side, from the Vatican to Neomalthusians, the general outline of the plan survived; it focuses primarily on the areas discussed below.

Human rights and well-being took center stage in the Program of Action; any population and sustainable development efforts must consider their implications. Though stabilizing population remained a goal, language regarding that goal was not strong enough to make much difference, at least in the short run. According to the Program of Action, population issues must be seen within the context of human-centered sustainable development plans.

For the first time in the history of these decennial population conferences, gender equity, equality, and the empowerment of women received considerable attention. The Program of Action sets the empowerment of women as one of the cornerstones of population and development policies.

The Program of Action was innovative also in its attention to reproductive rights and health, especially for women. The document stresses the need for incorporating family planning into the broader area of reproductive health; it also reaffirms the universal right of couples to decide for themselves how many children they should have and how those births should be spaced. Also for the first time at such conferences, abortion, though not forbidden, was recognized as a potential health risk.

Another major goal of the Program of Action is to encourage men to help bring about gender equity, sharing power in every arena—from development to reproductive behavior and parenthood. Equal opportunities for education and employment, equal wages, and the sharing of responsibilities in the home are all to be encouraged.

■■■ TABLE 7-5. Estimated Number of Married Women Practicing Contraception, by Region, for Selected Years from 1984-2025

Region	Women Practicing Contraception (millions)				
	1984	1990	2000	2010	2025
Africa	11	16	35	74	168
Latin America	32	41	58	73	90
East Asia*	132	160	180	184	171
South Asia	81	120	190	250	307
Developing countries	256	337	462	581	736

*Excluding Japan.
Source: John Bongaarts. (1986) "The Transition in Reproductive Behavior in the Third World," in Jane Menken, ed. *World Population and U.S. Policy: The Choices Ahead.* (New York: W.W. Norton and Company), table 7, p. 131. Used by permission.

spouses. Although exact figures are unavailable, there are thought to be perhaps 100 million couples or unmarried individuals in Third World countries who have an unmet need for contraceptive services. Satisfying this unmet demand, alone, could significantly reduce fertility, and morally it seems the right thing to do (Potts, 2001). Though family planning programs focus on providing contraceptives and knowledge about them, and though meeting unmet needs for contraception seems crucial for most developing countries, other considerations need to be made as well. For example, more attention needs to be given to unwanted sex, male sexual aggression, and male roles in families and households.

Before going on to family planning in the developed countries, one further note should be made about programs in the developing countries; funding for family planning is increasingly a problem. International aid has accounted for one-third of the funds, while the less developed countries themselves have supplied the remaining two-thirds of the funding. The developing countries will be expected to increase their financial contributions in the future, though economic problems in many of them may make that difficult. The contribution by AID in the 1990s was lower in real dollar terms than it was in the late 1970s. More significant, however, is the increasing demand and desire to practice family planning by people throughout the world. Third World countries continue to increase their commitment and support for family planning efforts, while support from the largest international donor, the United States, is waning. In the early 1990s, according to Ross, et al (1993), official development assistance to the developing world from the United States amounted to only 0.32 percent of GNP (the same as for Japan); by comparison, Norway provided 1.14 percent, Denmark, 0.96 percent, and the Netherlands, 0.88 percent of their respective GNPs.

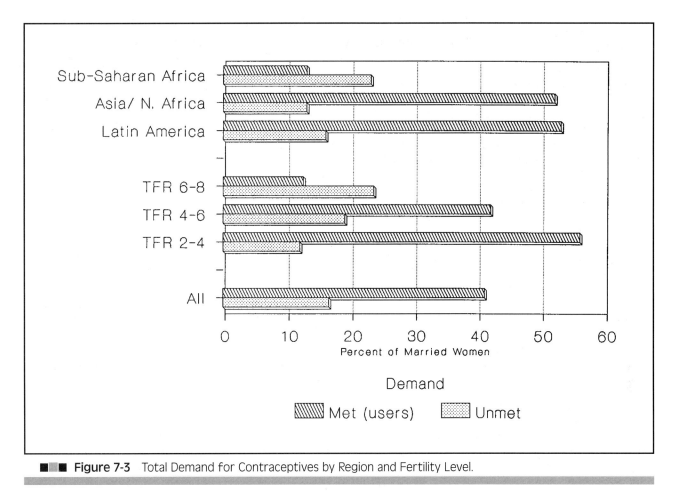

Figure 7-3 Total Demand for Contraceptives by Region and Fertility Level.

Source: Reprinted with permission of The *Population* Council, from John Bongaarts, "The KAP-Gap and the Unmet Need for Contraception," *Population and Development Review 17* No. 2 (June 1991): 310.

INDIA

Among the developing countries, India is generally believed to be the one whose government first favored lowering the rate of population growth. This antinatalist policy dates back to 1952, and it was incorporated into the first Indian Five-Year Plan. However, in the early years Indian family planning tended to be poorly financed. After a miserable start, followed by occasional lapses, the program was revitalized in 1965.

In 1965 the Indian Government established the Department of Family Planning, under the authority of the Ministry of Health and Family Planning. The Indian program is funded by the Central Government; however, most of the organization and operation of the program is left to the various states.

Until 1976, participation in the family planning program was voluntary. However, in 1976 compulsory sterilization was introduced in India, in the state of Maharashtra. From 1975–1977, then-Prime Minister Indira Gandhi suspended civil liberties and both the number

The Happening in Cairo (Continued)

Aside from those goals described above, the Program of Action also dealt with the reproductive health needs of adolescents, the family unit, migration (including urbanization and refugee flows), and the provision of resources of development projects. In every case the document reasserts the need for gender equity and the empowerment of women; in addition, the care of the elderly and the very young received attention in some categories, especially that of migration.

of sterilizations and abuse of the program reached its peak. "Public sector employees needed to show a vasectomy certificate to get a transfer, promotion, food license, or an allocation of a house or medical care," noted Singh (1990, 127).

Until the introduction of involuntary sterilization, the mainstay of the voluntary program had been sterilization, coupled with traditional contraceptives (especially IUDs and condoms). In the 1970s the Indian program attempted to improve communications and motivation; a need for convincing people of the advantages of smaller families was by then obvious to most observers.

A major element in Indian programs during the 1970s was a package of incentives. The major incentive was a financial award offered to men or women, with two or more children, who underwent a sterilization operation. The amount of the incentive was inversely related to the number of children. The maximum cash award was 150 rupees, more than a month's wages for the average agricultural worker. Beyond this cash incentive from the national government, the various states added other incentives, such as salary increases for sterilization, along with such disincentives as not providing maternity leave for children beyond the second one.

When the Janata Party assumed power in 1977, the family planning program was not dismantled. However, the program's emphasis was shifted considerably; only noncoercive methods were acceptable to the new government. Also, a new emphasis was placed on family welfare. Monetary incentives for sterilizations and IUD insertions were retained, with new rates of 100 rupees for a vasectomy and 120 rupees for a tubectomy, irrespective of the acceptor's current family size (Visaria and Visaria, 1981). Unfortunately, under the Janata Party people involved in the family planning program became demoralized and program achievements fell to low levels as well.

Between 1978 and 1981, however, the program gradually recovered. In 1979 the Working Group on Population Policy set a new long-range goal—India should reach replacement-level fertility by 1996 (a goal that has not yet been reached). In the same year the government planned for an increase in international family planning assistance for the 1980–85 period.

An analysis of data on contraception reveals some interesting changes. Approximately one-third (34.9 percent) of the total eligible couples in 1985–1986 were using effective contraceptive methods. Fifteen years earlier, in 1970–71, the comparable figure was 10.4 percent. The ineffectiveness of the program, however, becomes clear when birth rate statistics are analyzed. The decline in the birth rate during the same time period was very small, from 36.8 per thousand in 1970–71 to 33.9 per thousand in 1985.

The principal contraceptive method was sterilization, especially vasectomy, and it has proved ineffective. According to an analysis of a village program by Singh (1990, 128–129):

> The implementation of the sterilization program left much to be desired for reducing fertility. The pressure of meeting quotas

must have resulted in the sterilization of persons who were old or otherwise marginal to fertility reduction such as grandfathers, or unmarried older uncles. A father who voluntarily opts for sterilization usually does so after he has reached his desired family size, five or six children, with a sufficient number of sons.

According to Freedman (1990, 39), there is probably a consensus amongst most Indian and foreign observers that the program has been poorly executed and ineffective for several reasons:

1. An obsession with unreasonably high targets for the program has demoralized personnel;

2. There has been too much of an emphasis on sterilization linked to a controversial incentive program;

3. There has been little continuity of family planning program personnel;

4. Officials within the program have been out-of-touch with the realities of Indian village life.

According to Roberts (1990, 88): "In India, where states function relatively independently and represent a range of religious and ethnic groups, the national coercive effort to reduce birth rates proved impossible to implement." Despite the partial success of family planning programs, India recently became the world's second "demographic billionaire," not a distinction that drew an ovation from the struggling Indian government. Taken in March of 2001, the latest census of India reported a population of nearly 1.03 billion, an increase of 21 percent since 1991.

SOUTH KOREA AND THAILAND: TWO SUCCESS STORIES

Two family planning programs that have achieved considerable success are those in South Korea and Thailand. The national program in South Korea was adopted in 1961 and has been one of the most successful in the developing world. The achievements of the South Korean family planning program went hand-in-hand with impressive economic growth. Between 1962 and 1988, per capita income increased from $62 to $3,600, and the proportion of women who completed secondary school increased over 65 percent between 1965 and 1987 (World Bank, 1990).

Although these economic changes influenced people's views on family size, South Korea's family planning program also played a significant role in shaping people's perceptions of childbearing and family size. In the early 1960s family planning was practiced by less than 10 percent of the women at risk of pregnancy and the total fertility rate was 5.4. By 1991 the total fertility rate had fallen to 1.6

and the percent of married women using contraception had risen to 77 percent; today the total fertility rate remains well below replacement level.

The government family planning program that contributed to this impressive change in reproductive behavior in South Korea was, according to Donaldson and Tsui (1990, 40), ". . . based on a system of targets that established performance standards for the country's family planning outreach workers who were responsible for visiting women in their homes and encouraging them to begin to use contraception." Recent changes in the South Korean program have placed the emphasis on family planning as a way to improve both maternal and child health.

Thailand has also had a very successful family planning program. Between 1965 and 1991 the proportion of married women of reproductive age using a contraceptive increased from 15 to 68 percent. The total fertility rate also declined dramatically, from 5.7 in the mid-1960s to 2.0 in 1998.

The national family planning program was aimed at providing information about contraceptive use to all citizens, especially those in rural areas. Non-physicians, mostly secondary school graduates and auxiliary midwives, were trained to distribute a wide variety of contraceptives. Thailand was one of the first nations to permit the use of DMPA, an injectable contraceptive, and recently the contraceptive implant Norplant has been introduced.

The Thai program also successfully linked national and international resources, as well as bringing together both public and private sector programs. These efforts were strongly supported by the country's economic development and by the cultural environment in which the family planning activities took place (Knodel, et al. 1987).

Throughout Asia fertility levels are falling. Aside from those already mentioned, Singapore, Hong Kong, and Sri Lanka have all experienced notable fertility declines; other countries in the region are likely to follow (Leete and Alam, 1993).

FAMILY PLANNING IN DEVELOPED COUNTRIES

Because of the smaller populations in most developed countries, along with their lower rates of population growth, less emphasis is generally placed on discussions of family planning programs in developed countries. Though the nature of population problems in the developed countries differs from that of such problems in the developing countries, it still deserves, and is receiving, increasing attention (Berelson, 1974).

Earlier we pointed out that population policies in the developed countries were generally pronatalist in the past. However, current policies in most developed countries are more likely to favor lower fertility. Low fertility is characteristic of these countries, all of which

In some developing countries, materialism, working mothers, and pessimistic views of the future are starting to play a role in lowering fertility rates to below replacement level.

have more-or-less completed their demographic transitions, and the use of contraceptives is widespread (Figure 7-4). In some developed nations, especially Spain and Italy, total fertility rates are so low that demographers are no longer sure of what to say about them—nothing in the demographic transition model predicted total fertility rates that would drop as low as 1.2. Materialism, working mothers, and pessimistic views of the future are playing a role that no one foresaw.

Though the United States has no explicit population policy, it does have a concern with family planning services. In 1970 the Family Planning and Population Research Act was passed. Laws governing abortions have been liberalized as well, although recent changes supported by a powerful conservative agenda (spearheaded by the "religious right") have increasingly tried to restrict access to abortion, especially for the poor.

In Europe several nations have reached, or are close to reaching, zero population growth. Most of these countries have permissive policies with respect to contraceptives and abortions, but no official antinatalist policies. In some cases, for example France, even permissive policies are of recent origin. As Schroeder (1976, 24) noted:

> France—historically one of the most aggressively pronatalist countries of Western Europe—has recently made some sweeping changes in its posture toward family planning. These include a law passed in 1974 permitting distribution of contraceptives financed by the social security system, plus a comprehensive public information campaign and the provision of free medical examinations and family planning counseling.

■■■ TABLE 7-6. Contraceptive Use by Women, 15–44 Years Old: 1995

Contraceptive Status and Method	All Women[1]	Age			Race			Marital Status		
		15–24 Years	25–34 Years	35–44 Years	Non-Hispanic White	Black	His-panic	Never Married	Currently Married	Formerly Married
All Women (1000)	60,201	18,002	20,758	21,440	42,522	8,210	6,702	22,679	29,673	7,849
Percent Distribution										
Sterile[2]	29.7	2.6	25.0	57.0	30.2	31.5	28.4	6.9	43.2	45.1
Surgically sterile	27.9	1.8	23.6	54.0	28.5	29.7	26.3	5.7	41.1	42.5
Noncontraceptively sterile[3]	3.1	0.1	1.2	7.4	3.2	3.7	2.3	0.9	4.1	5.8
Contraceptively sterile[4]	24.8	1.7	22.4	46.6	25.3	26.0	24.0	4.8	37.0	36.7
Nonsurgically sterile[5]	1.7	0.7	1.3	2.8	1.6	1.8	2.0	1.1	2.0	2.2
Pregnant, postpartum	4.6	5.9	6.9	1.3	4.3	4.5	6.3	3.1	6.4	1.9
Seeking pregnancy	4.0	2.1	6.2	3.5	3.7	4.6	4.0	1.5	6.4	2.1
Other nonusers	22.3	44.4	13.3	12.6	21.1	23.1	26.3	46.8	4.7	18.4
Never had intercourse	10.9	30.8	3.4	1.4	10.4	8.9	12.1	28.9	—	0.0
No intercourse in last month[6]	6.2	7.0	5.3	6.5	5.7	7.2	8.6	11.5	0.5	12.7
Had intercourse in last month[6]	5.2	6.6	4.6	4.7	5.0	7.0	5.6	6.4	4.2	5.7
Nonsurgical contraceptors	39.7	45.0	49.1	26.1	41.2	36.1	35.1	41.8	39.7	32.4
Pill	17.3	23.1	23.7	6.3	18.8	14.8	13.6	20.4	15.6	14.6
IUD	0.5	0.1	0.6	0.8	0.5	0.5	0.9	0.3	0.7	0.4
Diaphragm	1.2	0.2	1.2	2.0	1.5	0.5	0.4	0.5	1.8	0.9
Condom	13.1	13.9	15.0	10.7	13.0	12.5	12.1	13.9	13.3	10.1
Periodic abstinence	1.5	0.5	1.8	2.0	1.6	0.7	1.3	0.6	2.3	0.7
Natural family planning	0.2	—	0.3	0.3	0.3	—	0.1	—	0.4	—
Withdrawal	2.0	1.6	2.3	1.9	2.1	0.9	2.0	1.5	2.3	1.8
Other methods[7]	3.9	5.6	4.2	2.1	3.4	6.2	4.7	4.6	3.3	3.9

— Represents or rounds to zero
[1]Includes other races, not shown separately.
[2]Total sterile includes male sterile for unknown reasons.
[3]Persons who had sterilizing operation and who gave as one reason that they had medical problems with their female organs.
[4]Includes all other sterilization operations and sterilization of the husband or current partner.
[5]Persons sterile from illness, accident, or congenital conditions.
[6]Data refer to no intercourse in the 3 months prior to interview.
[7]Includes implants, injectables, morning-after-pill, suppository, Today(TM) sponge and less frequently used methods.
Source: U.S. National Center for Health Statistics, Advance Data from Vital and Health Statistics, No. 182.

In the Commonwealth of Independent States population policy has been concerned mainly with population distribution. Higher fertility is being encouraged in selected areas, where growth rates are currently low. Both contraceptives and abortion, however, are readily available. Since the end of the Cold War fertility in Russia has fallen considerably; the total fertility rate was only 1.2 in 1998, well below replacement level. Furthermore, Russia's crude birth rate of 9 was well below the crude death rate of 18, resulting in a negative rate of natural increase of 0.5 percent annually. Fertility has also declined in most of the former Soviet satellites; Bulgaria, for example, has a total

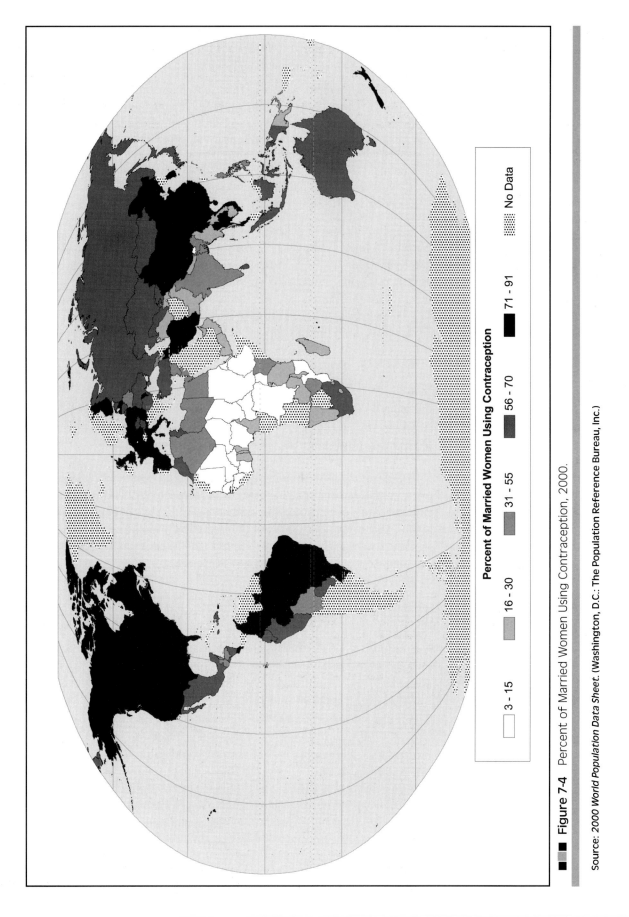

Figure 7-4 Percent of Married Women Using Contraception, 2000.

Percent of Married Women Using Contraception

☐ 3 - 15	▨ 16 - 30	▨ 31 - 55	▨ 56 - 70
■ 71 - 91	▦ No Data		

Source: *2000 World Population Data Sheet.* (Washington, D.C.: The Population Reference Bureau, Inc.)

fertility rate of only 1.2. In an intriguing study of low fertility in the former East Germany after the fall of the Berlin Wall, Witte and Wagner (1995, 394) concluded the following:

> Our finding that lower fertility is tied to individuals' own economic situation is an indication that some intimate aspects of behavior have already been changed through assimilation. Couples, knowing that unemployment is high, that their labor market value is relatively low, and that there are less generous maternity benefits, perhaps less flexible employers, and far fewer child care alternatives, respond rationally to socioeconomic change by limiting fertility.

In Asia, Japan, the largest developed country in the region, has a widely-accepted family planning program. Family planning services are available through more than 800 health centers throughout the country; abortions are readily available as well. The result is that Japan's total fertility rate has fallen to 1.3. According to Butler (1998) the cause is mainly a decline in marriages.

Throughout the industrialized world low fertility is common, even though explicit policies to encourage low fertility are not. A major element in low fertility has obviously been the effect of high and rising socioeconomic status, coupled with urbanization. Family planning programs often aid in the maintenance of low fertility levels.

EASTERN EUROPE: A MODERN DILEMMA

It is curious, to say the least, but in a world in which China and India, among others, struggle to alter birth rates downward through various policies and programs, the countries of Eastern Europe found that social and economic policies during the Cold War years led to undesirably low fertility levels. During the past thirty years fertility in Eastern Europe has followed a downward trend, in large part because of abortion.

In Eastern Europe, then, governments have often tried to encourage couples to have more children. "Prior to the adoption of explicitly pronatalist goals," according to David (1982, 32), "most of the countries already had a complicated structure of programs and policies that worked both directly and indirectly to ease the financial burden of large families." Included were paid maternity leaves, monthly payments to families with children, lump sum payments for the birth of a child, price supports for children's needs, tax breaks for parents, and a host of other features that would appear to encourage couples to have large families. But the desired results failed to materialize. David (1982, 32) pointed out that ". . . while these measures did lessen the economic burden of childbearing, they were not powerful enough to overcome the physical and psychological burden which childbearing posed to women holding full-time jobs—as socialist policy decrees they must— and also faced with inadequate housing and chronic shortages of consumer goods and services and childcare facilities." As van de Kaa (1987, 52) noted in a study of European population changes, ". . . collective and individual interests do not seem to coincide . . . Eastern European countries appear incapable of overcoming individualistic desires and raising fertility to replacement level."

Understanding fertility trends in Eastern Europe before the end of the Cold War requires an appreciation of the status of women in these formerly socialist countries. Although Marxist ideology gave women equal status with men, reality was not always in accord with ideology. David (1982, 39) noted that in Eastern European countries ". . . laws have not eradicated the vestiges of male chauvinism or changed the economic conditions which still make childbearing a particularly heavy burden for women, whatever financial incentives governments might devise to encourage them to bear children."

The recent democratic revolutions in Eastern Europe have brought monumental changes to the political and economic structure of these nations. It will be interesting to follow these changes and to see if they have an impact on demographic behavior. So far, fertility has dropped in virtually all of these nations, as it has in Russia.

LAW AND POPULATION POLICY

Policies designed to affect demographic variables are often politically and emotionally sensitive issues. They often involve serious changes in preexisting laws, especially laws that have controlled the dissemination of knowledge concerning contraceptive devices.

Gazing into the future, it seems that family planning programs are likely to expand, to reach an increasing number of couples. Laws governing the sale of contraceptives will most likely change in favor of increasing their availability. Governments are likely to increase their support for family planning programs and population policies designed to slow rates of population growth. If all goes well, then the decline in the world's population growth rate will continue, and the next century should see a world population that is nearly stable in size, though probably not before we reach 10 billion or so.

SUMMARY

Fertility control policies and family planning programs are of special interest to us because they are so closely related to the future growth of the world's population, as well as to the geographic distribution of that growth. Though population policies are also concerned with mortality and migration, both internal and international, our focus has been primarily on fertility.

Until recently everything seemed rather simple; we assumed that future fertility trends could be predicted, at least in general terms, from the demographic transition model. Given sufficient time and sustained development in a country, we could expect fertility to decline, though few demographers were very specific about when the decline would occur, how much decline there would be, or which developmental variables were most clearly related to changes in fertility. However, van de Walle and Knodel (1980, 38) commented, in their summary of literature on fertility declines in Western Europe, that ". . . there is only a loose relationship between the level of socioeconomic development and fertility decline." Economic development is no longer viewed as a necessary precondition for fertility decline.

Certainly some changes associated with modernization may facilitate fertility declines, as also may family planning programs. Reducing infant and childhood mortality; expanding educational opportunities, especially for women; improving the status of women; providing a more equitable distribution of the benefits of development; and raising the legal marriage age may all contribute to reducing fertility levels in the developing countries. All of these variables can be related to the decision to have a child, though specifying mathematically precise relationships is tenuous, if not impossible.

However, few researchers doubt that an essential prerequisite for lowering fertility is the motivation that couples acquire. Only as couples begin to see that lower fertility will benefit their lives are they likely to respond to programs that are designed to lower fertility. Until people are strongly motivated to have smaller families, no amount of propaganda and no supply of modern contraceptives will bring about a significant change. "Whether or not a family planning program will meet with success," wrote van de Walle and Knodel (1980, 39), "will be determined by how receptive couples are to the idea of reducing their fertility once the knowledge and means of birth control are available." Success is also going to depend on how societal changes take place, especially with respect to the changing role of women. As World Bank economist Lant Pritchett (1994, 42) so nicely put it:

> . . . reducing fertility is best seen as a broad problem of improving economic and social conditions, especially for women: raising their levels of education, their economic position, their (and their children's) health, and their role and status in society. That is a task altogether more difficult, but with more promise, than manipulating contraceptive supply.

Migration and Mobility

One of the most distinguishing characteristics of humans is their propensity to **migrate.** The distance of these movements and their frequency are distinctive. This mobility is evidenced by the linguistic, social, and nationalistic mixing of much of the world's population. Though the human population has always been mobile, its mobility has accelerated with economic and technological progress, particularly with progress in the fields of communications and transportation. Cultural adaptability has allowed humans to adjust to major ecological changes by employing mental abilities and technological skills. As Michael Parfit (1998, 11) recently observed, "In airports and seaports and railway stations, along forested borders, and even where steel and barbed wire make barriers that seem impenetrable, thousands upon thousands of people are on their way to somewhere new . . . they change the world."

Migration differs from fertility and mortality in several ways. First, the biological processes of birth and death are uniform, discrete, and distinct events; for individuals each is a one-time-only event (save that growing number among us who at least claim that they crossed over to the other side, only to be brought back to life). Because migration is not a biological event, it does not have a uniform process. There is no upper limit to migration, but there is an upper limit to the number of children a woman can have. Goldscheider (1971, 49) remarked that "The total migration of communities or societies would not imply their demise, but their removal to another location." For example, the ghost towns of the American West were created when people left after the inducement that drew them there in the first place, typically minerals such as gold or silver, disappeared.

One of the most distinguishing characteristics of humans is their propensity to **migrate.**

Migration differs from fertility and mortality in several ways.

Second, because migration involves leaving one place and entering another, two populations must be considered: the population of the area of origin and the population at the destination. Changing mortality and fertility affect an area in a relatively simple way, whereas migration always simultaneously impacts two areas; in order to fully understand the migration process, both areas must be studied.

Finally, births and deaths are universals; that is, societies must have reproduction and as much control as possible over longevity in order to survive. On the other hand, migration is not a universal. Not everyone migrates, though in modern societies most people do. Furthermore, migration is a selective process that effectively has no upper limit. Some societies experience a great deal of migration, whereas others experience very little. In addition, migration can be repeated and even reversed.

MEASURES OF MIGRATION

At first glance it may seem that defining the term "migrant" would be relatively simple. However, several definitional problems can be identified. For example, if a migrant is simply defined as someone who "moves," then several questions must be answered. What constitutes a move? Should the move be permanent or do transients classify as migrants? Many people, such as commuters, shoppers, and tourists, change their geographical position, but should they be considered migrants? There are not definitive answers to all these questions. According to demographer Ralph Thomlinson (1976, 267–268): "Demographers thus define persons as migrants if they change their place of normal habitation for a substantial period of time, crossing a political boundary." Geographers use the term **circulation** for activities such as shopping, commuting, and touring because each of these activities begins and ends at a person's place of residence; a considerable literature exists, for example, on commuting, the journey to work.

Demographers distinguish between movers and migrants according to a single criterion. Movers are persons who change their place of residence; migrants are those whose change of residence takes them into a new political unit. All migrants are movers, but some movers are not migrants. Thomlinson also pointed out that getting information for all moves for all persons is almost impossible without a national registration system; however, it is usually possible to obtain a record of moves where people cross political boundaries (from a census, for example). Also, crossing a political boundary can be significant, particularly if a person moves from one country to another, even if the distance involved in such a move is short.

Geographer Curtis Roseman (1971) provided a different and conceptually useful way of describing migration. He distinguished between partial displacement migration and total displacement migration. His differentiation was made by considering a household's *activity space*, the set of places such as schools, work, stores, and recreational facilities with which the household interacts on a regular basis (circulation systems). Partial displacement moves are those that disturb only a portion of a household's activity space. For example, a family moves to a different part of town; the husband and wife keep the same workplaces, but the kids go to a different school and the family shops in a different shopping center. In contrast, total displacement migration, which usually involves longer distances and is more disruptive, requires moving the entire activity space as well as the place of residence.

A FEW BASIC RATES AND CONCEPTS

A standard set of migration rates and concepts has gradually evolved. They are useful in testing various migration hypotheses and facilitate the collection of data. In order to understand the migration literature, we need to understand the terminology and concepts.

Gross Migration

The sum of all the people who enter and leave an area is considered **gross migration.** Thus, it measures the total volume of population turnover in a community.

The sum of all the people who enter and leave an area is considered **gross migration.**

Net Migration

During any specific period of time a region may be receiving migrants from one region and losing migrants to another region. The difference between the arrivals and departures is called **net migration.** When more people leave an area than move into that area, net migration is said to be negative. Positive net migration, on the other hand, occurs when more people enter than leave an area.

The difference between the arrivals and departures is called **net migration.**

Out- and In-Migration

Each migratory event involves two actions: leaving one place and arriving at another. Leaving the place of origin is referred to as **out-migration,** whereas arriving at the place of destination is referred to as **in-migration.**

Areas of Origin and Destination

The area of origin is the place from which a migrant leaves, whereas the area of destination is the place at which the migrant arrives.

Migration Rates

The relative frequency of migration is called the migration rate. It is the number of migratory events divided by the population exposed to the chance of migrating. Donald Bogue (1969, 758) defined the following four such rates:

$$\text{Out-migration rate: } \frac{O}{P} \times k$$

Leaving the place of origin is referred to as **out-migration,** whereas arriving at the place of destination is referred to as **in-migration.**

In-migration rate: $\dfrac{I}{p} \times k$

Net-migration rate: $\dfrac{I - O}{p} \times k$

Gross-migration rate: $\dfrac{I + O}{p} \times k$

where O is the number of out-migrants from an area,
    I is the number of in-migrants to an area,
    p is the average or mid-interval population of the area, and
    k is a constant, usually 100 or 1000.

These migration rates can be computed as specific rates if both the numerator and denominator refer to the same particular substratum of the population. There can be specific rates for age, occupation, income, race, sex, or any other particular subgroup for which data are available.

Migration Interval

Because **migration** is a time-specific process, it is necessary to specify the time **interval** over which migration is studied or observed. All other things being equal, the longer the time interval, the smaller the average size of the annual number of migrants, because a significant proportion of migrants return rather quickly to their places of origin. Even though migration data may have been computed on an average annual basis for two sample groups, if the migration intervals involved are not the same, then the data may not be comparable.

Streams and Counterstreams

Migrants who move from a particular origin to a particular destination over the same migration interval are considered part of a **migration stream.** Migrants who return to that origin during the same period are part of a **counterstream.** Keep in mind, however, that even if the stream of migrants between two places is exactly equal in numbers to the counterstream, the selectivity of migration is still likely to produce significant changes in both places. For example, the average age of persons in the stream may be younger than that of the counterstream; these age differences may in turn affect fertility, mortality, employment levels, incomes, and a host of other variables in both places.

Return Migration

Though often difficult to find data for, a certain number of people who make a move later return to their original location (though not necessarily to their original address). To some extent **return migration** is an index of dissatisfaction with moves; it could even be construed as a "failure" rate. Because we depend heavily on census data, however, we cannot differentiate between people who move into an area on the basis of whether or not they had previously lived in that area. Without doubt, some portion of most counterstreams is comprised of returnees.

Differential Migration

The study of the selectivity of migration and the differing rates between various social, demographic, and economic groups is referred to as **differential migration.** This term is also used when studying the differences in the population composition of migration streams.

Emigration and Immigration

A person who changes residence from one country to another is considered an **emigrant relative** to the country of origin and an **immigrant relative** to the country of destination.

International and Internal Migration

International migration involves a change of residence from one country to another, whereas internal migration refers to a change of residence within a country. Because some territories have characteristics similar to those of nation-states, the distinction between international and internal migration is not always clear-cut. For example, difficulties arise when trying to designate migration between Puerto Rico and the United States. The migration literature can not only be broadly categorized into international versus internal, however, but internal migration can be further categorized into rural-urban, inter-urban, and even intra-urban; other possibilities exist as well.

TYPES OF MIGRATION

Migration can take many forms. Though most people in the United States think of migration streams as comprised of individuals, or perhaps families, moving freely from place to place, this has not always been the typical migration pattern. Several types of migration have been defined, including

1. primitive migration

2. group or mass migration

3. free-individual migration

4. restricted migration

5. and impelled or forced migration

Most of these types were first recognized by a noted demographer, William Petersen (1958).

Primitive Migration

This type of migration is associated with groups that are unable to cope with natural forces in their physical environments. You see, one method of coping with the deterioration of the physical environment of an area is to move from it. This is generally done by a group of people, and many times it is related to hunting or gathering food. When there is a disparity between the produce of the land (carrying capacity) and the number of people that must subsist from that land, a primitive migration

The study of the selectivity of migration and the differing rates between various social, demographic, and economic groups is referred to as **differential migration.**

■■■

When there is a disparity between the produce of the land (carrying capacity) and the number of people that must subsist from that land, a **primitive migration** takes place.

■■■

takes place. According to Petersen (1975, 320): ". . . this can come about either suddenly, as by drought or an attack of locusts, or by the steady pressure of growing numbers on land of limited area and fertility."

Group or Mass Migration

Most of the major population movements, certainly until the seventeenth century, consisted mainly of the movement of groups of people. **Group migration** refers to the migration of a clan, tribe, or other social group that is larger than a family. Throughout history entire societies left their original domiciles and laid claim to or invaded other areas. In some cases armies invaded other areas and some of the soldiers settled in the region of conquest. After an invasion, the native population was then either assimilated or displaced.

Another form of mass or group migration is colonization. The early stages of the colonization process involve group migration; but during later stages, most migration is by individuals or families (Simkins, 1970).

Group migration refers to the migration of a clan, tribe, or other social group that is larger than a family.

Free-Individual Migration

Free-individual migration has been described by Fairchild (1925, 20) as follows:

> The movement of people, individual, or in families acting on their own individual initiative and responsibility without official support or compulsion, passing from one well-developed country (usually old and thickly settled) to another well-developed country (usually new and sparsely settled) with the intention of residing there permanently. A great deal of international migration since the seventeenth century has been characterized as this type, particularly migration to Australia, New Zealand and the Americas.

Migration from Europe to the United States has been particularly important and, since the founding of the early colonies in the United States, it has involved some 30 to 40 million people. Much of the subsequent growth of the population of the United States is attributable to these immigrants. For example, an estimate by Gibson (1975, 158) was that ". . . about 98 million, or 48 percent, of the 1970 population (203 million) is attributable to the estimated net migration of 35.5 million in the 1790–1970 period." Similar percentages would probably apply to the counts provided by the 1980 and 1990 censuses; the escalation of immigration in recent decades will in turn result in considerable growth of the United States population in the decades ahead.

Restricted Migration

Free migration has slowly been replaced by restricted migration. Since the turn of the century numerous laws have been enacted to **restrict the migration** of people between countries. In some cases the restrictions involve a complete ban on all movement of certain types of people, whereas in other countries migration quotas have been set up to curtail movements.

Since the turn of the century numerous laws have been enacted to **restrict the migration** of people between countries.

Restrictions on international movements to the United States in the twentieth century were primarily the result of increasing numbers of immigrants who desired to move to the United States, as well as the changing nature of the immigrants themselves—from Anglo-Saxon Protestants of northern and western Europe to Catholics of southern and eastern Europe. Immigration to the United States has been curtailed since 1921, when barriers were set up to restrict both the numbers and the types of migrants. A loosening of restrictions in movements to the United States began in the mid-1960s, resulting in a rapid increase in the annual number of immigrants entering the country since then. A similar situation exists in other countries, such as Australia and Canada, countries that originally opened up their doors to all immigrants but have since restricted immigration.

It was ironic that as the following words of Emma Lazarus' poem, "The New Colossus," were being inscribed and dedicated on the Statue of Liberty, the United States was about to impose racial barriers on migration and to establish a quota system:

> Not like the brazen giant of Greek fame,
> With conquering limbs astride from land to land;
> Here at our sea-washed, sunset gates shall stand
> A mighty woman with a torch, whose flame
> Is the imprisoned lightning, and her name
> Mother of Exiles. From her beacon-hand
> Glows world-wide welcome; her mild eyes command
> The air-bridged harbor that twin cities frame.
> "Keep ancient lands, your storied pomp!" cries she
> With silent lips. "Give me your tired, your poor,
> Your huddled masses yearning to breathe free,
> The wretched refuse of your teeming shore.
> Send these, the homeless, tempest-tost to me,
> I lift my lamp beside the golden door!"

Presently, restrictive policies in most countries present difficulties for prospective international migrants. These people must either stay where they are or find illegal methods of overcoming restrictions. According to Thomlinson (1976, 288): ". . . two results of all these confining ordinances are that an enormous potential is being pent up in certain areas, and inequalities in the distribution of scarce goods and the standard of living are being aggravated." Pressures for more international migration continue to increase, especially as the developed countries have low rates of population growth and the developing countries continue to grow at medium to high rates. While the ratio of earnings in the richest countries to those in the poorest countries keeps increasing, the fraction of the world's population enjoying that wealth continues to decline. No one can believe right now that containing international movements will continue to be possible, at least not without committing an ever-expanding pool of resources toward that end. In the United States, for example, an increasing number of people are urging such measures as the use of the military and national guards to aid in patrolling borders. Though resisted for the most part so far, such urgings have encouraged the Clinton administration to put more resources into the Immigration and Naturalization Service (INS).

Impelled and Forced Migration

When the state or some other political or social institution is the activating agent in migration, then that migration is referred to as **impelled or forced.** There is a simple distinction between these two types: with **impelled migration** the migrant holds some degree of choice, whereas with **forced migration** the migrant has no power or control over the situation. For example, the Nazi policy between 1933 and 1938 that encouraged Jewish emigration by various anti-Semitic laws and acts would be considered impelled migration, whereas the policy that followed, with the actual forcing of Jews to leave their homes and in many cases actual extermination, would be considered a forced migration policy (Petersen, 1975, 321). Neither seems justifiable, of course, and differentiating between them makes little sense.

Forced migration generally serves one of two purposes. First, it is a means whereby a potentially hostile group can be removed from a country, and second, it is a means of furnishing an unskilled labor force for certain areas. The slave trade was a forced migration. According to Bouvier, Shryock, and Henderson (1977, 16):

> Forced migration, the most tragic of group movements, has meant flight and often enslavement for untold millions of human beings. Yet, despite the involuntariness of the act, this represents a major form of migration that has been prevalent throughout much of history. . . . The millions upon millions of 20th century refugees represent today's variation of "forced migration," which can be expected to continue as long as mankind insists on waging war or exercising total ethnic domination.

Tempting though it is to believe that slavery and the slave trade is a thing of the past, confined to the pages of history books, that is hardly the case. Tyler (1991), for example, pointed out that there are more slaves in the world today than at any time in the past, perhaps as many as 200 million, though most of today's slaves are either bonded laborers or child laborers, not chattel slaves. Even the latter are still with us, however, as children are sold in slave markets in Thailand, the Sudan, and a few other countries (Tyler, 1991).

These children are in turn used for everything from sex slaves and child laborers to inexpensive fodder for "religious" sacrifices. A grim article in *Stern*, for example, suggested that as many as 10 million youngsters worldwide are involved in child-prostitution (Oberlander, 1992). Sachs (1994, 24) began an article about child prostitution with the following statement: "Once considered a universal crime, the trade in children's bodies is increasingly regarded simply as a business—with plenty of support from tour agencies, affluent travelers, and even governments." As Robert Scheer (1998, B7) bluntly noted, "Wherever foreign businessmen and tourists gather, they are readily serviced by sad-eyed economic refugees, often mere children, who

When the state or some other political or social institution is the activating agent in migration, then that migration is referred to as **impelled or forced.**

With **impelled migration** the migrant holds some degree of choice, whereas with **forced migration** the migrant has no power or control over the situation.

have been lured to the sparkling centers of commerce in a desperate attempt to escape the poverty of their homelands."

Even the United States is not above reproach. In 2000 the Central Intelligence Agency estimated that about 50,000 women and children are brought illegally to the United States by "people-traffickers" annually to work as prostitutes, servants, and sweatshop laborers. Primary source regions for traffickers include Thailand, Vietnam, China, Mexico, Russia, and the Czech Republic. Most of these women and children are initially lured by the promise of good jobs in America.

EXPLAINING MIGRATION

Geography is only one among many disciplines that share an interest in understanding and explaining migration patterns. From the economists, with their focus on the role of labor and labor markets, to sociologists, with their eye for institutional changes, all are hampered somewhat in their search for explanations of migration by the lack of adequate data and other constraints. As Richey (1976, 399) noted, for example: "A major problem hampering theory development in migration research is a simple methodological problem that undermines the utility of many studies—the time to which measures of influences and of migration relate." These times, of course, are often different, depending on what data were available for the variables in use.

Migration, particularly that between nations, is among the most striking changes that can occur in a person's life. According to Bogue (1969, 801):

> In a high proportion of such moves the person not only changes his national loyalties, but also forsakes his native language, his cultural heritage and customs, his relatives and lifelong friends, and his occupation. So drastic is the change that many adults never make a complete adjustment in their lifetime, and it is only their children or grandchildren who are fully integrated into the receiving society.

Yet, often in the face of considerable odds, people have always made such moves; today more people than ever are on the move, and high rates of international migration over the coming decades seem assured.

RAVENSTEIN, LEE, AND THE PUSH-PULL MODEL

The reasons why people migrate are varied, and several theories have been developed to help explain migration patterns. The pioneering effort in migration theory was an early study by Ravenstein in the 1880s and was set down in his "Laws of Migration." He studied population movements in Great Britain and related migration to population size, density, and distance. He had a minimum of records for migrants in England, but in a fairly short time he extracted the essentials from those records and published a series of generalizations, most of which still hold true today. Some of Ravenstein's generalizations were the following:

1. Most moves cover only a short distance.

2. Females predominate among short-distance movers.

3. For every stream there is a counter stream.

4. Movement from the hinterland to the city is most often made in stages.

5. The major motive for migration is an economic motive.

With respect to the latter, Ravenstein (1889, 286) stated the following: "Bad or oppressive laws, heavy taxation, an unnatural climate, uncongenial social surroundings, and even compulsion . . . all have produced and are still producing currents of migration, but none of these currents can compare in volume with that which arises from the desire inherent in most men to 'better' themselves in material respects. . . ." It is somewhat surprising (but a tribute to him) that, from Ravenstein's day until now, relatively few attempts have been made to extend his generalizations, or to devise additional ones, or to gather his generalizations into a more theoretical framework. That is not to say, however, that migration researchers have been sitting still; recent efforts have been reaching along new, if still unproven, lines.

One important conceptual framework, accompanied by a set of hypotheses about the volume, streams, and characteristics of migrants, was broached by Lee (1966). He began by classifying the elements that influence migration into the following groups:

1. factors associated with a migrant's origin

2. factors associated with a migrant's destination

3. obstacles between the two that the migrant must overcome, which Lee calls intervening obstacles

4. personal factors

People move for a variety of reasons, including job changes, marriages, divorces, graduations, retirements, and trouble with the law. They may move because of conditions in their area of origin, conditions at their destination, or some combination of the two. Not all people in a particular area perceive conditions the same way, thus they may respond differently to the same stimuli. Lee suggested that a good climate was almost universally attractive, whereas a bad climate was undesirable. However, not everyone agrees on exactly what is or is not desirable. Recently some researchers have focused on the role of environmental preferences and migration patterns.

Lee summarized his ideas in the schematic diagram shown in Figure 8-1. The circles representing the places of origin and destination have pluses, zeros, and minuses. The pluses indicate elements to which potential migrants respond favorably, whereas the minuses are elements to which they react negatively; zeros stand for elements to which potential migrants are indifferent.

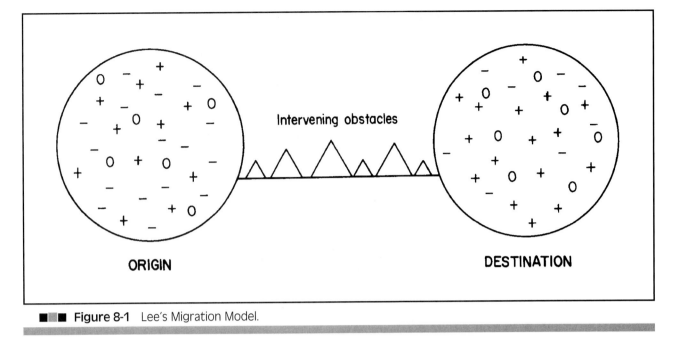

■▓■ **Figure 8-1** Lee's Migration Model.

Source: Everett S. Lee, "A Theory of Migration," *Demography,* Vol. 3, No. 1 (1966), p. 50. Used by permission.

Conceptually, a potential migrant adds up the pluses and minuses for both the origin and one or more possible destinations and then decides whether the balance of pluses and minuses favors moving or staying. **"Push-pull" models of migration** derive from the observation that some elements of an origin "push" people to migrate, whereas some elements of a destination "pull" migrants toward it. However, before the migrant decides to move, another set of circumstances must be considered—Lee's **intervening obstacles.** These include such things as the actual cost of making the move, which is obviously related to distance, and psychic costs, such as the necessity of breaking ties with family, friends, and community, especially when long distances are involved. Furthermore, many personal factors may also influence a migration decision; for example, potential migrants may consider such things as health, age, marital status, and number of children. Clearly, a person's decision to migrate depends upon many considerations. Some of these considerations require further elaboration.

Operationalizing push-pull models is more difficult than conceptualizing them. Researchers must consider which variables to measure at origins and destinations, what time periods to measure them for, and how to include some measure of intervening obstacles. Most migration studies have been cross-sectional, using variables at origins and destinations to explain observed migration flows during a specific time interval (between two census periods, for example). Most economic and social models of migration are actually specific cases of the generalized push-pull model.

"Push-pull" models of migration derive from the observation that some elements of an origin "push" people to migrate, whereas some elements of a destination "pull" migrants toward it.

■▓■

Some Reasons for Migrating

In a theoretical economic sense, migration is the mechanism by which human resources move to their highest-valued use; however, this ideal often goes unrealized. Among the major determinants of migration are distance, income differences, and information flows.

Repeatedly it has been shown that most migration takes place over short distances. The distance itself, however, is related to other factors that also inhibit migration. For example, distance is related to both the economic and psychic costs of moving. Furthermore, distance can determine the ease or difficulty with which one can collect reliable information about alternative destinations. Several studies have suggested that the psychic costs of leaving friends and relatives may be significant, though they are difficult to measure. Diverse approaches to explaining migration are considered in subsequent sections; each is described only briefly, though each has generated a literature and a following of its own—none so far has received unanimous support.

Two Micro Approaches from the Economists

The two economic models discussed in this section are micro-level in the sense that they focus on the rational behavior of individuals who are potential migrants rather than the macro-level system within which they are planning to migrate. Modernization theory and neoclassical economics have focused attention primarily on labor mobility; the basic argument is that migration flows are necessary and rational adjustments to geographical differences in the supply and demand for labor, as measured primarily by regional differences in real wage rates and unemployment rates.

In modernizing economies both of these models help explain the cityward movement of people as industrialization progresses. Higher wages in cities, coupled with diminishing opportunities for employment in agriculture, lead people, especially the young, to move steadily toward expanding urban areas. They also help explain why the globalization of the economy is attracting people in new directions as opportunities change. Large wage differentials between the developed and the developing nations do not go unnoticed, especially as global communications systems become increasingly common (think of the ubiquity of CNN, for example). The surplus of low-wage labor in the poor countries needs to come together with the large-scale capital accumulations in the rich ones. Two trends are then set in motion. One is a flow of capital outward from the developed countries into the poor countries, where it can take advantage of low wage levels. The other is the flow of surplus labor from the poor to the rich countries, where demographic stagnation and affluence are creating a notable demand for unskilled labor. Once confined primarily to agriculture, nations such as the United States now need labor in many service sectors—hotels and restaurants, for example—and in more industries as well—apparel and construction, for example. So long as large wage differentials persist, the economists will tell us that labor migration will follow (even if it means crossing borders that are otherwise marked "closed"). Migration is an integral part of the globalization of the economy.

The Factor Mobility Model

The **factor mobility model** of migration is based on wage rate differentials, which in turn reflect geographic variations in the supply and demand for labor, one of the major factors of production. We could argue that migrants move up the wage rate gradient; that they move from low-wage areas to high-wage areas, in other words. In a *perfect* world, migration would then serve as an equilibrium mechanism. As people left low-wage areas, the smaller remaining labor supply would drive wages upward. At the same time, the growing supply of labor in the high-wage area would tend to depress wages there. Theoretically, migrants would move between the areas until wage rates were equal in both places; however, imperfections exist in the market mechanism and people are not perfectly mobile. For example, many Appalachians have chosen to stay in Appalachia despite low wages and high unemployment rates, partly because they have had deep attachments to the region and partly because they have been unaware of opportunities elsewhere.

Some important earlier findings of a study by the United States Department of Labor (1977) are worth considering at this point.

1. The study found that household-level unemployment or job dissatisfaction does operate as a "push" factor in migration. People who are recent arrivals in an area, and who fail to immediately find employment, are especially likely to move again (often becoming return migrants).

2. Local economic conditions, such as unemployment rates, do affect outmigration, but only among those who are unemployed.

3. Unemployed individuals and others currently seeking work are more responsive to such economic determinants as family income, local wage rates, and expected earnings increases than are persons who are satisfied with their current jobs.

4. Families are much more likely to move in a given time period if they moved in the recent past, mainly because of a tendency for people to return to places that they recently left.

5. Families that have moved several times previously are more likely to move again than are families that have made only one or no moves recently, if the multiple moves were nonreturn moves. Such people are often referred to as *chronic movers*.

> The **factor mobility model** of migration is based on wage rate differentials, which in turn reflect geographic variations in the supply and demand for labor, one of the major factors of production.

■ ■

6. More important now than ever before, spouses are not passive, secondary migrants, but have a significant influence on the family's decision to move. Also, contrary to the results of some studies, families with working wives are not necessarily less likely to move than are families in which the wife does not work. These families will seek locations that maximize both incomes.

7. Surprisingly, the study found that age and education, typically found to be among the strongest correlates of the propensity to migrate, appear to be relatively unimportant in explaining the migration of married couples when other migration determinants, many of which vary with age and education, are held constant.

The Human Capital Model

First developed by Sjaastad (1962) and nicely summarized in Clark (1986), the **human capital model** is an alternative economic model. Because individuals seek to improve their incomes over the long run, movements take place when individuals or families decide that the benefits of a move outweigh the costs. The model is in many ways an improvement over the more simplistic factor mobility approach, though it is not unrelated to it conceptually. Clark (1986, 67) identified the following advantages of the human capital model over competing models:

1. Benefits occur over a period of time, which helps explain why migration rates drop with age.

2. Psychic as well as monetary costs and benefits can be included.

THE "NEW ECONOMICS APPROACH"

The "new economics approach," an unfortunate name, perhaps, is a useful expansion on the micro models discussed above. The primary difference is this: the **"new economics approach"** argues that larger social units, primarily the family, rather than individuals, must be considered in the making of migration decisions. Rather than income maximization, proponents of this approach believe that risk minimization is the motivating factor in migration decision-making.

For example, consider a farming family in Mexico. If a son decides to migrate to the United States because of higher wages there, his decision affects the entire family's welfare. If he goes, his labor is lost at home; however, if the family can manage the farm without him, then remittances from him will increase the family's overall income and welfare. The remittances also act as a counterbalance for low food prices at home or poor crops. The family, then, decides as a unit whether or not the son should seek employment elsewhere.

The "new economics approach" argues that larger social units, primarily the family, rather than individuals, must be considered in the making of migration decisions.

■■■

A STRUCTURALIST VIEW

The essence of the **structuralist view** of migration, primarily international migration, is derived from neo-Marxist ideology or from world systems theory. It adds another dimension to the above economic arguments, maintaining that the wage rate differentials may stimulate migration but they are maintained by dependency relationships between countries of the periphery and those at the core. Briefly, the argument that **structuralists** make is that uneven development (an outcome of global capitalism) generates not just wage differentials but also dependency relationships. Migration results from differences in social class position, this view tells us, in a world divided into a few economic centers or cores (Japan, Western Europe, and North America) and a host of countries at the periphery, many of them former colonies of developed countries. As Goss and Lindquist (1995, 322) noted, "The processes of underdevelopment create and sustain a dual labor market at the global level; . . . the Third World periphery provides for the reproduction of cheap labor that is selectively recruited by the core to counter falling rates of profit. . . ."

The argument that **structuralists** make is that uneven development (an outcome of global capitalism) generates not just wage differentials but also dependency relationships.

A STRUCTURATION APPROACH

Among the newest approaches to the study of migration is the **structuration approach,** based primarily on the theory of structuration set forth by Anthony Giddens (1982, 1984), who concerned himself primarily with the "structure-agency problematic." Goss and Lindquist (1995) proposed this approach as a means of explaining international labor flows—we can only sketch for you here their major ideas.

According to Goss and Lindquist (1995, 335):

> . . . we will attempt to develop an alternative approach to international labor migration, drawing up Giddens's conception of agency and structure, the notion of institutions as "sedimented" social practices, the effects of the operation of rules and resources or "modalities of interaction" within institutions, the strategic actions of individuals, and time-space distanciation. More specifically, we suggest that what has previously been identified as migrant networks be conceived as migrant institutions that articulate, in a nonfunctionalist way, the individual migrant and the global economy, 'stretching' social relations across time and space to bring together the potential migrant and the overseas employer.

The essential idea here is that in the complex world of today's global economy there are intermediaries (migrant institutions) which function to bring together potential migrants in Third World nations with opportunities offered within the developed or rapidly developing nations within the global economic system. As Goss and Lindquist (1995, 335) put it, "International labor migration can then be conceived as a process whereby individuals transcend the limits to presence-availability and negotiate their way across boundaries between locales in order to establish presence and control over resources in a distant place. . . ."

On their own, neither individuals in the Third World nor employers in wealthier countries are likely to succeed in matching their needs. Formal and informal organizations often, and increasingly, serve the function of "matchmaker" between the two. The authors use the Philippines as an example. Foreign employers are not allowed to directly recruit Philippine laborers; rather, they must go through licensed recruitment agencies—overseen by the Philippine Overseas Employment Agency—in that country. These recruitment agencies, then, are able to decide which potential migrants are going to be linked via migration to foreign employers in search of their labor. Regardless of where they live, these migrants ultimately depart from Manila.

Given the likely long-term persistence of structural differences in the global economy, it seems promising to look at the role of such migrant institutions in establishing and perpetuating streams of labor migration.

A Behavioral View

Among geographers, Julian Wolpert (1965) was one of the first to approach migration by concentrating on individual **behavior** rather than on aggregate data. He argued that understanding and explaining migration patterns was dependent on sorting out constants in migration behavior. Wolpert identified three central concepts of migration behavior:

1. place utility

2. field theory approach to search behavior

3. life-cycle approach to threshold formation

According to Wolpert (1965, 162), place utility ". . . refers to the net composite of utilities which are derived from the individual's integration at some position in space." Dissatisfaction with one's current location, then, is the major stimulus for beginning a search for another location. Place utility may be either positive or negative, depending upon how an individual perceives his or her location. Of considerable importance, then, is how the individual perceives the utility of his or her current place relative to the perceived utility of other places. According to Wolpert (1965, 162):

> The utility with respect to these alternative sites consists largely of anticipated utility and optimism which lacks the reinforcement of past rewards. This is precisely why the stream of information is so important in long-distance migration—information about prospects must somehow compensate for the absence of personal experience.

Subsequently, Wolpert (1966) suggested that the study of residential mobility should be viewed in terms of stresses in the current place of origin and the potential migrant's threshold for stress. Thus, the decision to move may be viewed as being dependent on both the alternatives available at a given time and a person's ability to cope with stress. Somewhat earlier, Rossi (1955) noted that residential complaints and dissatisfaction were important determinants of the decision to move. Wolpert proposed that a person's tolerance for stress could be measured by a "threshold function." Once a person's stress threshold is surpassed in a particular location, that person is likely to move.

Despite the many reasons for which individuals choose to change their places of residence, it is generally agreed upon by researchers that the study of movement behavior should focus on two major decisions:

1. the decision to seek a new residence

2. the search for, and selection of, a new residence (Moore, 1972). Roseman (1977) also emphasized the need for considering separately the decision to move and the decision about where to move.

THE ROLE OF INFORMATION

Wage rate differentials alone fall short of fully explaining most migration patterns, though the prediction of movement up the economic gradient tends to hold true. Information is also important in the decision to migrate, especially information about alternative destinations. People are hesitant to move to areas about which they know little or nothing. Nelson (1959) found that friends and relatives who had previously moved from one place to another were especially good sources of information about their current location. Often this "friends and relatives effect" leads to chain migration, reflecting a directional bias on movers who leave a particular place. What we find, then, is that the distribution of past migrants tends to be a good predictor of future migration patterns (Greenwood, 1975). Migration streams, once established, are likely to be perpetuated for a long period of time. With respect to international migration, Goss and Lindquist (1995, 321) recently commented that ". . . it is as important to examine the flow of information and the role of previously successful migrants in recruitment networks as it is to identify objective income differences in the explanation of migration."

GENDER AND MIGRATION

At least until recently, most migration studies, especially those that focused on national and international labor flows, paid little or no attention to gender. However, as Castles and Miller (1993, 31) remind us, ". . . the migrant is also a gendered subject, embedded in a whole set of social relationships." Because of their traditional roles in patriarchal societies—primarily wives and mothers—female migrant workers have often been offered lower wages than males (Phizacklea, 1983). They have often been seen as easier to control than males as well.

Not only are female migrants sought for employment in many traditionally low-paying sectors of modern economies—including the garment industry, motel and hotel workers, maids, and nannies—they have also often been sexually exploited as well. In an excellent study of Filipina migrant entertainers, geographer James Tyner (1996) considers the importance of perspective in migration studies. He deconstructs four controlling images of these Filipina migrants—the Other, the prostitute, the willing victim, and the heroine—and argues that these images in turn reflect the intentions, objections, and motives of the observers as they look at Filipina migrants. In concluding, Tyner (1996, 89) commented that:

> The discourse surrounding the causes and consequences of migration is equally informed by a set of images. These images reflect the observer's intention, objectives, and motives, as well as the political, social, economic, historical and geographical context of the encounter. . . . Only by transcending these representations, and uncovering the underlying sentiments associated with them, will meaningful dialogue—and hence policy formulation—be possible.

Another study of gendered migration focused on the importation of nannies into the United States (Hochschild, 2000), not just to care for children but also the elderly—in the process creating "global care chains." Paradoxically, many of the women who come to the United States to be nannies are themselves mothers, and they typically leave their own kids behind in the care of others. As the number of mothers in the work force in the United States has increased, so has the demand for people to take care of those children. Hochschild (2000, 35) comments that:

> At the first world end of care chains, working parents are grateful to find a good nanny or child care provider, and they are generally able to pay far more than the nanny would earn in her native country. This is not just a child care problem. Many American families are now relying on immigrant or out-of-home care for their *elderly* relatives. . . . But this often means that nannies cannot take care of their own ailing parents and therefore produce an elder-care version of a child care chain—caring for first world elderly persons while a paid worker cares for their aged mother back in the Philippines.

Though such migration streams may be rational responses to global capitalism's shifting opportunities, little is said of the "human costs" of such moves, when mothers leave their own children behind in poor countries to care instead for the children of working mothers in rich countries. Raising wages and respect for care-givers and encouraging males to contribute more to child care might improve the situation. Nonetheless, female migration from poor to rich countries is likely to continue to fill many care-giver niches, even though those women often leave behind children and elders that must themselves be cared for by others. As Hochschild (2000, 36) tells us, "Sadly, the value ascribed to the labor of raising a child has always been low relative to the value of other kinds of labor, and under the impact of globalization, it has sunk lower still. . . . And . . . the low market value of care keeps low the status of the women who do it."

RESIDENTIAL PREFERENCES AND MIGRATION

According to geographer Paul Schwind (1971, 150):

> Patterns of migration presumably are functions of, and should therefore contain information on, both the total magnitude of population systems and the aggregate preferences of migrants for relevant characteristics of regions. Migrants, however, probably respond not so much to factual statistical information about regions as to rather general perceptions of regional attractiveness.

Though geographers have accomplished a great deal in studies of place preferences, most of the work specifically relating to urban migration and city size preferences has been done by sociologists. Changing migration patterns in the United States since 1970 have stimulated researchers to seek explanations of migration patterns outside of the standard wage theory and its derivatives.

Studies of city size preferences are often concerned with the question of the need for a population distribution policy in the United States. Numerous proposals have been made, including those designed to aid depressed rural areas, such as Appalachia; to focus migrants on cities in the medium size range, thus decreasing in-migration to the most congested cities; and to guide the overall development of the American urban system. Fuguitt and Zuiches (1975, 491) noted that:

> An important element figuring in this discussion is concern about public preferences and attitudes on desirable places to live. A policy that provides community and housing options compatible with preferences should have a greater chance of success and could be expected to lessen any discrepancy between the actual and ideal distribution of the population.

Geographer Gundars Rudzitis (1991, 81) argued that: "Today, there is an increasing recognition of the need to incorporate noneconomic variables or surrogates into migration and regional growth models." He went on to note the importance of climate, trade-offs between climate and income, and the importance of what he called the "location-specific amenities model."

One criticism of many studies of city size preferences is that they failed to include an indication of the relative locations of cities, especially those in the smaller size categories. For many people a considerable difference in attractiveness exists between an isolated small town and a small town located within easy driving distance of a major city. Puzzled by the apparent contrast between survey results and actual migration patterns, Fuguitt and Zuiches (1975) added to a city size questionnaire a question asking those who preferred small towns whether they preferred those towns to be within thirty miles of a large city. What the distance-qualifying question showed was that a majority of those respondents who preferred small towns and rural areas said that they would like to live within thirty miles of a city of at least 50,000. Thus, we must be careful in interpreting the results of preference surveys that fail to include a distance-qualifying question or some relative location measure.

It appears that the perceived quality of life in small towns is their major attraction, however romantic the perception may be, whereas the lack of employment acts as a constraint on movement to them. As we have mentioned already, migration into small towns often destroys the very characteristics that attracted migration in the first place. Fuguitt and Zuiches (1975) concluded that if all people would move to the places they preferred, no mass exodus to remote areas would occur. Most people seem to prefer the best of both worlds, a quiet bucolic residential location and the opportunities and amenities that can only be provided by a metropolitan area.

More recently, Rudzitis (1991, 86) commented that there is a "...need for imagination rooted in regional historical reality if we are to better understand and promote the vitality of nonmetropolitan areas." He also suggested the need for geographers to consider such concepts as "sustainable development" in rural areas and small towns and the need for incorporating aspects of the relationship between local vitality and the natural environment in such areas. In concluding, Rudzitis (1991, 86) remarked that "A major shortcoming is the lack of studies of place, a comparative base from which to measure what might be expected. . . . A place, the world is what people take it to be, lying scattered openly on the surface, not as we social scientists often regard it, concealed beneath deceptive appearances." As geographer Peirce Lewis noted in his Presidential Address to the Association of American Geographers: "Good intellectual description provokes strong thought." (Lewis, 1985, 469) Many geographers are beginning to look more closely at the relationship between particular places and the migration streams that flow into and out of them. The relationship between mobility and attachment to place was explored by Bolan (1997).

SELECTIVITY OF MIGRATION

Migration is a selective process. Trewartha (1969, 137) described it in the following way:

> Assuming a sedentary population with an inducement to move, typically some individuals will leave and others remain where they are. But those who leave do not represent a random distribution of the biological and cultural characteristics of humanity in either the region of exit or entrance, for certain elements of the population tend to be more migratory than others. This is termed **migratory selection.**

Probably the most important and universally accepted migration differential is age. In movements within countries, as well as those between countries, it is older adolescents and young adults who predominate in most areas, as is apparent for the United States in Figure 8-2. Young people can generally adapt more easily to new conditions and, because they have only recently entered the labor

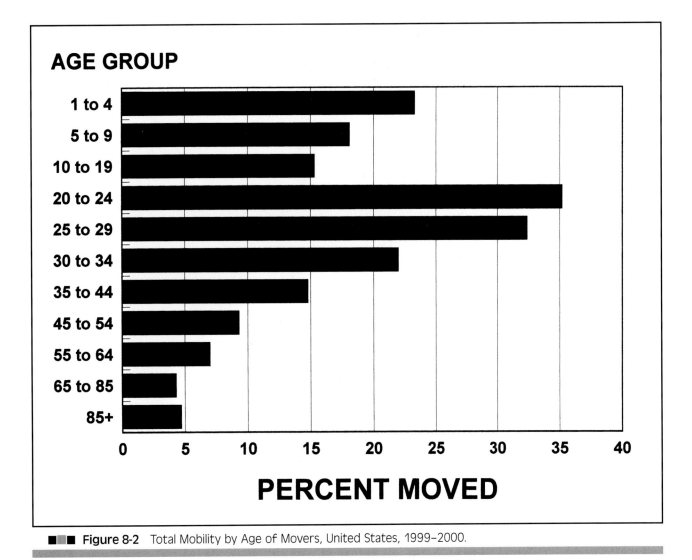

AGE GROUP

PERCENT MOVED

■■■ **Figure 8-2**   Total Mobility by Age of Movers, United States, 1999–2000.

Source: United States Bureau of the Census, *Geographical Mobility, March 1999–March 2000,* Current Population Reports, P20-538.

force, they can change jobs more easily. Because of the age selectivity of migrants, we generally find that areas that have experienced a great deal of in-migration have a young age structure, whereas areas that have experienced a lot of out-migration are generally older in their population composition.

Marital status also has a bearing on migration selectivity. In the developing areas of the world migration is usually undertaken by young, single adults. This was also probably true of the developed world in past times. However, in the developed countries today the married seem to be almost as mobile as the single.

Another facet of migration is sex selectivity, but whether the migration stream is male- or female-dominant depends upon a variety of factors. For example, in nineteenth-century Europe the movement from farm to city was primarily accomplished by young farm girls who went to work in the cities as domestic servants. On the other hand, the frontier towns in America were largely male-dominated. According to Trewartha (1969, 138):

> . . . there is some validity, no doubt, in the generalization that in less-developed countries migrants are predominantly male. . . . All the large commercial and industrial cities of India show a strong male predominance. The same is true in the newer settlements and mining towns of Negro Africa. . . .

Occupation and education are other variables that demonstrate migration selectivity. Unskilled workers are less likely to migrate than the skilled and semi-skilled; professional people are typically the most mobile occupational group. A considerable amount of research has been done on the relationship between education and migration selectivity. These studies have focused on the similarities and differences between the well-educated and the poorly-educated in regard to distances, rates, and direction of migration. Also, several studies dealing with education levels at the places of origin and destination of migrants have been undertaken.

In general, migrants have higher levels of education than nonmigrants, particularly if they have moved long distances. Folger and Nam (1967, 210) found that ". . . persons with higher levels of education are more migratory than persons with lower levels of education, and the difference in migration rates between poorly educated and well-educated persons increases for longer-distance moves." Similarly, data on migration in the United States between 1965 and 1970 pointed out the same pattern. For example, during this period, though the college-educated made up only 11 percent of the nation's total population aged 25 and over, they accounted for 25 percent of the people who migrated between noncontiguous states. Similar conclusions were reached with older age groups of the college-educated.

The above data and conclusions refer to relatively long-distance moves. An analysis of short-distance moves, such as from one house to another within the same county, reveals that in the United States the less educated are somewhat overrepresented; they tend to make more short-distance moves. They tend to change jobs more frequently, live in rental housing, and have more difficulty with the legal system.

With regard to changes in the educational status at the places of origin and destination, the data suggest relatively few significant changes at either place. In summarizing several studies, Folger and Nam (1967, 185) concluded that ". . . the effects of interregional migration on the educational attainment of the resident population are small."

Though there appears to have been relatively little effect on educational levels at places of origin and destination relative to internal migration in the United States, there has been a significant impact on the international scene. Educational self-selection, operating in the form of the so-called "brain drain," has become an international issue of concern. The developing countries feel that the brain drain is hindering their cultural and economic development. Perhaps the most significant aspect is the migration of trained medical personnel to the developed countries. One report by the United States Agency for International Development (1973, 112) found the following:

> . . . there are not enough medical and paramedical personnel in Haiti to meet the ordinary needs of the people. In relation to population, available personnel are equivalent to one physician for every 12,000 people, one nurse for every 12,000, and one paramedic for every 6,200. Nor is the medical situation improving. There has been an exodus of health professionals from the island. It is said that all medical school graduates in 1969 left Haiti because of the lack of economic opportunity.

There are those, however, who argue that the brain drain does not necessarily have an adverse effect on the poorer countries. For example, they argue that discoveries made in one area will eventually benefit all areas, and that scientists and other highly educated persons should be in those areas where they can be the most productive. Though this may be true in the long run, there is little doubt that most developing countries are in dire need of trained professionals if they are to proceed with economic and social development programs.

Interestingly, as the world economy becomes increasingly "globalized," we are likely to see increased movement among the developed countries (especially within Europe), as well as from the poor nations of the world to the rich ones. In an interesting study of Americans who are seeking (and increasingly finding) employment abroad, journalist Paul Gray (1994) found M.B.A.s and others seeking employment throughout both hemispheres. Among reasons cited by interviewees were better opportunities for advancement and more responsibility at a younger age—most were finding the challenge of working abroad a positive experience. As Gray (1994, 46) concluded:

> Such optimism is comforting and, in a global perspective, almost certainly correct. National boundaries become ever less important in the world's economies; a job is a job, whether it be in Budapest, Buenos Aires or Birmingham, Alabama. Still, certain ancient human emotions have not yet adapted to the new realities.

On the other hand, keep in mind the growing number of Japanese and German auto workers who are now working in plants in America, building Hondas and Mercedes-Benzes. An escalation of movement across international boundaries seems inescapable in the decades ahead; it also will inevitably toss up new challenges for the survival of democratic societies everywhere.

CONSEQUENCES OF MIGRATION

Migration creates many opportunities for geographic research. So far, we have focused primarily on the causes of migration, on the reasons why flows of migrants occur between places. However, migration can alter demographic, socioeconomic, and other conditions within societies. Because migration is a selective process, it produces changes in both the areas of origin and destination. The magnitude of these changes depends upon several variables. Because migrants are largely young, for example, areas of high out-migration will experience an aging of their populations. We see this in parts of the Great Plains area of the United States, as well as in parts of Appalachia. In turn, as the age structure of an area becomes older, death rates increase and birth rates decrease, altering local age structures even more. In receiving regions, the opposite impact results. Thus, the effect of migration is not just a matter of numbers; selectivity by age, gender, race or ethnicity, education, income, and even ideology can be profitably considered.

Migration and its concomitant mixing of peoples have a variety of effects. There is a complex pattern of interaction between geographic, demographic, social, and economic factors, the consequences of which are therefore varied. Of particular significance to geographers is the impact of migration on space and numbers, an impact that takes on two forms:

1. If destination areas absorb large numbers of people, their cities and towns will expand and, with the filling of the countryside, new lands open up, affecting the population distribution and density of the receiving area.

2. Conversely, the source areas for these migrants are likely to experience a decrease in population and a decline in density. Both population growth and decline produce far-reaching socioeconomic consequences.

Good examples of the impact of migration are the changes that took place in the United States and Ireland since the first third of the nineteenth century. Ireland has probably been more affected by emigration than any other country. Currently, the combined population of the Republic of Ireland and Northern Ireland is approximately 4.5 million people; in 1841 the Irish population was 8.2 million. Since that time, almost 5 million Irish have migrated to the United States and millions more have settled in Australia, Canada, and England.

An analysis of the Irish situation by Meenan (1958) pointed out that emigration has rarely been viewed as a beneficial force. Meenan (1958, 80) also noted some self-perpetuating features of the Irish situation and made the following statement:

> In each generation it is a movement of the younger sons and daughters. It is not an emigration of whole families and it never has been, except perhaps at the time of the famine. If it had been such, it might well have come to an end decades ago when the removal of families had led to a redistribution of farm holdings and the establishment of a new rural equilibrium. As things are, emigration enables those who are left to marry and rear families whose members will in their turn seek a living abroad. Certainly this is not all the story: A vicious circle is created whereby emigration causes underdevelopment, which in turn leads to further emigration.

Many researchers believe that emigration can be advantageous to areas of dense populations. Studies of Italy's post-World War II

Courtesy of Photo Disc

Ireland has probably been more affected by emigration than any other country. Since 1821, almost five million Irish people have migrated to the United states and millions more have settled in Australia, Canada, and England.

emigration have shown that the economically disadvantaged regions, where most of the population was engaged in agricultural pursuits, provided most of the overseas emigrants. Population transfers relieved pressure on resources in these regions and made it possible for disadvantaged areas to increase their per capita incomes. Another factor noted in the study of Italy was that Italians living abroad sent money back to their homeland, thus reducing Italy's adverse trade balance. Similar circumstances have existed in other countries as well and may have helped to diminish the differences between the poor and the prosperous nations.

Consequences of migration can be studied at many scales, from the above international example to that of the neighborhood or household. Changing metropolitan neighborhoods have long been studied by social scientists, though not always with a direct linkage to the migration processes that are altering them. Alba, et al (1995) is an excellent recent example of such studies; selected neighborhoods in the Greater New York metropolitan area are their focus. These neighborhoods, impacted considerably in the last two decades by immigration, are experiencing two different trends. On the one hand, the authors found that the entire area was become increasingly complex in its racial and ethnic characteristics; on the other hand, they found that segregated neighborhoods—primarily either African American or Latino, were becoming more numerous. The linkages between immigration and localized transitions in population composition deserve far more attention. Alba, et al (1995, 653–654) noted that:

> . . . the growing diversity of many neighborhoods may make it easier for their residents to ignore the problems besetting those who are left in all-minority neighborhoods. People, especially of the majority group, tend to judge diversity from their experiences in

their own neighborhoods; seeing diversity among their neighbors, they may fail to realize that it leaves out a large part of a minority group. As our analysis has shown, this issue particularly affects blacks, a disproportionate number of whom do not reside in diverse neighborhoods.

At the household scale researchers have begun to look more closely at such things as how households adjust when one or more members migrate—what happens to those left behind? Often, males migrate, leaving behind females who must then become household heads. Clearly, whether we focus on households in developing countries, from which males have emigrated to take advantage of jobs in developed countries, or households in metropolitan areas, from which males have moved (often simply abandoned), we are going to find changing roles for women and children. If remittances are not made to these households from those who left, then we can expect to find household incomes lower, and the fate of children within them will become more precarious. Everywhere we look—from the richest to the poorest nations—children are experiencing increasingly difficult roads to adulthood.

INTERNATIONAL MIGRATION

There are many ways to classify the complex and diverse movements of human populations, but perhaps the most common division is to divide movements into two basic categories, international migration and internal migration. As Clark (1986, 74) has noted:

> International migration has a two-part structure: On one hand, it is composed of the labor streams seeking jobs in foreign countries; on the other, it is composed of the increasing flows of refugees from war and political disruption. However, neither of these flows takes place in a vacuum. The labor streams are inextricably woven together with changing urban development and there are increasingly dramatic social and economic impacts from global refugee migrations.

Each of these deserves attention here.

THE CHANGING NATURE OF INTERNATIONAL MIGRATION

Though international movements of people are as old as history, there was little useful information on the volume and nature of such movements before the nineteenth century. Archaeological evidence suggests that even the earliest ancestors of humans were wanderers. A progenitor of modern humans, *Homo erectus*, has been traced through fossil evidence to such widely separated places as China, Africa, Europe, and Java (Howell, 1965). The first "Americans" were most likely people who migrated from Siberia to Alaska over a land bridge that existed at various times during the Pleistocene. A

Biblical example of migration was the movement of the Children of Israel out of Egypt to the Promised Land under the leadership of Moses. Many other examples of early migrations exist, and there is little doubt that all of human history has been dramatically affected by such movements.

One of the largest migrations in human history was the emigration from Europe that began in the sixteenth century and continued into the early twentieth century. Estimates are that over 60 million Europeans were involved in this migration. The primary places of origin for migrants were Germany, Russia, Poland, Spain, Austria, Hungary, Italy, Portugal, Sweden, the Netherlands, and the British Isles. The primary destinations for the emigrants were the United States, Canada, Australia, New Zealand, Brazil, Argentina, the British West Indies, and South Africa.

The flow of migrants out of Europe was relatively slow until the nineteenth century. Because of declining death rates in Europe during the early nineteenth century, the European population was growing rapidly. Also, Europeans began to feel that life was perhaps more rewarding abroad and that the overseas lands looked attractive. At the same time new technological improvements, such as the steamship, made migration easier. This is not to say that emigration from Europe was not fraught with dangers; geographer Beaujeu-Garnier (1966, 172) pointed out that

> . . . the exiles, often very poor, made lengthy journals on the deck or in the hold of a wretched vessel. Of the starving Irish who quitted their country in the famine of 1845, 6 percent died at sea, and counting also those who died soon after arrival, one in five of these unhappy creatures failed to make a home on the welcoming continent.

International migration from Asia was relatively small compared with the massive movement out of Europe. For example, Table 8-1 gives the estimated numbers of expatriate Chinese in Asia, amounting to a total of over 21 million people who emigrated from China to other Asian regions. These data can be compared with data for other areas of the world, such as Africa, which received only 94,000 ethnic Chinese; Oceania, which received 99,000; the Americas, which received 643,000; and Europe, which received 89,000.

Migration from Africa was significantly different from movements out of other regions of the world. The huge exodus of Africans was primarily involuntary. Slaves were taken from Africa and brought to the New World from the sixteenth century to the nineteenth century. Approximately 10 million slaves were imported into slave-using areas between 1451 and 1870. The number of slaves actually taken from Africa has been estimated at over 11 million, since perhaps 25 percent of the slaves died enroute to the destination. In terms of both the numbers moved and distance this was the greatest slave migration in history.

■ ■ ■ TABLE 8-1. Estimated Numbers of Expatriate Chinese in Asia, About 1969

(Outside the People's Republic of China, Taiwan, Hong Kong, and Macao)		
	Numbers (000s)	% of Total Population
Thailand	4,930	14
Indonesia	4,800	4
Malaysia	3,602	34
South Vietnam	2,661	15
Singapore-Malacca	2,200	98
Cambodia	2,000	30
Burma	500	2
Philippines	376	1
Laos	61	2
Japan	54	—
Korea	26	—
India	16	—
Pakistan	3	—
Ceylon	1	—
Total	21,229	14

Source: Leon F. Bouvier with Henry S. Shryock and Harry W. Henderson, "International Migration: Yesterday, Today, and Tomorrow," *Population Bulletin*, Vol. 32, No. 4 (September, 1977), p. 7. Courtesy of the Population Reference Bureau, Inc., Washington, D.C.

INTERNATIONAL MIGRATION SINCE 1940

Recent international migration has taken several forms. Refugee movements have been a primary facet of migration since 1940, and there have been two major movements:

1. population transfers within the Indian subcontinent between India and Pakistan

2. the uprooting of Eastern and Central Europeans that represented a drastic solution to the problems of ethnic minorities in these regions prior to World War II

These forced migrations were not of a selective nature because entire communities were uprooted and moved.

Not all of the post-World War II migration movements were as dramatic or tragic as these forced migrations. The movement of Europeans from 1956 to 1970 is shown in Table 8-2. These data show that almost 7 million Europeans moved to Australia, New Zealand, Canada, the United States, South Africa, and Latin America. In some countries, like South Africa and Australia, the data reflect the special efforts of those countries to attract Europeans. The large decline in migrants moving to Latin America reflects the increase in job opportunities for Southern Europeans in countries closer to home, as well as the measures taken by Latin American countries to slow down immigration. Recent migratory movements within Europe are reflected in Table 8-3. In total, the United Nations estimates that Western and Northern Europe gained 2.6 million people through

■■■ TABLE 8-2. Immigrants from Europe to Principal Overseas Destinations, 1956–1970[a]

Country or Region	(In thousands)			
	1956–1960	1961–1965	1966–1970	1956–1970
Australia	524.9	588.8	764.8	1,878.5
New Zealand	70.9	86.1	70.4	227.4
Canada	672.9	365.2	589.9	1,628.0
United States	700.1	531.6	600.6	1,832.3
South Africa	45.2	78.6	149.7	273.5
Latin America[b]	490.0	226.3	131.0	847.3
Total	2,504.0	1,876.6	2,306.4	6,687.0

[a]Based on statistics of receiving countries.
[b]From Italy, Portugal, and Spain only, based on statistics of the sending countries. Migrants from these three countries constituted about 95 percent of all European migrants to Latin America.
Source: United Nations, *International Migration Trends, 1950–1970*. Conference Background Paper, E/CONF .60/CBP/18, May 22, 1974, Table 8.

■■■ TABLE 8-3. Migration Balances Between Principal Sending and Receiving Countries of Europe, 1960–1970

Sending Countries	(In thousands) Receiving Countries					
	Total	West Germany	France	Switzerland	Belgium	Netherlands
Greece	250	240	[a]	[a]	[a]	[a]
Italy	500	230	100	150	12	[a]
Portugal	510	40	450	[a]	[a]	[a]
Spain	640	150	330	100	40	20
Yugoslavia	490	400	55	25	[a]	[a]
Total	2,390	1,060	940	280	70	40

[a]Less than 10,000
Source: United Nations, *International Migration Trends, 1950–1970*, Conference Background Paper, E/CONF .60/CBP/18, May 22, 1974, Table 4.

migration from Southern Europe from 1960 to 1970. More recent studies of European migration by geographers include Salt (1985) and White (1985)

In summary, the balance of international migration since the end of World War II has increasingly shifted toward North America and Oceania. According to the data shown in Table 8-4 these were the only two major regions of the world to record a net immigration during the 1960 to 1970 decade. Also, emigration from Europe dropped significantly, from 5.4 million between 1946 and 1957 to 300,000 between 1960 and 1970. Detailed estimates of historical net migration in selected countries of Europe, North America, and Oceania for the 1950–1970 period are shown in Table 8-5.

TABLE 8-4. Estimated Net Migration Balances for Major World Regions, 1946–1957 and 1960–1970

	(In millions)	
Region	1946–1957	1960–1970
Africa	+0.5	−1.6
Asia	−0.5	−1.2
Europe	−5.4	−0.3
Latin America	+.09	−1.9
North America	+3.4	+4.1
Oceania	+1.0	+0.9

Source: United Nations, *International Migration Trends, 1950–1970,* Conference Background Paper, E/CONF .60/CPB/18, May 22, 1974, Table 5.

In recent years labor migration has become increasingly more important in Asia, as well as from Asia to the Middle East and Oceania. Rapid economic development within these regions has altered the demand for labor, encouraging international flows from the poorer to the wealthier. The composition of these migration streams has altered over time; major construction projects in early stages of development attract primarily male laborers, for example, whereas the need for domestic workers—as wealth increases in a region and more women enter the labor force—attracts more females. Longitudinal studies of such streams are likely to find that changing sex ratios are common over time, even if the number of migrants remains the same.

REFUGEES: A SPECIAL CATEGORY

"A **refugee** is defined as a person who is outside the country of his nationality . . . because he has or had well-founded fear of persecution . . . and is unable or, because of such fear, is unwilling to avail himself of the protection of the government of the country of his nationality."

The Protocol Relating to the Status of Refugees—agreed to by the United Nations in 1967—defines a **refugee** as a ". . . person who is outside the country of his nationality . . . because he has or had well-founded fear of persecution . . . and is unable or, because of such fear, is unwilling to avail himself of the protection of the government of the country of his nationality." Estimates of the number of refugees worldwide vary from 15 to 20 million, though no one knows with any degree of certainty; there may be an additional 25 million people who are "refugees" within their own countries. What is known with more certainty, however, is that in recent decades, as a result of wars and famines (Bosnia and Rwanda are excellent examples), refugees have increased in numbers. At the same time, geographers have shown an increased interest in refugees in recent years, and in one article Kliot (1987) noted that since the beginning of the twentieth century more than 100 million people have become refugees. As with other forms of migration, refugees affect both the region they leave and the one to which they move. Refugees often flee one form of conflict only to precipitate another.

■■■ TABLE 8-5. Estimates of Net Migration in Selected Countries of Europe, Northern America, and Oceania, 1950–1970

Region and Country	Net Migration (000s)	Region and Country	Net Migration (000s)
Europe	−3,028	*Northern Europe*	−698
Western Europe	+8,748	Denmark	−32
Austria	−103	Finland	−214
Belgium	−211	Ireland	−558
West Germany	+4,780	Norway	−10
France	+3,258	Sweden	−297
Luxembourg	+22	United Kingdom	−181
Netherlands	−50		
Switzerland	+630	*Northern America*	+8,698
		Canada	+1,802
Southern Europe	−7,301	United States	+6,896
Greece	−651		
Italy	−1,958	*Oceania*	+1,857
Malta	−81	Australia	+1,712
Portugal	−1,952	New Zealand	+145
Spain	−1,377		
Yugoslavia	−1,282		
Eastern Europe	−3,777		
Bulgaria	−178		
Czechoslovakia	−174		
East Germany	−2,488		
Hungary	−161		
Poland	−526		
Romania	−250		

Source: United Nations, *International Migration Trends, 1950–1970,* Conference Background Paper, E/CONF .60/CBP/18, May 22, 1974, Table 6.

Kliot (1987) argued that among the major causal agents of current and future refugee movements are the following:

1. ethnic, racial, and religious problems within countries

2. local and regional wars, from South Africa and the Middle East to the breakup of what only a short time ago was Yugoslavia

3. changing racial policies of many governments in Africa and Asia

In another study, geographer Richard Black (1991, 281) importantly noted that "... geographers must draw links between this field and their own longstanding interests in issues such as migration, 'natural' and other disasters, and the politics of conflicts which are often the immediate cause of refugee flows."

An excellent exposition of refugees in the world today is Zolberg, Suhrke, and Aguayo (1989). The authors begin by criticizing the accepted definition of refugees (cited above); instead ". . . we point out only that disagreement on these issues is unavoidable, because defining refugees for purposes of policy implementation requires a political choice and an ethical judgment." (Zolberg, Suhrke, and Aguayo 1989, 4). Their major fear is that acceptance of an earlier definition gives it credence and focuses research on ongoing practices. After thirty years, they suggest, perhaps the definition itself needs reconsideration because of the changing world system. The authors, after several detailed discussions, suggest that refugee studies need to focus more on the *root causes* of flight—included among them are poverty and social change—and the development of policies that might mitigate them. They note that the burden of asylum has fallen heavily on the developing countries and argue the following (Zolberg, Suhrke, and Aguayo 1989, 282):

> . . . the richer states must, at a minimum, accept a greater financial obligation to assist the countries of first asylum in the South. . . . Refugee policy must be held up against the negative yardstick that at least it should not contribute to greater refugee flows in the future.

Though we have looked only briefly at refugees in this section, geographers are now studying refugees in many places. Rogge (1991), for example, has looked at the uncertain future for refugees in Southeast Asia and Everitt (1991) considered the refugee problem in Middle America since the 1970s. Luciuk (1991, 197) ended his study of refugees in Afghanistan by saying that "Since even the most optimistic estimates suggest that many years of economic, political, and social rehabilitation will be required before the land and its peoples recover from what has been described as a genocidal war, it seems evident that Afghanistan will remain a deeply troubled region of the world, and, as such, a source area of refugees and displaced persons, well into the foreseeable future." In surveying the refugee situation in Africa, Kenzer (1991, 200) concluded that the situation was ". . . horrible at best." This was said, prophetically, long before the debacle that has all but completely destroyed Rwanda. Elsewhere on that continent other devastating events are probably not far off, perhaps in Burundi, or Nigeria, or Mozambique.

IMMIGRATION: THE UNITED STATES

Immigration has played a key role in the social and economic development of the United States. If there had been no immigration to the United States during the nineteenth and early twentieth centuries, then population growth would have been much slower and the ethnic composition of the United States' population would be substantially different.

Immigration history in the United States can be divided into four distinct periods: colonial, old immigration, new immigration, and restricted immigration. The first three periods could be considered

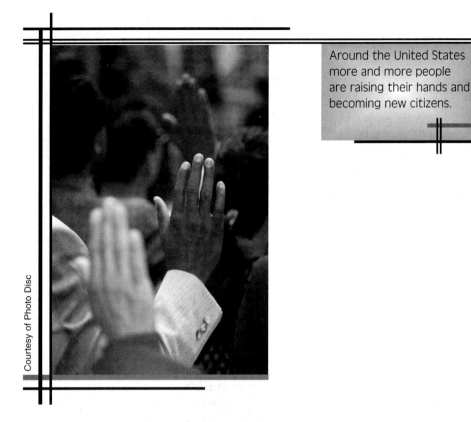

Around the United States more and more people are raising their hands and becoming new citizens.

times of free-individual moves, whereas the last period involved significant restrictions on immigration.

The immigrants who came to the United States during the colonial period were confronted with a situation that was significantly different from that of later arrivals. Most were of British ancestry, though other ethnic groups such as the Dutch, Swedes, Germans, French, and Spanish were also represented.

The period of old immigration was approximately 1800 to 1880 and involved a considerable increase in immigration. The overwhelming majority, about 95 percent, of these immigrants came from western and northern Europe. Though the largest proportion of these immigrants came from England, large numbers of them were from France, Germany, Scotland, and Ireland.

The period of new immigration started around 1880 and was marked by a shift in the source of immigrants from western and northern Europe to eastern and southern Europe. This new wave of immigration was much larger than the old, and in the first decade of the twentieth century approximately 9 million immigrants entered the United States. A breakdown of total immigration from Europe is outlined in Table 8-6.

Since 1921 immigration to the United States has no longer been free. Restrictions on both the number and types of migrants are in effect. The origin of immigrants has shifted dramatically, however. Before 1900 immigrants came almost exclusively from Europe. Today they come mostly from Asia and Latin America.

■■■ TABLE 8-6 U.S. Immigration from Europe, 1820–1976			
Country of Origin	Number of Immigrants (000s)	Country of Origin	Number of Immigrants (000s)
Germany	6,960	Spain	248
Italy	5,278	Belgium	201
Great Britain	4,863	Romania	168
Ireland	4,722	Czechoslovakia	137
Austria and Hungary	4,313	Yugoslavia	109
U.S.S.R.	3,362	Bulgaria	68
Sweden	1,270	Finland	33
Norway	856	Lithuania	4
France	744	Luxembourg	3
Greece	638	Latvia	3
Poland	506	Albania	2
Portugal	422	Estonia	1
Denmark	363	Other Europe	55
Netherlands	357		
Switzerland	347	Total	36,033

Source: Leon F. Bouvier with Henry S. Shryock and Harry W. Henderson, "International Migration: Yesterday, Today, and Tomorrow," *Population Bulletin* Vol. 32, No. 4 (September, 1977), p. 17. Courtesy of the Population Reference Bureau, Inc., Washington, D.C.

The attributes of places of origin and destination hold true for **undocumented immigrants** as well as for legal immigrants.

Undocumented Immigration

Since the establishment of migration quotas there has been relatively little concern about the origin of present-day legal immigrants to the United States. There is, however, increasing concern over the presence of immigrants who have entered the United States illegally. Many of the factors that help explain legal migration also apply to illegal migration. The attributes of places of origin and destination hold true for **undocumented immigrants** as well as for legal immigrants. The undocumented immigrant is confronted by unfavorable conditions in his place of origin and tries to overcome these difficulties by emigrating.

The problem of undocumented immigration is not just associated with the United States. Similar problems exist in Latin America, Asia, and Europe. In a study of immigration in Europe, King (1974, 82) noted the following:

> Increasingly restrictive immigration policies in Europe had made it more difficult to migrate legally to Western Europe, with the result that by the summer of 1973 it was estimated that there were more than half a million foreigners illegally working or living in Europe, mainly engaged in activities such as road construction, building, agriculture, hotels, and the public services.

More recently France, Germany, and Austria have experienced rightward political shifts as voters have become increasingly concerned with the presence of immigrants, especially those who have, or are at least perceived to have, entered illegally.

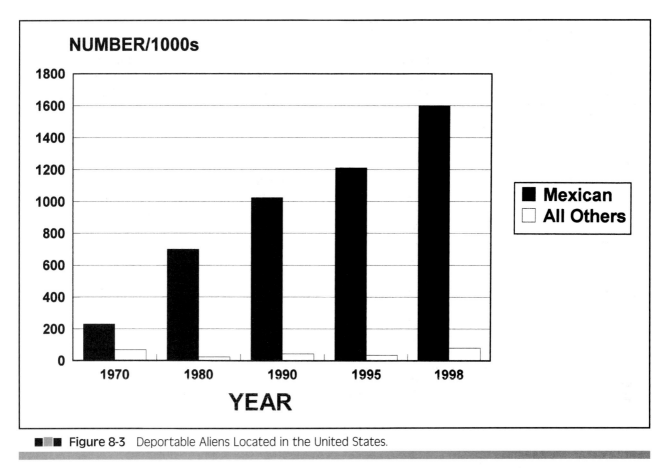

NUMBER/1000s

■■■ Figure 8-3 Deportable Aliens Located in the United States.

Source: *Statistical Abstract of the United States, 1995,* Table 331, and *INS-Fiscal Year 1998 Statistical Yearbook,* Table 57.

Presently the United States attracts the largest number of illegal migrants. There is no actual count of the number of illegal migrants in the United States, but estimates range from 5 to 10 million. Although no exact count of illegal aliens exists, there are data on the number of illegal aliens who have been apprehended. Data from 1970 to 1998, shown in Figure 8-3, suggest approximately a 500 percent increase in the apprehension of illegals during that period. It has been estimated that only one out of every three or four persons attempting to enter illegally was caught. The largest proportion of undocumented immigrants come from Mexico, as is shown in Figure 8-3. Of the 1,199,000 illegal aliens located in the United States in fiscal year 1993, 1,169,000, or about 97 percent, were from Mexico. Economic conditions in Mexico provide a "push" to migrate to the United States for employment, and the estimated growth of the Mexican labor force at about one million annually during the 1990s is a virtual guarantee that this particular migration stream will continue.

In the past many migrants stayed close to the border after illegal entry. However, recent data show that large numbers are now scattering throughout the nation, to New York, Houston, Chicago, Miami, San Antonio, and Los Angeles. The problem of illegal aliens is complex. Many interest groups are involved with both sides of the issue.

The United States has spent a great deal of time and money trying to make illegal border crossings from Mexico more difficult. Unfortunately, they have had little success. President Bush in 2001 made it clear that he wanted significant changes in immigration laws.

Labor unions are fearful that jobs that could be held by United States citizens are being taken by illegal aliens. Those who favor limited or zero population growth see the influx of illegal aliens as a stumbling block in achieving a stable population. On the other hand, there are those who are against efforts to curb immigration, including agricultural interest groups that benefit economically from the cheap labor supply of illegals. Also, some Hispanic rights groups, as well as the United States Catholic Bishops, fear increased discrimination against Mexican Americans.

The Immigration Reform and Control Act of 1986 (IRCA)

After years of political bickering, the Congress of the United States finally passed legislation designed to deal more realistically with immigration and the problems resulting from illegal immigration. This bill contained two key provisions. First, there was an amnesty program for aliens who had lived and worked in the United States since before January 1, 1982. Second, there were sanctions for employers who knowingly hire unauthorized workers.

In order to qualify for amnesty under this act a person must have proved that he or she entered the United States before January 1, 1982, and had resided in the United States continuously since that time of entry. Proof of eligibility required such records of long-term residence as health and school records, employment records, and utility receipts. The deadline for applications ended early in May, 1988, at which time approximately 1.5 million people had applied.

In an attempt to reduce the number of illegal aliens residing in the United States, employers are required to screen applicants.

Special provisions were made for agricultural workers so they could be given temporary status if they had worked for at least 90 days between May 1, 1985, and May 1, 1986. After two years of temporary status they can then apply for permanent residency. After November 6, 1993, all applicants who successfully qualified for amnesty will begin to qualify for citizenship.

Employers are to be enforcers of the new program because of their responsibility for not hiring illegal aliens. Severe penalties will be given out to those who violate the new law. Employers are required to screen applicants, who must provide documentation of their status in the form of a passport from the United States, certification of United States citizenship, an alien registration receipt card, or a foreign passport with a work authorization stamp. If they have none of these things, then they must have either a birth certificate showing that they were born in the United States or a social security card and, in addition to one of those things, proof of identification in the form of a valid driver's license or state identification card. Though employers cannot be held responsible for the authenticity of each of the above items, they must keep them on file for a period of at least three years once they have been submitted.

One unplanned result has been the development of a cottage industry in document production in the border states. Another effect has been a change in the traditional pattern of circular and repeat migration from Mexico to the United States. As Vernez and Ronfeldt (1991) noted, "Mexican immigration to the United States can no longer be characterized by the persistent image of the

Mexican immigrant as a temporary worker staying here for a short period of time and leaving his or her family behind, if it ever could be so characterized." This change will have considerable impact on states in the southwest, especially California and Texas. Bean, Edmonston, and Passel (1990) provided coverage of many of the implications of undocumented immigration since IRCA for those who would like to pursue this topic further.

In the new millennium, Americans still remain divided over many immigration issues. Nonetheless, Washington spent much of the 1990s trying to make illegal border crossings from Mexico more difficult. Remote video surveillance sites, thermal infrared imagers, night vision goggles, long-range infrared systems, or fiber optic borescopes, and intelligent computer-aided detection systems all are employed to help Border Patrol officers detect illegal immigrants. In response, immigrant routes shifted to less-guarded interior sites.

The Immigration Act of 1990

Congress passed legislation to increase the number of legal immigrants that would be accepted each year. The result of that legislation was the **Immigration Act of 1990.**

Despite polls that regularly show that Americans are concerned about the level of immigration (Page and Shapiro, 1992), Congress passed legislation to increase the number of legal immigrants that would be accepted each year. The result of that legislation was the **Immigration Act of 1990.** The impact will be to increase legal immigration from a level of some 500,000 annually to around 700,000 annually, with some adjustments to be made after 1993, when a required review of immigration will be made.

Part of this revised legislation is designed to increase the immigration of skilled and educated people, as well as more wealthy people. Up to 10,000 slots per year will be available to investors who are willing to put at least $1,000,000 into an American business that employs ten or more workers. Only half that amount is required if the investment is made in a rural or depressed area. Another 40,000 slots were set aside in a lottery for potential immigrants from 34 countries, mainly European, that felt neglected under previous legislation. Of those, 40 percent were set aside for Irish immigrants.

The Illegal Immigration Reform and Immigrant Responsibility Act of 1996 (IIRIRA)

After continued discussions of immigration, especially illegal immigration, in the early 1990s, Congress passed the **Illegal Immigration Reform and Immigrant Responsibility Act in 1996.** In part, at least, this act was sparked by concerns about terrorism in the United States. IIRIRA approved an expansion of border guards by 1,000 per year for five years, bringing the total number of border patrol agents to around 10,000 in 2000. It also introduced a pilot telephone verification program so that employers could more easily check on the immigrant status of job applicants and social service agencies could do the same for applicants for various benefits, and altered several requirements for residents who sponsor immigrants.

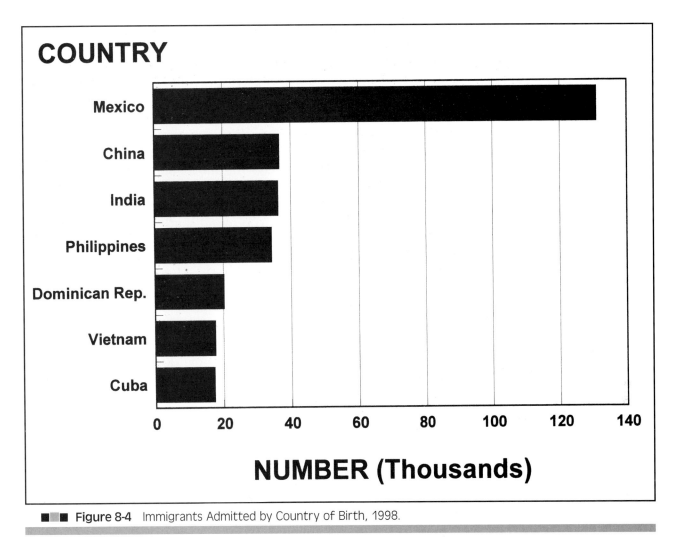

COUNTRY

Mexico

China

India

Philippines

Dominican Rep.

Vietnam

Cuba

0   20   40   60   80   100   120   140

NUMBER (Thousands)

■■■ **Figure 8-4** Immigrants Admitted by Country of Birth, 1998.

Source: INS-Legal Immigration, 1998, Table 2.

In subsequent years Congress enacted further immigration laws, including the Nicaraguan Adjustment and Central American Relief Act in 1997 and the Haitian Refugee Immigration Fairness Act in 1998. In 2001 the immigration issue continued to get attention from Congress and from President Bush. Whatever the outcome of this current discussion, immigration is likely to remain a contentious issue well into the future in the United States.

Refugees

Since the end of its involvement in the Vietnam War in the mid-1970s, the United States has admitted numerous refugees from Southeast Asian countries under a special refugee act. Figure 8-4 shows the source regions for refugees into the United States for 1998; the predominant sources are still countries in Asia and Southeast Asia as well as Latin America.

Originally the government of the United States encouraged a refugee distribution policy that spread Southeast Asian refugees around the United States. However, subsequent migration patterns for these groups has redistributed their numbers considerably, and, like other Americans, they have been moving to the sunbelt. As a result, concentrations of Asian refugees have developed in many areas, especially in Southern California, where Long Beach has the largest settlement of Cambodians outside of Cambodia and where nearby Westminster has gained the nickname "Little Saigon" as a reflection of the concentration of Vietnamese in that city. The redistribution of Indochinese refugees within the United States was carefully studied by Desbarats (1985), who found that refugees were concentrating in states that offered some combination of high public assistance benefits, lenient eligibility requirements for receiving those benefits, mild climates, reasonable employment opportunities, and large Asian communities.

Figure 8-5 shows just how dramatically immigration has been growing in recent decades. This growth in turn is fueling debates in the United States about immigration policy (and concern about both legal and illegal immigration in some quarters). Growing anti-immigrant rhetoric in the United States during the 1990s (exemplified by the passage of the anti-immigrant Proposition 187 in California) has resulted in some rethinking of refugee admission policies. Refugees from Haiti were sent back, and the long-standing policy of accepting all Cuban refugees has given way to a more restrictive policy.

INTERNAL MIGRATION

A large proportion of migrants never cross national borders; they move only within their own countries. Such internal migration is often more difficult to assess, and only those countries that keep a complete dossier on each citizen can accurately reconstruct the nature of internal migration. However, in countries such as the United States, census information—supplemented by periodic surveys, social security data, and information on driver's license exchanges—can be used to trace internal migration patterns.

GENERAL PATTERNS OF INTERNAL MIGRATION

Some generalizations can be made about the geographic patterns of internal migration. For example, one of the major movements in modern times has been the movement of people from rural to urban areas. This movement has been the result of many complex interactions, especially the shift of employment away from rural agricultural pursuits and toward urban industrial and service occupations. As geographer Wilbur Zelinsky (1971, 236) once noted, "The onset of modernization . . . brings with it a great shaking loose of migrants from the countryside."

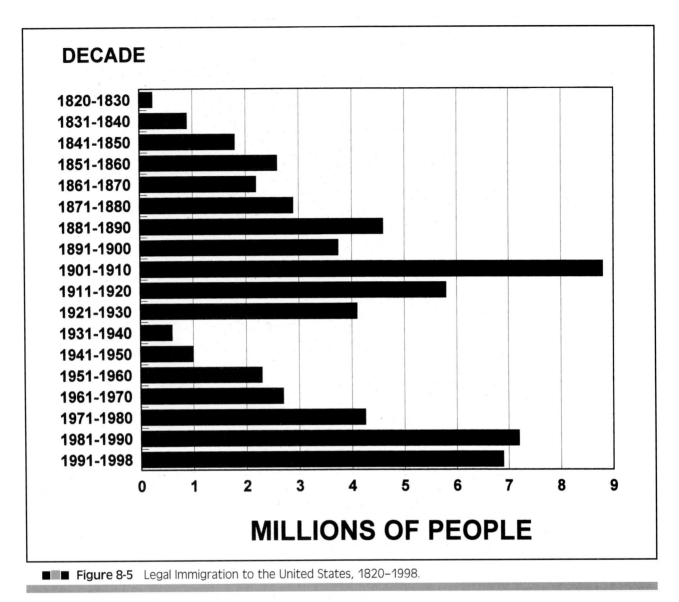

DECADE

MILLIONS OF PEOPLE

■■■ **Figure 8-5**   Legal Immigration to the United States, 1820–1998.

Source: Immigration and Naturalization Service, 2000.

The conquest of new territory has been an important force in the settlement of many countries. The movement of the frontier in the United States, from the original thirteen colonies to the Pacific Ocean, has its parallels elsewhere. Examples include the westward push of the pioneer fringe in Brazil, the northwestward expansion across Canada, the northward movement in Australia, and the movements of peoples from the Andes to the coastal plain and interior forests of South America.

Nomadism is another type of internal (and sometimes international as well) movement. Even at the start of the twenty-first century there are still people who are constantly moving about, often with little regard for international boundaries. Many nomads are still found today in the Sahel region of Africa, for example, and in the high latitudes of both Europe and North America.

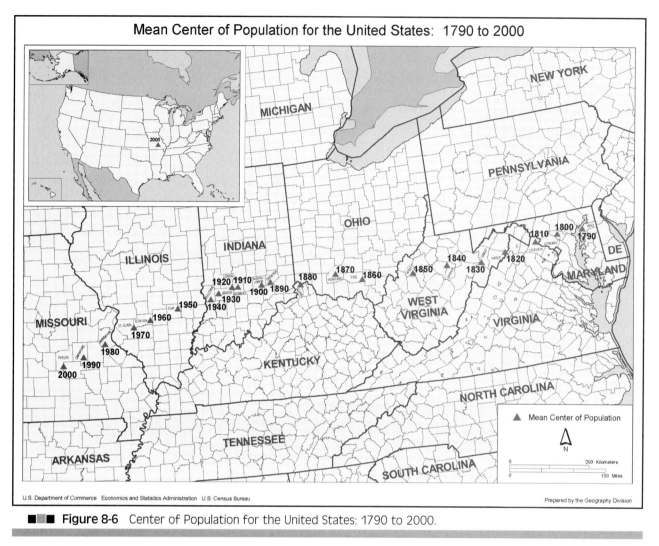

■■■ Figure 8-6 Center of Population for the United States: 1790 to 2000.

Source: United States Bureau of the Census.

MIGRATION WITHIN THE UNITED STATES

Approximately four million Americans lived in a narrow belt along the Atlantic Ocean in 1790; only a hundred years later the United States occupied an area that extended to the Pacific Ocean. Movements during the twentieth century have been primarily westward and coastward, with a loss of population in rural areas and a concomitant concentration of people first in cities and then in suburbs, though recently there has been a revival of growth in nonmetropolitan areas. Because of the large gains in population in the western states, as well as in Florida, Texas, and other states in the southwest, the center of population for 2000 was farther west and somewhat south of its 1990 location, as is apparent in Figure 8-6—you can expect this to continue. The United States is indeed a nation of migrants, as well as a nation of immigrants. Since the annual mobility sample studies were first begun in 1947, about one-fifth of the American population has changed residence each year (about 17 percent in recent years).

The United States has entered a new era with respect to both the distribution of population and economic activities (Fosler, et al, 1990). Capital and people are increasingly being attracted away from the old industrial heartland and toward the South, Southwest, and West. Forces not only in the national economy, but increasingly in the global economy as well, are constantly at work changing locational advantages, altering economic landscapes, and encouraging consequent patterns of internal migration. The essence of American mobility was captured in the following comment by geographer Paul Simkins (1978, 204): "Although it does not appear as such in any of the state seals or flags, the wheel in many ways may be considered a symbol of America; the wagon or train wheel which early carried the nation westward, or the automobile wheel that moves it now increasingly to and from metropolitan centers." Americans have always sought to better their lot in life, whether they were pioneers seeking "elbow room" or folks in twentieth-century rural Iowa seeking the perpetual sunshine of Southern California.

A recent governmental survey of mobility in the United States (Schacter, 2001) found that more than 43 million Americans (15 percent) changed residence in a one-year period between March 1999 and March 2000. Of those, however, 56 percent moved within the same county, whereas only 19 percent moved to another state. This reflects, of course, the generalization that most moves occur over short distances.

Some salient features of these recent movers give us a better sense of American mobility at the turn of the millennium. For example, long distance moves have been increasing, even though the overall frequency of moves has slowed somewhat since 1990. Also, the most mobile age groups were 20–24 and 25–29-year-olds, and the least mobile racial/ethnic group was non-Hispanic whites. Other observations include the following: married people moved less than single and divorced people; one-third of renters moved, and lower income people moved more often than those in higher-income groups. People of different educational levels had similar mobility rates, though movers with a college degree tended to move longer distances than those without.

At the forefront of motives for migration, of course, has been the hope of material gain, though the recent rise of the Sunbelt as a center of attraction for migrants suggests that other motives are also important, especially such amenities as mild climates. Many seek to escape the long cold winters of the cities of the Northeast, as well as their smog, traffic congestion, and high crime rates. Though an overview of migration patterns can deal only with aggregates of people, we need to remember that the motives for migration differ for various age groups and for various other characteristics of people as well.

Despite the rural-urban migration that characterized the redistribution of people throughout the United States during the twentieth century, Americans have never been in unanimous agreement that cities, especially large ones, were desirable places. Industrialization concentrated people in cities, primarily for economic reasons. Urban locations often meant lower costs for industries, as well as better access to service needs and to local markets for goods. However, from the worker's viewpoint, industrial concentration in urban areas meant more jobs and higher wages, so to the cities they moved.

Once in the cities, however, most groups have tried to work their way outward again, giving rise first to suburbanization, then increasingly to exurbanization. Many problems in the cities today are at least in part related to suburbanization because it, like other migration processes, is selective—those economically able to move outward tend to do so. Though suburbanization was long considered an almost exclusively white process, Latinos and African Americans are increasingly moving toward the suburbs as well. In Los Angeles, for example, newly-arrived immigrants in formerly black communities such as Watts and Compton have accelerated the movement of African Americans toward such suburban locations as Bellflower, Lancaster, and Palmdale within Los Angeles County and to suburban parts of neighboring Riverside and San Bernardino counties.

Two noticeable and interesting migration trends that emerged in the 1970s are still of special interest and may have possible long-range implications for the redistribution of people, jobs, and wealth in the United States:

1. movement to the "Sunbelt"

2. the "Rural Renaissance"

Though movement to the Sunbelt continued throughout the 1980s and on into the 1990s, the Rural Renaissance seemed more ephemeral because during the 1980s metropolitan areas again grew faster than nonmetropolitan areas. Nonmetropolitan growth has subsequently accelerated again, indicating that the 1980s were only a blip in an ongoing process of decentralization.

Movement to the Sunbelt

The Sunbelt states are those of the South and Southwest, along with California, though there is not complete agreement from one list of Sunbelt states to the next. Colorado, for example, is sometimes included because of its growth pattern rather than its climatic claims. Furthermore, within the Sunbelt itself there is considerable variation in the growth and attractiveness of different states. Mississippi, for example, has hardly had the same experience during the 1980s that Texas or Florida has had. Since 1970 a dramatic shift of population away from states in the North Central and Northeast and toward the Sunbelt states has occurred. A disproportionate share of recent population growth in the United States has occurred in only three states: California, Florida, and Texas. In the 1980s the Sunbelt states with the highest rates of growth were Nevada, Arizona, Florida, New Mexico, Hawaii, and Texas. Not only people, but also industry, are increasingly attracted to the Sunbelt.

Regional growth during the 1970s and 1980s was highest in the West, though economic difficulties during the early 1990s slowed the process considerably. Reductions in population growth rates were occurring in many states in the Midwest and Northeast. People are increasingly moving away from the older areas of the North Central and Northeastern states, partly in response to environmental preferences and partly as a result of aggressive economic development in many of the Sunbelt states. Many people are also being attracted to the Rocky Mountain states, though new energy development no longer offers as many job opportunities as it did in the 1970s, and to the Pacific Northwest, especially to Portland and Seattle, although early in the 1990s opportunities were limited in those cities and unemployment was high. The following comment by Biggar (1979, 27) is probably still correct:

> . . . the Sunbelt migration, then, is part of a larger radical change in the South and the West. It is closely related to the industrial development which typifies most states in the "Southern Rim." It is a reflection of changing American lifestyles. It is part of the equalization of U.S. regions which has brought the South more in line with the rest of the nation.

The Rural Renaissance

During the 1970s, for probably the first time in American history, more people were leaving metropolitan areas than entering them. The long-established pattern of rural-urban migration underwent a reversal during the 1970s. As you might imagine, not all nonmetropolitan areas were equally pleased with the prospect of increased population growth and "rural urbanization." Conflicts often arise between growth advocates and those who seek to limit growth. Land use conflicts are also common. Many small-town residents fear that growth will destroy the very features that attract people to small towns. For example, they point to increasing crime rates, traffic congestion problems, and even smog as inevitable results of small-town growth. It is interesting to note that quite often it is the newcomer who is most vocal about limiting further growth and change; visit with ex-Californians in Oregon or Idaho, for example, and see if they invite you to stay.

The sudden reversal of long-term urbanization in America caught many people, including demographers and geographers, by surprise. Many had thought that most of the nonmetropolitan growth was just spillover from expanding metropolitan areas; however, even nonmetropolitan areas far from urban centers were participating in the new trend. Spillover, of course, has been commonplace in recent decades, but a new pattern seemed to be at hand. Among its major causes were the following:

1. changes in communication and transportation technology that have taken away, or at least mitigated, the necessity of urban concentration

2. the expansion of highways that allow easy access to urban areas

However, regional variations in the rates of both metropolitan and nonmetropolitan growth rates may be observed.

The growth of retirement and recreational communities has also been a factor in nonmetropolitan growth. Early retirement, coupled with better benefits for retirees, frees an increasing number of people to seek locations based on personal preferences rather than economic necessities. Wherever the elderly choose to go, of course, services are required; thus jobs are created and further growth is stimulated. Not only Florida and the Southwest, but also the Ozarks, the Texas hill country, and California's "Gold Country," in the western foothills of the Sierra Nevada, are experiencing growth from substantial influxes of retirees. In the 1990s retirees may be looking in new directions as well. From Hanover, New Hampshire, to Sequim, Washington, retirees are seeking smaller, less expensive, less crime-ridden areas, and they may be willing to tradeoff some sunshine for more peace and quiet.

Just as social scientists were getting comfortable with the idea of the rural renaissance, it seemingly came to an end. By the early 1980s metropolitan areas were again growing faster than nonmetropolitan ones. For example, Richter (1985, p. 260) wrote that ". . . the turn-around from negative to positive net migration in nonmetropolitan areas was sustained throughout the 1970s . . . nevertheless a slow-down in the growth of nonmetropolitan areas (occurred) in the late 1970s." Richter (1985, p. 261) went on to suggest that ". . . the validity of the turnaround and the evidence for a slowdown in nonmetro growth by the end of the decade may be incorporated into new theories of urban-rural migration." These theories need to explain changing mobility patterns in "post-industrial" societies. They must include the effects of innovations in communication and transportation technology, and increasingly are going to have to grapple with the geographic effects of a globalizing economy.

The Canadian experience has been similar, as noted by Keddie and Joseph (1991). Though they found less concrete evidence of a rural turnaround during the 1970s in Canada, they did find that ". . . the data for the early 1980s point toward a return to the urban-dominated growth scenario prevalent in the 1960s." (Keddie and Joseph, 1991, 378) They also noted that changes in rural population trends generally were a good indicator of the economic fortunes of different regions.

Although during the 1980s and early 1990s nonmetropolitan areas grew more slowly than they did in the 1970s, many of them were still attracting migrants. Studies of the patterns that emerged in the late 1970s and 1980s will help to better illuminate underlying causes of changing mobility patterns in the United States. By the mid-1990s nonmetropolitan area growth had again increased, leaving some social scientists shaking their heads (and teaching us all to beware of short-term trends).

It is worth noting here briefly also that national and regional trends often obscure underlying patterns that may be seen for particular subgroups of the population, from retirees to ethnic groups. These various patterns are important as well, and do receive considerable attention in the migration literature. Examples of such studies include Robinson (1986), who wrote about the reversal in the 1980s of the previous pattern of African American net migration out of the South, and McHugh (1989), who studied the internal redistribution of Hispanics within the United States.

Places Left Behind

Like economic development and globalization, migration creates winners and losers, reshaping America's demographic landscape impassionately and inexorably. Of the places left behind in this vast land, none stands out so clearly as the Great Plains, where a broad swath of population loss stretches northward from Texas and Oklahoma to the Dakotas, eastern Montana, and western Minnesota. As Jeff Glasser (2001, 19) remarked, "Up and down the Great Plains, the country's spine, from the Sandhills of western Nebraska to the sea of prairie grass in eastern Montana, small towns are decaying, and in some cases, literally dying out. The remarkable prosperity of the last decade never reached this far."

Many counties in this vast interior landscape have reverted back to "frontier" status, as their population densities have declined to below six persons per square mile. The population of the Great Plains actually peaked during the 1930s, and has declined since then, albeit unevenly in both space and time. Most of the declines have come in places that specialized in some combination of wheat farming, cattle ranching, and mining (including oil production). Without subsidies for farming and energy production, declines probably would have been even faster.

Farming in much of the Great Plains is in a depressed state, stung by low grain prices, rising production costs, and competition from imports. Additionally, young people are less interested than ever in continuing to live and work on farms—they go off to college then on to careers in places that offer them jobs. The Great Plains remains a harsh land of great vistas but declining opportunities, a place where some will always remain transfixed but others are more than willing to give it back to the buffalo. Poet and resident Kathleen Norris (1993, 110) has perhaps described it best:

> Where I am is a place where the human fabric is worn thin, farms and ranches and little towns scattered over miles of seemingly endless, empty grassland. . . . But some have come to love living under its winds and storms. Some have come to prefer the treelessness and isolation, becoming monks of the land, knowing that its loneliness is an honest reflection of the essential human loneliness. The willingly embraced desert fosters realism, not despair.

In a nation where most people embrace "reality TV," cell phones, and instant gratification, however, the Great Plains is likely to continue its decline as its young leave for cities and its old gradually die. Contraction feeds on itself—as small towns decline, property values shrink, schools and businesses close, and people have even fewer incentives to stay.

RURAL-URBAN MIGRATION AND URBANIZATION: REGIONAL COMPARISONS AND CONTRASTS

Rural-urban migration drives the urbanization process.

■▮■

An analysis of **urbanization** by geographer Ray Northam suggested that it could be viewed chronologically as a three-stage process, as shown in Figure 8-7.
1. Initial Stage
2. Acceleration Stage
3. Terminal Stage

■▮■

Rural-urban migration drives the urbanization process. The population of the world as a whole has been increasing rapidly, but the increase of the urban population has increased even more rapidly than the overall rate in recent decades. According to Berry (1993, 399):

> . . . transnational urbanward migration is an important force that redistributes urban growth across the globe. The sharply-rising rate of this migration reflects radical increases in the underlying migration propensities: a sevenfold increase 1900–80; and fourteenfold 1860–1980. . . . Significant differences have emerged in the relationship across levels of development since World War II. Transnational migration remains a potent source of urban growth in a number of low- and middle-income countries, at the expense of other countries at these levels of development.

An analysis of **urbanization** by geographer Ray Northam suggested that it could be viewed chronologically as a three-stage process, as shown in Figure 8-7. In the Initial Stage the population is primarily rural, engaged in agricultural pursuits, and of a dispersed nature. The second stage, referred to by Northam as the Acceleration Stage, is a period in which an increasingly large share of the population lives in urban centers. During this stage there is a marked redistribution of the population, with the urban component rising from less than 25 percent to over 70 percent of the total population. The Acceleration Stage also involves a basic restructuring of the economy; there is a rapid shift from agriculture to industry, with a notable concentration of economic activity in the city. Large numbers of people are employed in manufacturing industries as well as in service and trade activities.

The final stage in the urbanization process is referred to by Northam as the Terminal Stage. During this stage the urban population is approximately 60 to 70 percent of the total population (occasionally even reaching 80 percent). Once the urbanization curve goes above 70 percent, it tends to flatten out. For example, the urbanization curve for England and Wales since 1900 has tended to flatten out after it reached the 80 percent level.

Today we might add another stage to Northam's model; a stage in which deconcentration begins in highly advanced "post industrial" economies. Some time ago, Berry (1980, 13) noted that, "Urbanization, the process of population concentration, has been succeeded in the United States by counterurbanization, a process of population deconcentration characterized by smaller sizes, decreasing densities, and increasing local homogeneity, set within widening radii of national interdependence."

A worldwide regional analysis shows that the world's nations are in various stages of urbanization. As a general rule, the developed

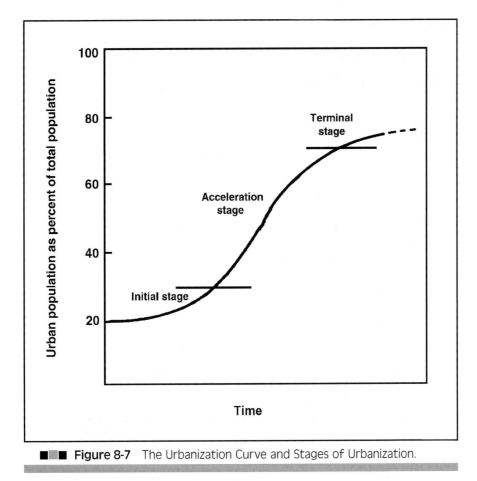

■■■ **Figure 8-7** The Urbanization Curve and Stages of Urbanization.

Source: Ray M. Northam, *Urban Geography* (New York: John Wiley and Sons, Inc., 1975), p. 53. Used by permission.

nations might be considered as being in the Terminal Stage of urbanization. On the other hand, the developing nations are primarily in the Acceleration Stage, though some nations that are largely agrarian in nature could still be considered in the Initial Stage. Figure 8-8 shows the world pattern of urbanization.

URBANIZATION IN DEVELOPING COUNTRIES

The growth of large cities is a well-recognized characteristic of developed nations, but urbanization is equally evident in developing countries. The twentieth century is the century of the urban transition; by the end of this century nearly three billion people, almost half of the world's projected population at that time, will live in cities. Two-thirds of those people will live in the less developed countries of Africa, Asia, and Latin America (United Nations, 1991). From crowded Cairo to Mexico City, from Rio to Bombay and Beijing, poor cities, teeming with residents, are going to have to grapple with problems that seem hard to even imagine. Already, in places such as

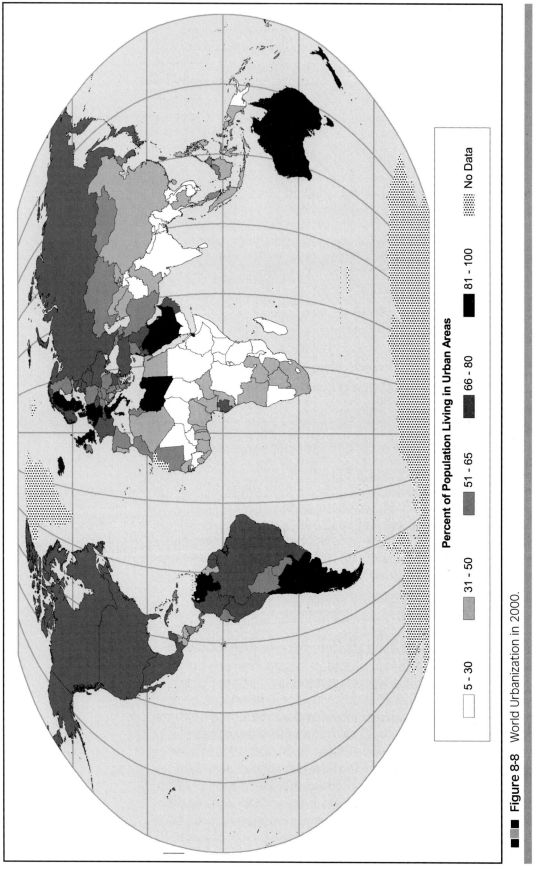

Figure 8-8 World Urbanization in 2000.

Percent of Population Living in Urban Areas

5 - 30   31 - 50   51 - 65   66 - 80   81 - 100   No Data

Source: *2000 World Population Data Sheet.* (Washington, D.C.: The Population Reference Bureau, Inc.)

Calcutta, people go through the streets early each morning removing those who failed to survive the night.

The movement of people from farms to cities has had enormous consequences. Because of overcrowding and unemployment in rural areas, families leave the poverty-stricken countryside with the hope of making a livelihood in the city. Unfortunately, jobs are often equally difficult to find in the cities, resulting in unemployment and the growth of urban slums. Robert S. McNamara, when he was president of the World Bank, commented as follows on the problems of the developing countries with regard to urbanization:

> To understand urban poverty in the developing world, one must first understand what is happening to the cities themselves. They are growing at a rate unprecedented in history. Twenty-five years ago there were 16 cities in the developing countries with populations of one million or more. Today there are over 60. Twenty-five years from now there will be more than 200. (Population Reference Bureau, 1976, 17).

Of course, rapid urbanization is partly a function of rapid population growth in the developing countries, but it is more than simply that. An analysis of population growth shows that though the total population in these countries is growing at about 2.5 percent, the population of urban areas is growing at nearly twice that rate. Approximately half of that urban growth is due to migration from rural areas and the remainder is due to natural increase.

These figures imply that cities in the developing countries are going to find it increasingly difficult to provide decent living conditions and employment for the hundreds of millions of new urbanites. According to Gugler (1988, 1):

> Rapidly growing urban populations have to find employment in urban labour markets characterized by widespread unemployment and underemployment, increase demands for urban housing and services already considered inadequate, and, for better or worse, exacerbate popular pressures on political systems.

Current Patterns

Prior to 1975 the majority of the world's urban population was located in what the United Nations calls the "more developed" nations. However, since 1975, according to United Nations estimates, there has been a shift; at the present time the majority of urban residents live in the developing countries. If current trends continue, this pattern will become more accentuated. It has been estimated that between 1950 and 2000 the urban population in the developing countries increased eightfold, whereas the urban population of the developed countries increased only by 2.4 times. This means that by the end of the last century approximately two-thirds of the world's 3.1 billion urban residents lived in the less developed countries, compared with the 1950 population, where only one-third of the urban population resided in the developing countries. According to Beier (1976, 3):

> From 1950 to 1975 the cities of the developing world absorbed just over a third (35.3 percent) of global population increase. Between 1975 and 2000 they will have to absorb over half (53.4 percent) of the increase—1.2 billion people. At the end of the century, the urban population of less developed countries will be close to 41 percent of the total population, compared to less than 16 percent in 1950.

In the countries that are now highly urbanized—mainly the leading industrial countries of the developed regions—urbanization proceeded slowly at the beginning of the development process, then rose sharply in the beginning stages of industrialization, and finally tapered off when a saturation point was reached. Because of the relatively low rate of urbanization in Europe, the emergence of new political, social, and economic institutions kept pace with the urban influx. Urban growth in the developing countries is taking place under much more difficult conditions. Several important factors that have distinguished the rapid and dramatic urban growth in the developing countries were discussed by Beier (1976) and they are summarized below.

First, the increase in population growth during the present century is the single most important factor distinguishing present from past urbanization. Population growth rates in Europe during its period of urbanization were around 0.5 percent a year, whereas in the developing countries today annual population growth rates are sometimes as high as 3.0 percent or more. In 2000 Africa had a rate of natural increase of 2.4 percent. Significantly higher growth rates mean that the developing countries have both larger natural increases within their cities and larger population movements to their cities. This combination is straining the very fiber of many Third World cities.

Second, because of widespread communication facilities, populations in the developing countries are provided with more information about urban amenities and opportunities, thus increasing the "pull" of the city on rural populations. At the same time the cost of migration is lower because of better transportation facilities.

Third, unlike the urbanization experience of most of the developed countries, urbanization in the developing countries is generally confined within fixed territorial boundaries. Little opportunity exists for free migration of the surplus population to other countries.

It appears that the developing countries will find it more difficult to urbanize than did the developed countries, and that they will have to accomplish that task in a shorter period of time. Though there will be country-by-country differences, approximately one-half of the new residents of cities will be native born and the remainder will be newcomers. These new urbanites will most likely be unskilled laborers; they are also liable to be relatively poor, undereducated, and probably illiterate—women and children will suffer the most. Only if they benefit materially from their urban experiences can we expect significant declines in their fertility, and such declines are essential in most

Third World countries if those countries are to avoid having death rates control their population balances.

India is a good example for comparing urbanization in the developing countries with the earlier European experience. In 1951 India was only 11 percent urbanized, a level reached by many European countries between 1850 and 1900; at this level of urbanization slightly over 50 percent of the population of European countries derived their livelihood from agriculture, whereas 65 to 70 percent of India's population were agriculturalists at that same level of urbanization. Approximately 10 percent of India's population was employed in manufacturing, whereas about 25 percent of Europe's labor force had been so employed. Though a common thread of rapid urban growth runs through the developing countries, widespread differences exist among them in long-term urban growth prospects and in the necessary resources to support such prospects.

What remains to be seen, of course, is the degree to which urbanization and the growth of large cities everywhere is actually sustainable. The litany of needs is considerable: health services, jobs, shelter, better schools, and incentives to keep urban sprawl under control are examples. Beyond that, as Brockerhoff (2000, 39) wonders out loud for us:

> What is much less certain is whether the horrific scenarios envisioned by some scholars will come to pass. Will earthquakes and hurricanes kill millions of people in big cities that are unable to prepare for or cope with such disasters? Will large and dense populations become breeding grounds for devastating new infectious diseases? Are ghettos a permanent and worsening aspect of the urban landscape in even the richest of countries? If cities swell with youth, but not with jobs, will violence erupt? Do increasingly volatile global financial movements impose an insurmountable barrier to informed urban planning?

Brockerhoff argues that just talking about these and other issues of large-scale urbanization is probably healthy. He ends his discussion on a positive note, writing that (Brockerhoff, 2000, 39), "These factors and others suggest that sustainable urban development, even under conditions of extreme population growth, is an attainable goal." We hope he is right.

CITIES IN DEVELOPING COUNTRIES

More than a billion people live in the cities and towns of the developing countries. Their numbers are rapidly growing in almost every Third World country and by the end of the twentieth century seventeen of the twenty-three largest metropolitan areas, with populations over ten million, will be in the Third World (United Nations, 1991).

Individual cities in the developing countries are already as large or larger than the major cities of the developed regions, and in a few years some may be considerably larger. Some interesting contrasts are also evident. The United Nations projects that in 2015 Tokyo will be the world's largest urban agglomeration, with 28.7 million residents, followed by Bombay (27.4 million), Lagos (24.4 million), Shanghai (23.4 million), and Jakarta (21.2 million).

There will be a continued concentration of people in the larger cities of both the developed and the developing countries. Cities with more than 5 million inhabitants will undergo the most rapid increases, particularly in the developing countries of Asia and Latin America.

Most of the increase in urban populations will occur through the expansion of existing cities. There will be few new cities created, and those are probable only where new resources are discovered. Because most of the choice urban locations are already usurped by urban centers, and because it is less costly to expand these areas than to develop new cities, it is unlikely that many new urban centers will appear.

MIGRATION IMPACT

As we noted previously, rural-urban migration is a major aspect of urban growth; in most countries today it is the driving force, as high fertility and high rates of unemployment make rural areas less and less desirable. Its significance varies over time and from country to country. In Latin America and other areas that already have a significant proportion of their populations living in urban areas, there will most likely be a decline in rural to urban migration in the next few years. However, in nations that are largely rural, like many in sub-Saharan Africa, migration will continue to play a key role in urban growth and distribution patterns for a long time.

The reasons why people migrate are many and varied, but the economic motive, the desire to better one's self and family, still seems to predominate. In most cases the migrants' expectations of economic improvement are met, as is suggested in the following comment by Yap (1975, 3):

> Not surprisingly, we find that migrants who stay in cities do seem to be better off, on average. Not everyone is employed immediately, but a large fraction of migrants find jobs in a reasonably short period. Incomes are higher, even for the unskilled; and many who start out in relatively undesirable jobs manage to find better ones in time.

Migrants come from varying cultural and socioeconomic environments and, according to Berry (1973, 82), they ". . . comprise a large and disparate array of social types both before and after migration."

Migration flows are important in explaining why large cities in the developing countries are growing relatively faster than the medium-size and smaller cities. Migration has often occurred in steps—from rural areas to small local cities, and then to the larger urban areas—especially during the early stages of urbanization (Shaw, 1975, 45–46). The evidence suggests that migrants are increasingly bypassing the smaller cities and moving directly to the larger urban areas. There are several reasons for this changing pattern, including the concentration of important economic opportunities in the larger cities and improved communication and transportation facilities.

URBANIZATION IN THE UNITED STATES

Urbanization in the developed countries has already reached Northam's Terminal Stage and, as we have suggested, it has even begun to go beyond that stage in some places to a new counterurbanization stage. Because the latter is best recognized in the United States, and because the pattern of urbanization is quite typical of the developed countries, we consider it in some detail rather than providing a more general overview of urbanization in the developed countries.

Since the first census was taken in the United States in 1790 there has been an increasing concentration of Americans in urban areas (Figure 8-9). According to Berry and Dahmann (1977, 433), "In 1800, the United States was 6 percent urbanized; today, more than 75 percent of the population reside in urban areas." Many of the original urban centers in the United States were forerunners of present-day urban agglomerations. They were located on the eastern seaboard and in peripheral areas and included Boston, New York, and Philadelphia. During the eighteenth century more cities were formed on the eastern seaboard, but there were also large numbers of cities founded in the Mississippi and Ohio River valleys as well as other riverene sites. This was also the time when the first major western seaboard cities were founded.

All sections of the country had some sort of urban agglomeration by the end of the nineteenth century, and most of the previously founded cities continued to grow. Cities formed during the nineteenth century were concentrated in the Midwest, primarily on navigable rivers and the Great Lakes. Approximately 40 percent of the United States population lived in urban centers by the end of the nineteenth century.

Rural-urban migration was a major component of urban growth in the United States in years past, but this pattern ended in the 1980s, perhaps even a few years earlier. According to Berry (1990, 99), "We are a completely urbanized society, with very few true rural residents left amidst large-scale factory style agriculture. Urban growth now is a product of natural increase and of foreign immigration." The latter, however, has certainly increased in importance since 1980, especially in cities such as Miami, New York, and Los Angeles. Changing population compositions within these cities are altering everything from local architecture to local politics.

The nation's largest cities, and the suburban areas around them, have grown continuously since the nation's inception. Between 1910 and 1988 the national population grew by 167 percent, whereas the metropolitan population grew by 449 percent. Metropolitan areas with at least one million residents grew by 630 percent. In contrast,

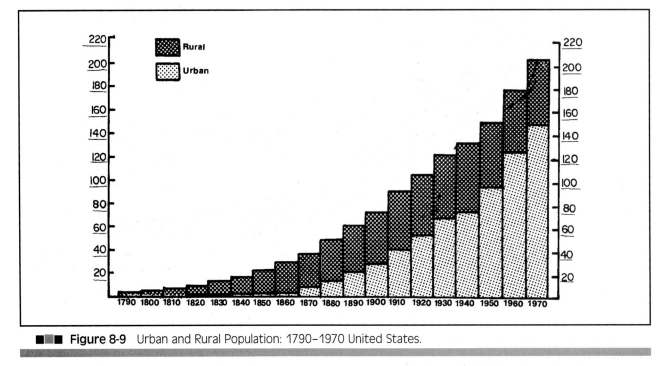

■■■ Figure 8-9 Urban and Rural Population: 1790–1970 United States.

Source: Ray M. Northam, *Urban Geography* (New York: John Wiley and Sons, Inc., 1975), p. 63. Used by permission.

throughout this 78 year period the nonmetropolitan population stayed within a narrow range between 56 and 67 million (Frey, 1990, 5).

The "look" of American cities is also changing. According to journalist Joel Garreau (1991), the majority of metropolitan Americans live and work in areas that look nothing like our old downtowns. Garreau calls these new multiple urban cores "edge cities," and he argues, quite convincingly, that they represent an important change in metropolitan development in this half century. Beyond the edge cities, around the peripheries of most metropolitan areas, continued expansion converts rural landscapes to urban land uses, a process geographer John Fraser Hart (1991) calls "the metropolitan bow wave," a process that is likely to continue inexorably in most regions of the nation. As Hart (1991, 50) commented:

> As population increases, the built-up areas of urban centers will continue to expand outward. In the vanguard of the expansion will be a bow wave of competitive and contradictory activities where the least intensive urban uses of land are steadily displacing the most intensive rural ones. This urban-rural fringe of greenhouses, nurseries, truck farms, and dairy farms will shift outward before the built-up area. The urban-rural fringe for one generation becomes a suburb for the next, and new fringes arise farther out.

Figure 8-10 shows the level of urbanization in the United States by state. The states with the highest proportions of people living in urban places are found in the Middle Atlantic and New England regions, which generally have over 80 percent of their populations

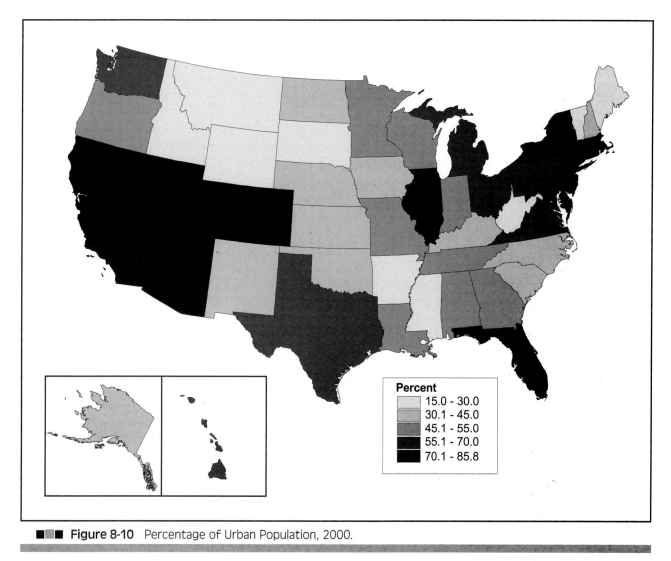

■■■ Figure 8-10 Percentage of Urban Population, 2000.

Percent
	15.0 - 30.0
	30.1 - 45.0
	45.1 - 55.0
	55.1 - 70.0
	70.1 - 85.8

Source: U.S. Census Bureau, 1990 Decennial Census.

urbanized. High levels of urbanization are also found in the Southwest and in California (the nation's most urbanized state despite its vast agricultural production). The states with the lowest levels of urbanization are found in the northern Great Plains and in parts of the Deep South. The states with the smallest proportions of urban populations are West Virginia, Louisiana, and North Dakota.

FUTURE MIGRATION TRENDS

What will happen to migration patterns in the United States in the future is a matter for speculation; however, some current trends will in all likelihood continue. The volume of migration will continue to increase, and the westward and southwestward movements will continue as well. Movement between cities will continue, as will the movement from central cities to suburbs.

Cities that are service centers, or that perform political or commercial functions, will grow more rapidly than manufacturing cities. Also, amenities such as climate and scenery will become increasingly important factors in people's decisions to migrate.

Elsewhere, the nature and extent of future migration will be primarily a function of environmental quality and population growth. There are no longer any large open areas or "safety valves" available. With a combination of scarce resources and increasing population pressure, there most likely will be continued migration in even larger numbers in search of shelter, food, and a better lifestyle.

At the international level, free migration is no longer possible in most instances because of conflicts between the interests of a society and the rights of the individual. Continued population growth, the exploitation of resources at an unprecedented rate, vast differences in living standards from place to place, and political instability are all related to migration processes, however, and we can rest assured that the restless movements of people at all scales will continue as individuals and groups seek ways to better their lives in this increasingly complex and interrelated global society in which we find ourselves.

Better transportation and communication systems have always impacted the volume and direction of migration streams. Among the newest communication technologies the Internet, via the World Wide Web, is becoming influential. As Tyner (1998, 340) recently concluded, "Web-related recruitment has the potential for revolutionizing patterns and processes of international labor migration."

Broader questions about the survival of nations, democracy in multicultural societies, and the meaning of international borders are going to be asked ever more frequently. Castles and Miller (1993, 275) observed the following:

> National states, for better or worse, are likely to endure. But global economic and cultural integration and the establishment of regional agreements on economic and political co-operation are undermining the exclusiveness of national loyalties. The age of migration could be marked by the erosion of nationalism and the weakening of divisions between peoples. Admittedly there are countervailing tendencies, such as racism, the "fortress Europe" mentality, or the resurgence of nationalism in certain areas. . . . But the inescapable central trends are the increasing ethnic and cultural diversity of most countries, the emergence of transnational networks which link the societies of emigration and immigration countries and the growth of cultural interchange.

Population and the Environment

So far we have considered primarily the demographic side of population geography; here and in the next chapter we turn to linkages between the demographic system (especially growth) and two other systems, the environment and food supply systems, which are interrelated to a considerable extent. The health of both of these systems affects us all; they are essential if the human species is to sustain itself. Unfortunately, as Abernethy (1992, 235) pointed out, "In the short run, both high fertility and an extractive, exploitative approach to the environment work well."

The interrelationships among population growth, resource depletion, and the environment have been topics of concern for many scholars, including geographers (Park, 1993). As with many other complicated issues, dealing with the relationship between population and other variables engenders considerable disagreement among the so-called experts. As Repetto (1987, 3) noted, "Both with regard to the past and outlook for the future, the connections between population growth, resource use, and environmental quality are too complex to permit straightforward generalizations about direct causal relationships. Further, many of these connections are manifested only at the local level and may be missed when the perspective is regional or global, where much of the debate has focused." Similarly, de Sherbinin and Kalish (1994, 1) noted:

> It seems self-evident that human numbers and human demands on the environment are driving ecological change in such areas as global warming, ozone depletion, deforestation, biodiversity loss, land degradation, and pollution of land, water, and air. But it is a trickier matter to figure out how much of the resulting environmental impact is from increasing numbers of people, and how much from human behavior, including consumption habits.

Smokestack industries are one important source of pollutants but automobiles collectively cause far more problems in most urban areas.

Courtesy of Photo Disc

On the one hand, there are those who feel that the root cause of most of our ecological problems is population growth—the Neo-Malthusians are perhaps the best example, led by biologist Paul Ehrlich. They argue that population growth is indeed detrimental to the environment, especially when it is accompanied by affluence (so that a birth in India, for example, is going to do far less environmental harm over that child's lifetime than will a birth in the United States). Another view (more complementary than contrary) is that many variables other than population are critical elements in environmental deterioration, and the use or abuse of technology plays an especially significant role in environmental degradation—Barry Commoner has long supported this view. People's impact on the environment can be traced far back into history, and their attitudes toward the natural environment have varied considerably, both from time to time and from place to place. A third view (contrary to either of the above, especially to the former) is that of the most optimistic economists, led by Julian Simon. In their view population growth is hardly harmful; in fact it can be good because it provides more brains, increases the ability of humanity to innovate, and creates markets for goods and services.

The two extremes in the debate about population and the environment are represented by biology and ecology (Ehrlich) and economics (Simon). As demographer Nathan Keyfitz (1994, 24) wryly noted, "Worldwide communication has never been easier, yet biologists and economists who eat in the same faculty club do not seem to communicate very effectively." He goes on to note the following:

> The central focus of modern economics is on growth: the idea that increasing productivity brings employment and prosperity. With sufficient economic growth, population growth is not to be feared. The resulting wealth, economists maintain, will in turn counteract the incidental ill effects of growth. Biologists, on the other hand, see the economy as embedded in a fragile ecosphere, upon which growth acts in dangerous, inscrutable ways. (Keyfitz, 1994, 24)

One geographer, Crispin Tickell (1993, 224), has gone so far as to suggest that "... demography has much in common with economics: it is a combination of social studies, algebra and necromancy."

Whether you feel that is too harsh or not, geographers can appreciate the nature of this debate better than most others because we have long grappled with our own disciplinary split personality—cultural/human and physical. Geographers are likely to have had training both in the social and natural sciences, as well as the humanities. Out of this historic development, geography has become an integrative discipline. This ability to see the "big picture," or think wholistically, often criticized by those in other disciplines, may give geography considerable importance as we struggle with growing global problems of every sort in the decades immediately ahead. For example, as geographer Dean Hanink (1995, 383) recently noted, "Location, it seems, matters in the analysis and solution of many environmental problems." Furthermore, geographer David Sauri-Pujol (1993, 9) noted that, "The environment is constantly transformed, degraded or improved by human action and so an understanding of human agency upon the Earth, as provided by human geography, remains essential if we are to understand and overcome our current environmental problems."

IMPACT OF EARLY CIVILIZATIONS

In order to understand the relationship of people to their physical environment in the modern world, it is helpful to trace the evolution of the people-environment interface. Several questions concerning this interface may be asked. What was the impact of early people on their environment? Have humans ever been part of an ecological community subject to control by ecological factors in the same way as other animals, or have humans always been able to influence their environment to a greater degree than any other animal?

Archaeological evidence strongly supports the thesis that early humans had at their disposal potent ecological devices with which they could alter their environment, though as Hern (1990, 10) has noted, "Small scale human assaults on the environment had little or no lasting impact during the early Pleistocene, although local and regional impacts began to be seen in the late Pleistocene and Neolithic. . . ." Humans have utilized fire for at least half a million years, for example, and it is a significant tool for modifying the environment, particularly in semi-arid and arid areas. Early humans also were able to modify natural vegetation and capture or kill animals with the aid of implements that they had made.

Perhaps even more significant was the development of farming techniques, which provided humans with the ability to alter the natural environment considerably, and the concomitant diffusion of agriculture. The relationship between population and land was discussed by Chinese writers at the time of Confucius, and Plato estimated that the ideal population for a Greek state would be 5,040 landholding households (Keyfitz, 1972, 12). As Eckholm (1976, 17) noted:

> Over the course of ten thousand years humans have successfully learned to exploit ecological systems for sustenance. Nature has been shaped and contorted to channel a higher than usual share of its energies into manufacturing the few products humans find useful. But while ecological systems are supple, they can snap viciously when bent too far. The land's ability to serve human ends can be markedly, and sometimes permanently, sapped.

Population growth was one response to improved agriculture; it further increased the need for food and made it impossible for agricultural people to avoid altering the ecosystem. For example, the development of agriculture in the wooded or parkland environment of the Near East and its eventual spread across Europe eventually brought about the decimation of forests in order to open up more land for food production. Deforestation in turn frequently led to ecological disasters of one form or another.

The collapse of numerous ancient agricultural systems has been traced to ecological imbalances. For example, in the semiarid regions of North Africa and the Near East this imbalance took the form of increased desiccation. It was not climatically induced; rather, it was a result of continued overgrazing and misuse of the land; some of these areas still remain derelict today. Another example of ecological imbalance brought about by population pressure can be seen in Mesopotamia, where efforts to extend agriculture into drier areas with the use of irrigation brought about disastrous results, including salinization and wind erosion.

The pollution of air and water is not just a product of modern society either; evidence suggests that early cultures also were confronted with it. When Juan Cabrillo visited California in 1542, he noted that while anchored in San Pedro Bay he could see the mountain peaks in the distance but not their bases. A thermal inversion in the area trapped the smoke from Amerindian fires, bringing about air pollution as well as human health hazards.

The increased concentration of people in urban areas is also associated with water pollution problems. There was, at least in some urban cultures, a need to avoid water pollution problems in order to insure safe drinking water. The contamination of surface water was sometimes recognized as a problem and led to the development of wells and water storage tanks. Canals were built in many areas in an effort to bring safe drinking water into urban areas. For example, the Aztecs under King Ahuitzotl brought in spring water through a stone pipeline.

In the previous chapter we noted Brockerhoff's optimism about the future of cities. A less sanguine view of urbanization, related to its environmental consequences, was suggested by McNeill (2000, 281-282):

> Twentieth-century urbanization affected almost everything in human affairs and constituted a vast break with past centuries. Nowhere had humankind altered the environment more than in cities, but their impact reached far beyond their boundaries. The growth of cities was a crucial source of environmental change. . . . For 8,000 years cities had been demographic black holes. In the span of one human generation they stopped checking population growth and started adding to it: a great turning point in the human condition.

McNeill recognizes that cities are complex, and that their support draws from a considerable area (often the entire world). He and others wonder if we can sustain urban behemoths with 20 or 25 or 30 million residents—cities that the United Nations projects will soon exist in several locations. As McNeill (2000, 287) commented, "Urban impacts extended beyond city limits to hinterlands, to downwind and downstream communities, and in some respects to the whole globe." Even as greater quantities of food, water, and materials moved into cities, greater volumes of garbage, human waste, and an assortment of pollutants moved out.

ENVIRONMENTAL DEGRADATION AND POPULATION GROWTH

Abundant evidence suggests that population growth and increased pollution went hand-in-hand in early civilizations. As the human population has increased in numbers and become more geographically concentrated, there has been an increased potential for disrupting the earth's ecosystems.

SOME DIMENSIONS OF THE PROBLEM

The problem of **environmental degradation** can be broken down into several key factors, including the following: pollution, crowding and violence, global warming and ozone depletion, deforestation, decreasing ecological diversity, and overgrazing. As discussed in this section, these factors have been among the consequences of human activity for several thousand years; many are becoming more critical as population and industrialization continue. We are inclined to agree with the following comment by Tickell (1993, 220): "We have the misfortune to be perhaps the first generation in which the magnitude of the global price to be paid is becoming manifest."

The problem of **environmental degradation** can be broken down into several key factors, including the following: pollution, crowding and violence, global warming and ozone depletion, deforestation, decreasing ecological diversity, and overgrazing.

■■■

Pollution

One important natural function of the earth's ecosystem is the absorption of waste material. Although the waste from one organism can be an important input to other organisms, when waste increases to the point that it can no longer be accommodated by the ecosystem, it becomes **pollution.**

We humans are currently polluting our planet at an unprecedented rate. The reasons for this are many and complex but, as economist Kenneth Boulding neatly summarized it, "Our desire to conquer nature often means simply that we diminish the probability of small inconveniences at the cost of increasing the probability of very large disasters." (Boulding, 1966, 14)

Pollution's primary forms are either biological or chemical (though filth, noise, and many other irritants may also be considered pollutants to many). Human population density often leads to increased biological pollution, as evidence from early civilizations has pointed out. Today, the crowding of large numbers of people into small places has brought about increased pollution almost everywhere. As a population increases, so does the accumulation of its human organic waste, for example. City water supplies may be contaminated as it becomes increasingly difficult to dispose of large volumes of waste. According to a study by Hardoy and Satterthwaite (1984, 49) for the International Institute for Environment and Development, only 209 of India's 3,119 towns and cities have even partial sewage systems and treatment facilities, whereas 114 towns and cities dump raw sewage into the Ganges, that nation's "holy river." In the Bogota River, downstream from the city of Bogota, the average fecal bacteria coliform count of 7.3 million stands in sharp contrast to the safe drinking water limit of 100 and the safe swimming count of 200. Heavy metal contamination of water supplies is also common around many Third World cities.

Chemical pollution is another by-product of a rapidly growing population, coupled with modern technology. Many streams and lakes have been polluted by the addition of toxic chemicals. One of the most notable—and most tragic—cases occurred in Minamata Bay, Japan, where industrial waste containing mercury was dumped into the fishing waters. Local fishermen (who knew nothing of the growing threat) continued to eat their catch; the result was several thousand cases of a debilitating disease now referred to as Minamata disease, a form of mercury poisoning.

Pollution problems that were usually only local in scope within earlier civilizations are now becoming global concerns, crossing borders indiscriminately and threatening international relationships in the process. Chemical wastes that break down very slowly can ultimately reach the oceans, which are constantly used as dumping grounds. As the population increases on the continental margins, the oceans become common sinks for industrial wastes and garbage. Compounds such as DDT have been carried to virtually all parts of the oceans (don't forget, no matter what we might name them, there is only one ocean), and they continue to be used.

Air pollution is also affected by population growth and the geographic concentration of population, as geographers Charles Collins and Steven Scott (1993) noted in their study of air pollution in Mexico. Climatic changes are being induced by concentrations of people in urban areas, frequently with a concomitant increase in airborne pollutants. As Repetto (1987, 28) pointed out:

> Estimated global emissions of carbon dioxide from the use of fossil fuels and burning of biomass have nearly tripled since 1950. Emissions of the most active chloroflourocarbons have increased from negligible amounts to almost 700,000 metric tons per year in 1985. . . . Samples of air trapped in ice cores suggest that the methane concentration has doubled.

In addition to lead, the urban atmosphere also contains high counts of microorganisms; gases, such as carbon monoxide and oxides of nitrogen; and a variety of other chemical compounds, including sulfur dioxide. Particulates and smog in the atmosphere cut down on visibility and produce a number of effects on people, from eye-irritations to much more serious conditions.

The United States Office of Technology Assessment (1984) estimated that the effects of air pollution may cause 50,000 premature deaths in the United States each year—about 2 percent of annual mortality. Especially vulnerable are the millions of people who are already suffering from asthma, emphysema, and other chronic respiratory disorders (National Research Council, 1985). The Environmental Protection Agency (EPA) and the American Lung Association estimate that approximately 150 million Americans live in areas whose air is considered unhealthy. This may lead to as many as 120,000 deaths each year (French, 1991, 97).

Although the Clean Air Act of 1970 has been successful in reducing air pollution, Ehrlich and Ehrlich (1990, 138) estimated that if population had not grown, air pollution would now be only a little more than half the 1970 level. Even Los Angeles (the butt of many a late-night comedian's jokes about smog) has cleaned up its act during the past two decades, though it remains the smoggiest city in the United States (Lents and Kelly, 1993). Clean industries and cars have become the rule, the establishment of an Air Quality Management District (AQMD) has helped fight major polluters, and new rules promise even stricter controls ahead. Nonetheless, the Los Angeles Basin continues to fill up with more and more people, driving hither and yon to work and play. Though, in speaking about Los Angeles, Lents and Kelly (1993, 39) point out that ". . . residents and business people seem to recognize the need to solve the serious air-pollution problems." Neither these authors nor the AQMD, however, suggest the contribution that sustained population growth has made to the problem.

With the increases in the human population and in industrial output that have occurred over the past century, persistent contamination of the biosphere has become a serious global problem. Continued population growth and the concomitant new demand for food, shelter, and other goods and services will make it ever more difficult to bring pollution under control. As you might imagine, China alone, as it rapidly develops, will run into numerous environmental problems, including pollution, according to Hertsgaard (1997).

Crowding and Violence

Though we have placed crowding and violence in the same category, we are not arguing that there is a clear-cut cause-and-effect relationship between the two, nor between them and population growth. However, it seems clear that population growth at the very least exacerbates such problems by constantly increasing the pressure on resources.

The pioneering work of anthropologist Edward Hall, dealing with how different cultures perceive and evaluate personal space, has been of considerable interest to geographers and other social scientists (Hall, 1969). As the population of an area grows, there will be more people per square mile; nothing can change this. Similarly, personal space becomes smaller and smaller. Evidence suggests that individuals and cultures can tolerate different levels of crowding. What, then, is the effect of crowding on specific populations?

Studies with animal populations (including rats, mice, and deer) suggest what happens when overcrowding occurs in those various populations. In his classic rat studies Calhoun pointed out the disastrous consequences of overcrowding (Calhoun, 1962). He found that high density resulted in the disruption of the rats' nesting patterns because of changes in their normal social behavior. Crowding intensified social interaction and the competition for resources. Infant mortality increased with the disturbance of nesting patterns, and some of the young were even consumed by other rats. Aggressive attacks became more frequent.

The relevance of these animal studies to the human situation has been questioned. For example, Freedman (1975) pointed out the following:

> . . . it is both difficult and risky to generalize directly from the behavior of one animal to that of another. It would be a mistake to conclude that dogs act a particular way just because cats do or that monkeys act the same way as lions; and it is of course much more difficult to conclude anything about humans from other animals. Humans are more intelligent, have language, and have an extremely complex social structure, are much more flexible and innovative than other animals . . . there is enough difference between humans and the rest of the animal world to make it difficult to conclude anything about humans from what other animals do.

Of course, as Freedman also noted, ". . . work on animals is not only extremely interesting but can also be a source of ideas and suggestions about how humans behave." (Freedman, 1975, 41) A knowledge of animal behavior may suggest hypotheses about human behavior, but before accepting them we need to test them by observing people. Even though some of the animal studies may not be directly transferable to people, we cannot help but wonder about the quality of life that can be maintained as ever more people are crowded into cities.

Attempts have been made to correlate high crime rates with population density, partly because researchers have suggested that aggression increases with **crowding** and partly because it has been shown that urban crime rates are higher than rural crime rates. For example, in the United States the crime rate per 100,000 residents is over five times greater in the largest cities than it is in rural areas, whereas intermediate crime rates prevail in suburbs and small towns. Still, large populations do not necessarily mean high population densities, at least in a relative sense. To date there have been few well-designed studies of the effect of population density on crime. After reviewing the results of several of those studies, Freedman (1975, 69) concluded that ". . . there are a great many reasons why people commit crimes, [and there are] many factors in modern, complex society that cause crime, but there is no evidence that crowding is one of them." Part of the problem with most of the studies is the difficulty of isolating the effect of crowding from other variables such as poverty, ethnic composition, and educational levels.

In an attempt to analyze the effects of cramming more and more people into cities, the organization, Zero Population Growth, devised an "urban stress test" and applied it to all 192 United States cities with populations over 100,000 people. The test looked at eleven criteria associated with urban blight and found that the twenty-two cities with the best scores averaged 116,000 people, with about 3,700 people per square mile and the twenty worst cities averaged 1,154,000 people with 8,200 people per square mile. According to Ehrlich and Ehrlich (1990, 156), "The message seems clear: measured either by social and environmental indicators together or by environmental indicators alone, more people mean more problems in American cities."

Regardless of the exact nature of the effect of crowding on human behavior, there is growing evidence that the quality of people's lives in many places is being diminished by population growth and the increased densities that often accompany it. For example, consider the following statement by Brown, McGrath, and Stokes (1976, 42):

> Aerial photographs of Java reveal that people are actually moving into the craters of occasionally active volcanoes in their search for land and living space. Periodic evacuations and loss of life result. In Bangladesh, people are driven by population pressure into floodprone lowlands and onto low coastal islands previously uninhabited because of the danger of tidal waves and typhoons. *The New York Times* of November 15, 1970, reported more than 168, 000 people killed by a tidal wave that swept the coastal area. Described as one of the worst natural disasters of the century, this loss of life is more accurately attributed to overcrowding than to any "natural" phenomenon.

Attempts have been made to correlate high crime rates with population density, partly because researchers have suggested that aggression increases with **crowding** and partly because it has been shown that urban crime rates are higher than rural crime rates.

Beyond the likely, if difficult to "prove," relationship between **crowding** and quality of life, population growth may be increasing levels of violence in other ways as well. For example, rapidly increasing populations in many developing countries are likely to intensify violent conflicts because of increasing resource scarcities; as Homer-Dixon, Boutwell, and Rathjens (1993, 38) noted:

> . . . scarcities of renewable resources are already contributing to violent conflicts in many parts of the developing world. These conflicts may foreshadow a surge of similar violence in coming decades, particularly in poor countries where shortages of water, forests and, especially, fertile land, coupled with rapidly expanding populations, already cause great hardship.

Aside from the obvious problems associated with population pressure on nonrenewable resources, the authors argue that growing problems over renewable resources are likely as well. They argue that human actions can generate scarcities of renewable resources in at least three different ways:

1. Use or degradation can occur because resources are used faster than they can be renewed.

2. Common resources such as water may have to be divided among too many users.

3. Changes in the way resources are distributed within a society, mainly concentration in a few hands.

Strong political and economic systems in Third World countries could certainly mitigate many potential conflicts, but developing countries are often lacking in those very strengths. Help from the developed countries is going to be needed, yet most of them are increasingly fighting their own battles to survive in a globalizing economic system that is creating new sets of winners along with vast numbers of losers. So far, the euphoria of having unfettered free markets diffused around the world has not led many political or corporate leaders to consider global capitalism's downside, those too poor and voiceless to fend for themselves.

An even broader view of the Third World (and parts of the rest as well) suggests that the very social and political fabric of societies is being shredded by overpopulation, resource scarcity, increasing crime, and the rapid spread of contagious diseases such as AIDS. Unsafe streets are common in many American cities at night, but they are nothing compared to the streets of West African cities after the sun goes down. In a thought-provoking article, journalist Robert Kaplan (1994) begins with a discussion of the breakdown of order in West Africa, then moves through other parts of the world, arguing that anarchy and chaos are becoming increasingly common as authorities lose control of growing populations. Kaplan sees a world increasingly confronted with more refugees, increasing interethnic conflicts (severely straining multiethnic nation-states), powerful international

Beyond the likely, if difficult to "prove," relationship between **crowding** and quality of life, population growth may be increasing levels of violence in other ways as well.

drug cartels, private armies, and security firms. Among his conclusions, the following (with blatant geographic appeal) is worth considerable reflection:

> Imagine cartography in three dimensions, as if in a hologram. In this hologram would be the overlapping sediments of group and other identities atop the merely two-dimensional color markings of city-states and the remaining nations, themselves confused in places by shadowy tentacles, hovering overhead, indicating the power of drug cartels, mafias, and private security agencies. Instead of borders, there would be moving "centers" of power, as in the Middle Ages. Many of these layers would be in motion. Replacing fixed and abrupt lines on a flat space would be a shifting pattern of buffer entities, like the Kurdish and Azeri buffer entities between Turkey and Iran, the Turkic Uighur buffer entity between Central Asia and Inner China . . . and the Latino buffer entity replacing a precise U.S.-Mexican border. To this protean cartographic hologram one must add other factors, such as migrations of populations, explosions of birth rates, vectors of disease. . . . This future map—in a sense, the "Last Map"—will be an ever-mutating representation of chaos. (Kaplan, 1994, 75)

Pessimistic? Yes, but sobering as well. Before this article was published the debacle in Bosnia was well underway; however, the horror in Rwanda came later.

Global Warming and Ozone Depletion

There is increasing evidence that people have changed the climate of large areas of the world. Climatic changes induced by human occupance are far-reaching and have affected food production, human health, and living patterns. Climate change is also likely to alter the course of future development.

Before going on, let's consider some recent changes that at least strongly suggest that our global atmosphere has been warming up:

- Spring has been arriving about one week earlier in the Northern Hemisphere.

- The length of the growing season in some places has increased by a week since 1980.

- The amount of vegetation in North America, Europe, and Siberia may have increased by as much as 20 percent since 1960.

- The carbon dioxide content of the atmosphere has increased steadily since monitoring began in 1958.

- Human activities contribute more than 3 billion tons of carbon dioxide to the atmosphere annually.

- After Hurricane Andrew in 1992 cost them some $16 billion, many insurance companies began to examine possible implications of global warming for their business.

- A study in Queensland, Australia, found that the number of days with frost had been declining since 1900.

- The Aletsch Glacier in the Swiss Alps has been retreating since 1865.

- Worldwide, glaciers are retreating almost everywhere, and the rate of retreat seems to be accelerating—even the ice on Mt. Kilimanjaro is retreating rapidly.

- The 1990s were the warmest decade since records started being kept around the world.

All of these observations and many more are at least compatible with, if not substantial confirmation of, the idea that the earth's atmosphere has been warming for more than a century now. Less well documented, but still likely, is that humans are helping the atmosphere to warm, mainly by burning fossil fuels and adding other "greenhouse" gasses to the atmosphere, including methane and chlorofluorocarbons. As Fagan (2000, 209) nicely describes it:

> We live on a benign planet, protected by the heat absorbing abilities of the atmosphere, the so-called "greenhouse effect." Energy from the sun heats the surface of the earth and so drives world climate. The earth, in turn, radiates energy back into space. Like the glass windows of a greenhouse, atmospheric gases such as water vapor and carbon dioxide trap some of this heat and re-radiate it downward. . . . But the effect is no longer purely natural. Atmospheric concentrations of carbon dioxide have now increased nearly 30 percent since the beginning of the Industrial Revolution; methane levels have more than doubled; and nitrous oxide concentrations have risen by about 15 percent. These increases have enhanced the heat-trapping capabilities of the atmosphere.

At the local level, temperature and rainfall patterns are affected by human occupance. Agricultural and urban-induced dust pollution and increased carbon dioxide in the atmosphere due to the burning of fossil fuels have led to changes in local rainfall and temperature regimes; urban heat islands and downwind rainfall plumes have been identified in a variety of places. Dust particles in the atmosphere act as condensation nuclei and thereby increase rainfall. It has also been hypothesized that increasing quantities of airborne dust act as insulation by reflecting the sun's rays away from the earth and thus lowering temperatures (National Academy of Sciences, 1975).

On the one hand, atmospheric dust has a cooling effect; but on the other hand, another atmospheric pollutant, carbon dioxide, seems to have a warming influence because it traps the earth's heat. Chlorofluorocarbons (CFCs, also known as halocarbons) and methane gas are also effective atmospheric heaters (via the greenhouse effect). At the local level the results of increased population, coupled with the increased burning of fossil fuels, will maintain urban areas as heat islands (Bryson and Ross, 1972, 61).

One environmental issue that has received a great deal of attention in recent years is the idea of **"global warming."** An increase in atmospheric carbon dioxide, in conjunction with increases in other heat-holding gases (methane, nitrous oxide, etc.), has led to an overall temperature increase for the planet. There is still debate among scientists about the precise nature and extent of this warming, but most scientists agree with the United States National Academy of Sciences that the warming over the next century will be from 1.5 degrees C to 4.5 degrees C. According to Stephen H. Schneider (1990, 30), head of Interdisciplinary Climate Systems at the National Center for Atmospheric Research, "The earth has not been more than 1 to 2 degrees C warmer during the 10,000-year era of human civilization. The previous ice age, in which mile-high ice sheets stretched from New York to Chicago, was only 5 degrees C colder than now."

International concern for global warming led to formation of the Intergovernmental Panel for Climate Change (IPCC) in 1988. Its purpose was to review the state of knowledge with respect to human-induced climate change and to assess possible responses. Composition of the IPCC is international, with hundreds of scientists making contributions to its work. IPCC's first report appeared in 1990, and it served as an important basis for discussions of climate change at the 1992 Rio conference.

IPCC's second report was published in 1995. It suggested that the world was likely to experience a warming of 2 degrees centigrade by 2100, along with a rise in sea level during that time of between .13 and .94 meters. It also for the first time went on record as implicating humans in the global warming process.

IPCC's third report was published early in 2001. Compared to the 1995 report, the 2001 report suggested a somewhat larger average temperature increase was likely by 2100 (between 1.4 and 5.8 degrees centigrade), but with a somewhat smaller rise in sea level (between .09 to .88 meters) than the panel had predicted in 1995. The panel also suggested that long-term patterns of changing temperatures and sea levels were likely to continue for thousands of years.

Future scenarios are just that—scenarios. They include the possibility of increased storm intensity, more coastal floods, shifting agricultural and forest belts, more tropical diseases in higher latitudes, intensified monsoon rains, ecosystem damage, and crop failures in many locales.

The result of global warming at such an unprecedented scale could be catastrophic. Two of the most significant changes would be a rise in sea level, which could flood low-lying river deltas throughout the world (especially in developing countries, which could not afford to erect dikes and other barricades), and changes in agricultural productivity that might be associated with shifting climate patterns. Christopher Flavin (1991, 82) believes that if global warming is permitted to continue, it may ". . . soon affect economies and societies worldwide. Indeed, it can be compared to nuclear war for its potential to disrupt a wide range of human and natural systems."

Global warming refers to an increase in atmospheric carbon dioxide, in conjunction with increases in other heat-holding gases (methane, nitrous oxide, etc.), has led to an overall temperature increase for the planet.

■▪■

The National Academy of Sciences (NAS), in response to a request from the United States Congress, released a study in 1991 on greenhouse warming. In that study (National Academy of Sciences, 1991) it linked population growth with global climatic change. Although the linkages between population growth and greenhouse warming were called "complex and not well understood," the NAS panel concluded that:

> More people create greater demand for food, energy, clothing, and shelter. Producing such products emits greenhouse gases. . . . Global population growth, which will largely take place in developing countries, is a fundamental contributor to increasing emissions of greenhouse gases . . . Even with rapid technological progress, slowing global population growth is a necessary component for the long-term control of greenhouse gas emissions.

On July 23, 2001, in Bonn, Germany, the 180-nation U.N. Climate Change Conference reached a milestone compromise agreement to reduce greenhouse-gas emissions that affect global warming. After its rejection of the Kyoto Protocol, the United States chose not to be part of the new accord, relinquishing leadership on the issue to the European Union. Nonetheless, the Bush administration continues to argue that it does take the issue of global warming seriously—a position hardly taken seriously by the other 179 nations that supported the new agreement. President Bush's expressed concern is that participation in the agreement would be harmful to the U.S. economy.

After major concessions were made to Japan, the final agreement (approved by 178 nations) was a considerably softened version of the 1997 Kyoto Protocol. Along with Japan, Canada, Russia, and Australia were also holdouts to some degree before final agreement could be reached. It seems that this new agreement, full of compromises, will at best have only a modest influence on global warming— its importance seemed mainly to provide a framework for future conferences, with the hope that the United States sooner or later would provide needed leadership on the issue.

Control of population growth has the potential to make a major contribution to raising living standards and to easing environmental problems like greenhouse warming. The United States should resume full participation in international programs to slow population growth and should contribute its share to their financial and other support.

Often confused with global warming, the **breakdown of the ozone layer** is really a separate issue, though the CFCs that are responsible for it also contribute to global warming. The ozone layer is located in the stratosphere; it protects the earth from much of the ultraviolet radiation that reaches the stratosphere. **CFCs,** primarily the chlorine in them, interacts with ozone to break it down into normal oxygen molecules and unstable chlorine-oxygen compounds.

CFCs, once popular as spray can propellants, are used primarily for refrigeration and airconditioning units and polyurethane foam. As observations in the 1980s detected a growing hole in the ozone layer

CFCs, primarily the chlorine in them, interacts with ozone to break it down into normal oxygen molecules and unstable chlorine-oxygen compounds.

Courtesy of Photo Disc

Deforestation sometimes leads to reforestation, as seen here. Many places in the United States have more forests now than they did 100 years ago, though many others have suffered from severe deforestation. Young forests can take a long time to mature.

over Antarctica (and more recently over the Arctic as well), the developed nations decided to act (unusual, we might note). The result was an agreement (to limit, and gradually phase out, CFC production) known as the Montreal Protocol; even if substitutes are found and production of CFCs ceases, however, their effect on the ozone layer will continue for many years.

In what could be more bad news for atmospheric pollutants, global warming, and the ozone hole, scientists in 2001 reported that the supply of hydroxyl radicals in the air may be declining. These short-lived molecules rapidly combine with other gasses in the atmosphere, helping to cleanse the air of undesired pollutants, so their decline would mean that more pollutants would remain unchanged in the atmosphere. They are particularly efficient at destroying hydrocarbons, which come from paints, solvents, and petroleum. Whether the decline in hydroxyl radicals is temporary or a new trend remains to be seen—more research is required.

Deforestation

The increase in the world's population over the centuries has brought about a concomitant decrease in forested areas, as we suggested earlier. Trees have been cut down for a variety of purposes such as home building and firewood. The primary causes of **deforestation,** clearing land for agriculture and the gathering of wood for fires, are directly related to population growth. In some areas of the world, such as Algeria, Tunisia, and Morocco, forests at one time covered over 30 percent of the total land area, whereas today they cover only about 10 percent.

The primary causes of **deforestation,** clearing land for agriculture and the gathering of wood for fires, are directly related to population growth.

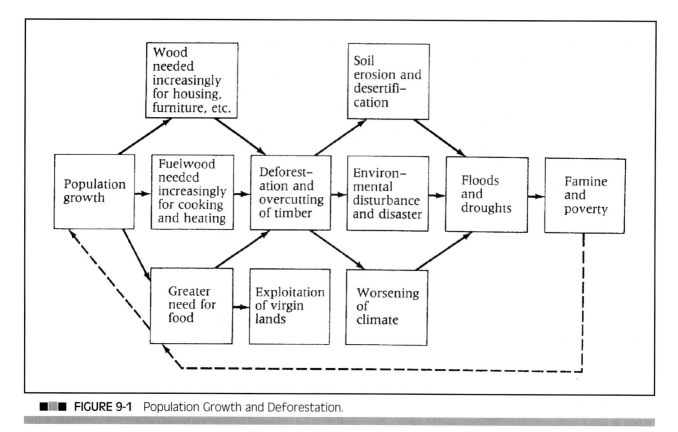

■■■ **FIGURE 9-1** Population Growth and Deforestation.

At the present time forests are being cut down at a faster rate than they are being replanted. It has been known for centuries that deforestation results in floods, local changes in climate, and heavy soil erosion, but little has been done about it. Certain areas that were once densely settled, such as the Middle East, were long ago deforested. At the present time many poor countries in other areas of the world are passing through the same stages of forest destruction, but at a much more rapid pace. Hines, in discussing China (1973, 2), stated that:

> The Yellow River, which flows through an area of great population, rides high on silt from the surrounding lands and periodically floods the region through which it travels. Through the ages, overcutting of forests and exploitative farming practices have reduced China's cultivable land so severely that today the despoiled areas, together with those climatically and physically unsuited to agriculture, constitute 80 percent of its total land—all now useless for farming.

The continuous growth of China's population and the clearing and burning of forest tracts in order to enlarge land for cultivation has placed considerable stress on China's forest resources. According to Jing-Neng Li (1991, 256), a vicious circle (Figure 9-1) appears to characterize the relationship between deforestation and population growth, and ". . . in the final analysis the most important thing the

Chinese government can do to break the vicious circle of overpopulation and deforestation is to promote the practice of family planning and to strictly control population growth." Geping and Jinchang (1994) agree that population growth is the major restraint that China faces in solving its growing environmental problems.

Throughout many tropical regions shifting cultivation takes its toll of forest lands. The impact of agriculture, along with those of logging and ranching, seriously threatens to destroy the tropical forest ecosystem (Richards, 1973, 67). There has been a clear downward trend in per capita forest production. Almost a billion hectares of forests and woodlands have been cleared since 1850. Although much of this deforestation took place in temperate forests, there has been a shift since World War II, with the locus of deforestation currently in tropical areas. According to FAO estimates, more than 11 million hectares, an area the size of the State of Pennsylvania, has been deforested each year in tropical regions during the 1980s. Population growth and energy price increases have intensified the pressures on fuelwood and timber supplies in developing countries. The open savannas of the Sahel region of northern Africa are also being rapidly stripped of their remaining trees. Annual fuelwood consumption alone exceeds annual tree growth by an estimated 30 percent (Repetto, 1987, 17). The worldwide estimate by the FAO is that 1.5 billion of the 2 billion people who rely mostly on fuelwood are cutting wood faster than it is growing back (World Resources Institute (WRI) and International Institute for Environment and Development (IIED), 1987, 272).

Tropical rainforests are being cut down at an alarming rate throughout the world. Although these forests only cover around 7 percent of the earth's land area, they play a disproportionate role in the biosphere. Rainforests are home to one-half of all the species found on earth and are a gigantic storehouse for carbon. Although exact figures are hard to find, tropical deforestation and burning probably contribute between 7 and 31 percent of the carbon dioxide released into the atmosphere each year (Durning, 1991, 169).

Current estimates are that somewhere between one-third and one-half of the world's original tropical rainforests are now gone (Muul, 1989, 29). The remaining area is slightly larger than the United States. In many instances population pressure has forced people to move to marginal lands where they clear and burn the forests in order to provide agricultural land. Unfortunately, this provides only a temporary solution as land stripped of forests erodes rapidly and soils are quickly depleted of their nutrients. Saving the rainforests will involve efforts at both the national and international level (Table 9-1).

Though periods of deforestation have occurred throughout much of human history, it is clear that, as McNeill (2000, 236) tells us:

> The late twentieth century was a great age of deforestation, like that of the Roman Mediterranean, Song China, or North America in the railroad age. The scale in the late twentieth century was larger, the ecological effects quite different, and the technologies employed radically different, but the motives much the same:

At the international level
- Support sustainable development programs that attack root causes of poverty, including maldistribution of farmland.
- Support family-planning programs in poor nations.
- Reduce international debt in exchange for forest protection policies.

At the national level
- Recognize and defend indigenous people's rights, including the right to exercise effective control over traditional lands and natural resources.
- Launch aggressive family-planning and land reform programs; step up participatory poverty alleviation programs.
- Reform land tenure, credit, subsidy, and tax policies to eliminate antiforest bias.
- Carry out through landuse planning based on ecological capacities of soils and biological diversity to demarcate and enforce biological reserves, extractive reserves, sustainable forestry areas, and agricultural land.
- Develop and disseminate tree-garden techniques based on indigenous models.

Source: Reprinted from THE WORLD WATCH READER ON GLOBAL ENVIRONMENTAL ISSUES, Editor: Lester R. Brown. By permission of W.W. Norton & Company, Inc. Copyright © 1991 by Worldwatch Institute.

arable or grazing land and marketable timber. Alarm about the obliteration of forest peoples, the lost ecosystem services, and the contribution of deforestation to greenhouse gas accumulation, while considerable after about 1980, weighed little in the balance against these motives. The political power of the beneficiaries of deforestation was far too great.

The good news with deforestation, as with many other environmental problems, is that we can, if given the will, probably reverse at least some of the damage. As writer Bill McKibben (1995) has pointed out, reforestation has been a success story in the eastern United States. Evidence of success includes more than just trees, as McKibben (1995, 64) makes clear in the following comment:

> The proof that what is happening is significant lies in the recovery not only of the forests themselves but of much of the life they always supported. Perhaps 40,000 black bears roam the East. In 1972, thirty-seven wild turkeys were introduced into western Massachusetts. By now the population exceeds 10,000.

Nonetheless, the eastern United States is not the wet tropics, and worldwide losses are likely to continue. More generally, wealthy countries can more easily afford environmental improvements.

Decreasing Ecological Diversity

As population growth forces people to seek out new areas for farming, mostly on increasingly marginal lands, habitats are being destroyed. Where they are not obliterated, natural habitats are often separated into "islands" cut off from each other and no longer parts of a whole ecosystem. Especially in the wet tropics, ecosystems are rich in species, many of which have yet to be identified and cataloged.

In addition to habitat destruction and habitat formation, further threats to **ecological diversity** include overkill and the introduction of exotic species. Even the extinction of one species in an ecosystem may in turn lead to other extinctions as the ecosystem adjusts.

Biologists argue that habitat destruction is resulting in an escalation of extinctions; we could be losing as many as 100 species a day according to some, though no actual number is really known. In fact, we don't even know how many species of life there are on the planet, and we probably never will. The major inference, however, is that if we destroy ecosystems, we are going to destroy species as well (Wilson, 1993).

In addition, threats to species may come from such environmental changes as global warming and increased ultraviolet radiation (because of depletion of the ozone layer). Concerns among biologists are aroused when similar biological changes are noted in disparate regions of the world. Two such changes deserve mention here: declining amphibian populations and declining human sperm counts.

Amphibians, especially frogs and toads, are in trouble on every continent (Blaustein and Wake, 1995). In California alone, foothill yellow-legged frogs, Yosemite toads, Cascade frogs, and leopard frogs are disappearing. According to Phillips (1994), the suspected cause of the decline in amphibian populations is the increased stress caused by exposure to ultraviolet-B radiation. Are frogs the proverbial "canaries in the mine?" More research is needed, of course, but the frog's problems should make thoughtful humans pause and reflect a bit.

Declining human sperm counts—we're putting you on, right? Again, evidence is accumulating from around the world, as are hypotheses to explain this rather disturbing (though it could be a benefit for population control, if not for *machismo*) observation. After sorting through a ream of studies, journalist Lawrence Wright (1996) found that not only were sperm counts declining around the globe but that a high proportion of sperm were damaged or misshapen as well. Urban living, stress, and environmental chemicals that masquerade as estrogen are among possible suspects as causes. As Wright (1996, 55) concluded:

> The most important unanswered question among the many theories that purport to explain the falling sperm count is whether the decline is permanent and irreparable. The hope, of course, is that some modern condition or habit has somehow waylaid the production of sperm, and that the cause needs only to be discovered and removed for the count to rebound. Unfortunately, the truth about the sperm count is that it is under attack from many different sources. . . . It is as if manhood itself were waging a losing campaign against forces as yet unknown but frighteningly overwhelming.

Overgrazing

The **overgrazing** impact of a growing population is similar to that of deforestation. As the human population increases, particularly in the poorer countries, there is a growing demand for livestock. The animals serve as food, security, repository of family wealth, and as power to

In addition to habitat destruction and habitat formation, further threats to **ecological diversity** include overkill and the introduction of exotic species.

■▪■

As the human population increases, there is a growing demand for livestock. Unfortunately, increasing numbers of livestock can quickly denude a landscape of its natural grass cover. Denudation in turn will increase runoff, accelerate erosion, and increase siltation.

pull agricultural implements. Increasing numbers of livestock, like cattle or goats, can quickly denude a landscape of its natural grass cover. Denudation in turn will increase runoff, accelerate erosion, and increase siltation. There are many historic examples of overgrazing. North Africa is presently largely unproductive and barren, although at one time it was considered the granary of the Roman Empire. The Tigris-Euphrates Valley, known as the Fertile Crescent, probably supports fewer people today than it did during the pre-Christian era. According to Brown, McGrath, and Stokes (1976, 41), overgrazing is not new, but its scale and rate of acceleration are. Damage that formerly took place over centuries is now being compressed into years by the fateful arithmetic of population growth. Populations are, in effect, outgrowing the biological systems that sustain them.

The expansion of arid areas, desertification, is at least partly due to overgrazing (Goudie, 1986, 46-50). As the size of the world's livestock herds increases, degradation of rangeland occurs. When the size of the livestock herds surpass the carrying capacity of perennial grasses on the range, the plant cover starts to diminish, leaving the land exposed to the ravages of both water and wind. According to Postel (1991, 27), "In the most severe stages, animal hooves trample nearly bare ground into a crusty layer no roots can penetrate, causing erosion to accelerate. The appearance of large gullies or sand dunes signals that desertification can claim another victory." Even if overgrazing were not a problem, growing herds of cattle pose another environmental concern—they generate significant quantities of methane gas, which is one culprit in global warming.

THE EHRLICH-COMMONER DEBATE

One of the most widely discussed topics among ecologists and other scientists concerned with population growth is the exact nature of the relationship between the size of a human population and its effect on the ecology of an area. Foremost among advocates of two different views on this question are Paul Ehrlich of Stanford University and Barry Commoner of Queens College in New York City. **Ehrlich** proposes that environmental deterioration is a direct consequence of population growth. **Commoner,** on the other hand, believes that although population plays a role in environmental deterioration, it is not the major determinant of the environmental crisis. He believes that other variables, primarily technology, play a much more significant role in the ecological crisis. Thus, those concerned with the relationship between environmental problems and population growth are served well by the confrontation of viewpoints held by Ehrlich and Commoner.

The Ehrlich Viewpoint

Ehrlich's basic argument is centered around five theorems that he believes provide a realistic framework for analysis. These five theorems originally were published in *Science* (Ehrlich and Holdren, 1971, 1212), and they are that:

1. Population growth causes a *disproportionate* negative impact on the environment.

2. Problems of population size and growth, resource utilization and depletion, and environmental deterioration must be considered jointly and on a global basis.

3. Population density is a poor measure of population pressure, and redistributing population would be a dangerous pseudosolution to the population problem.

4. Environment must be broadly construed to include such things as the physical environment of urban ghettos, the human behavioral environment, and the epidemiological environment.

5. Theoretical solutions to our problems are often not operational and sometimes are not solutions. Ehrlich believes that each person has a negative impact on his environment. In order to meet his needs, man simplifies ecological systems through the establishment of agriculture and is involved in the utilization of nonrenewable and renewable resources.

In a restatement of these ideas, Ehrlich and Ehrlich (1990, 39) state that the "population connection" is the key to understanding the root causes of our environmental problems. The number of people in an area relative to the carrying capacity of that area is used to determine if an area is over populated. If the long-term carrying capacity of an

> **Ehrlich** proposes that environmental deterioration is a direct consequence of population growth. **Commoner,** on the other hand, believes that although population plays a role in environmental deterioration, it is not the major determinant of the environmental crisis. He believes that other variables, primarily technology, play a much more significant role in the ecological crisis.

area is clearly being degraded by its current human occupants, then that area is overpopulated. By this standard, the Ehrlichs believe that virtually every nation today is overpopulated.

According to Ehrlich and Ehrlich (1990, 58):

> The impact of any human group on the environment can be usefully viewed as the product of three different factors. The first is the number of people. The second is some measure of the average person's consumption of resources (which is also an index of affluence). Finally, the product of those two factors—population and its per-capita consumption—is multiplied by an index of environmental disruptiveness of the technologies that provide the goods consumed . . . In short,

$$\text{Impact} = \text{Population} \times \text{Affluence} \times \text{Technology, or } I = PAT$$

The Ehrlichs believe that this I = PAT equation is the key part of the population connection and useful in understanding our environmental crisis. Under this formulation, environmental problems are found throughout the world because rich nations with relatively small populations and poor nations with large populations have a significant environmental impact (Espenshade, 1991, 332).

The Commoner Viewpoint

Commoner, Corr, and Stamler (1971) suggested two ways to test the validity of Ehrlich's assumption that population growth is the crucial variable. The first way is to quantify the variables in the equation. The second is to analyze a specific environmental problem and determine the exact nature of the impact of population growth.

The time period between 1946 and 1968 was chosen for analysis because many of the present-day environmental problems—for example, pollution from detergents, photochemical smog, and pollution from synthetic pesticides—began after World War II. In addition, many new production techniques were introduced during this period. Pollution levels in the United States between 1946 and 1968 increased 200-1000 percent, while the increase in the U.S. population for the same time period was approximately 43 percent. Commoner, Corr, and Stamler argued that population growth alone could not account for such large increases in pollution levels and that the population component was not large enough to balance Ehrlich's equation.

Their next step was to see whether economic growth and increased per capita consumption could account for the increases in pollution levels. However, the income data for that period show that income alone did not account for the increased pollution levels.

After analyzing changes in per capita consumption of a variety of selected products, Commoner concluded that increases in the consumption of some products were counterbalanced by decreases in consumption of others. This led him to conclude that the most important factor was the nature of technologies used to produce various goods, and the impact of those technologies on the environment.

The largest increases in per capita consumption were related to products that turned out to be important causes of pollution. New technologies were the major culprits in rising pollution levels, not an increase in population. In a recent restatement of his position, Commoner (1990, 14) developed his own equation to measure total pollution as follows:

> The total amount of pollution generated can . . . be expressed by multiplying the pollution per unit good (technology factor) by the total amount of good produced. Finally, the latter figure can be broken down into the product of two factors: good produced per capita (the affluence factor) multiplied by the size of the population. In this way, the total amount of pollution can be expressed numerically in the form of an equation:
>
> Total Pollution = Pollution per good x
> good per capita x population.

With this equation the total amount of pollution can increase when any of the factors increase. Thus, total pollution can be increased because of population increases (Ehrlich) or because of high pollution technologies (Commoner).

Commoner went on to look at these three factors relative to post-1950 production technologies and concluded (1990, 151) that:

> . . . the data both from an industrial country like the United States and from developing countries show that the largest influence on pollution levels is the pollution-generating tendency of the system of industrial and agricultural production, and the transportation and power systems. In all countries, the environmental impact of the technology factor is significantly greater than the influence of population size or of affluence.

This debate between Ehrlich and Commoner outlines two of the major viewpoints regarding the role of population in environmental degradation. Probably the truth, as often is the case in such complex areas, embodies elements from both sides.

In another study of the importance of population growth and its role in environmental degradation, Ridker concluded that three generalizations can be made about this interrelationship (Ridker, 1980, 116):

1. Most of the environmental effects of changes in assumptions about population growth are relatively small.

2. The resource and environmental impacts of changes in per capita income are significantly larger than those of an equal-percentage change in population in early years, but the latter impacts grow over time, so that by the year 2025 they are roughly equal in magnitude.

3. Other determinants, such as the extent of recycling, technological changes, changes in availability of resources, and specific policies directly aimed at reducing the emission of a particular pollutant, generally have larger impacts than either population or economic growth rates.

POPULATION AND RESOURCES

The interrelationship between population and resources has been the subject of study for many natural and social scientists. With population growth and the concomitant increase in demand for goods and services, the pressure on resources also increases.

In a recent evaluation of the need to balance population and resources, Hinrichsen (1991, 27) concluded that, "It is increasingly evident that many developing countries, struggling with rapidly growing populations and dwindling stocks of natural resources, must evolve strategic development plans that incorporate population and resource concerns." There are different opinions, however, about the impact of population growth on resource depletion.

THE POPULATION-RESOURCE REGION

Geographers have always been concerned with regions and have spent a considerable amount of time and energy examining the problems of resource adequacy and population growth. If we try to define a **population-resource region,** then the concept of technology must be examined. Thus, the state of a country's technology is an important measure of the availability of its resources. In general, the greater a country's level of technology, the greater is its ability to exploit its resources.

A useful regional classification, based on population-resource relationships and technology, was developed by Ackerman (Ackerman, 1967). He divided the world into five categories:

1. European Type—technology-source areas of high population-potential resource ratios

2. United States Type—technology-source areas of low population-potential resource ratio

3. Brazil Type—technology-deficient areas of low population-resource ratios

4. India-China Type—technology-deficient areas of high population-resource ratio

5. Arctic-Desert Type—technology-deficient with few potential food-producing resources

Ackermann's discussion of population-resource regions points out some interesting observations. More than half of the world's population lives in areas that could be considered technologically deficient and have high ratios of population to potential resources. The remaining population is rather evenly divided among three of the other population-resource types. The technology-deficient countries of low population-potential resource ratios, including much of Latin America and Africa, have about one-sixth of the world's population. The Western European countries and Japan, which have another sixth of the world's population, are examples of regions where "industrial organization and technology permit them to extend their resource base through world trade, thus effectively meeting the deficiency of their low domestic per capita resource production" (Ackerman, 1967, 87). The remaining portion of the world's population lives in technically advanced societies with low ratios of population to potential resources. These countries include the United States, Canada, Australia, and Russia.

THE LIMITS TO GROWTH

One of the most ambitious attempts to bring together forecasts of resource depletion and its relationship to population growth was the publication of *The Limits to Growth* (Meadows, et al., 1972). Through the use of system dynamics, the authors developed a model with which they could examine the following five factors: population, agricultural production, industrial production, natural resources, and pollution.

One of the principal objectives of *The Limits to Growth* study was to examine the long-term outlook for the world. The various levels of human concern are depicted in Figure 9-2. All human concerns can be found somewhere on the graph, depending upon how far in time we wish to go forward and the amount of geographical space we are interested in. The concern of the vast majority of humankind would be concentrated in the lower left-hand corner of the graph. That is, most people are concerned with their immediate family on a day-by-day basis. However, the concern of the authors is depicted in the upper right-hand corner of the graph. They examined the long-range future, of the children's lifetime, of the entire world.

> One of the principal objectives of **The Limits to Growth** study was to examine the long-term outlook for the world.

In order to examine this long-range future, a formal mathematical model was developed. The advantages of such a model are twofold:

1. All of the assumptions of the model are written out in a precise form and are thus open for inspection and criticism.

2. After all of the assumptions have been analyzed and revised to agree with the most current ideas, a computer can be used to determine the nature of the complex interactions.

The construction of the model followed four principal steps (Meadows, et al., 1972, 98):

1. Professionals in many fields (economics, demography, geology, and nutrition) were contacted and the literature was searched in order to identify the important causal relationships among the five levels of the model.

2. Using global data where it was available, each relationship was quantified as accurately as possible. If global data was not available local data was used.

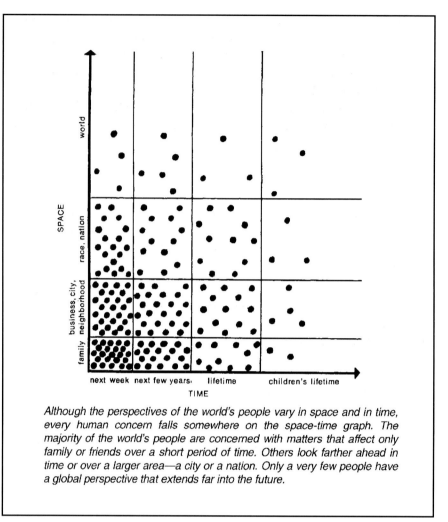

Although the perspectives of the world's people vary in space and in time, every human concern falls somewhere on the space-time graph. The majority of the world's people are concerned with matters that affect only family or friends over a short period of time. Others look farther ahead in time or over a larger area—a city or a nation. Only a very few people have a global perspective that extends far into the future.

■■■ **FIGURE 9-2** Human Perspectives.

Source: From THE LIMITS TO GROWTH: *A Report for THE CLUB OF ROME'S Project on the Predicament of Mankind,* Second Edition, By Donella H. Meadows, Dennis L. Meadows, Jorgen Randers, William H. Behrens III. A Potomac Associates Book published by Universe Books, New York, 1974. Graphics by Potomac Associates. Used by permission.

3. The computer was used to calculate the simultaneous operation of all the relationships over time. The effects of the numerical changes were then tested in the basic assumptions in order to determine the principle and critical determinants of the system's behavior.

4. The fourth and final step was to test the effect of various policies that were currently being proposed on the global system. In order to apply their model to a single nation, or even a smaller areal unit, each relationship in the structure had to be quantified with numbers characteristic of that nation. In order for the model to represent the world, data on world characteristics had to be used.

After quantifying all of the above variables and their relationships, the computer was used to make graphs of the changes that would occur in each of the five factors from 1900 to 2100. The standard world model assumed that there were no major changes in the economic, physical, and social relationships that have historically determined the development of the world system. The mode of behavior for the world system in the standard run is that of overshoot and collapse. According to the authors of *The Limits to Growth* (Meadows, et al., 1972, 129):

> Food, industrial output, and population grow exponentially until the rapidly diminishing resource base forces a slowdown in industrial growth. Because of natural delays in the system, both population and pollution continue to increase for some time after the peak of industrialization. Population growth is finally halted by a rise in the death rate due to decreased food and medical services.

In addition to the standard model, several other assumptions about the world system were also developed, and models of the system under those assumptions were run through the computer. Included were the following assumptions:

- doubling of natural resources;

- unlimited resources;

- unlimited resources and pollution controls;

- unlimited resources, pollution controls, and increased agricultural productivity;

- unlimited resources, pollution controls, and perfect birth control; and

- unlimited resources, pollution controls, increased agricultural productivity, and perfect birth control

Based on their analysis of the world model under varying assumptions, the authors of *The Limits to Growth* arrived at the following conclusions (Meadows, et al., 1972, 29):

1. If the present growth trends in world population, industrialization, pollution, food production, and resource depletion continue unchanged, then the limits to growth on this planet will be reached sometime within the next one hundred years. The most probable result will be a rather sudden and uncontrollable decline in both population and industrial capacity.

2. It is possible to alter these growth trends and to establish a condition of ecological and economic stability that is sustainable far into the future. A state of global equilibrium could be designed so that the basic material needs of each person on earth would be satisfied and each person would have an equal opportunity to realize his individual human potential.

We're Flat Being Overrun

The Colorado Plateau of America's Southwest is a land of spectacular beauty with a rich cultural heritage. It includes a vast area of national forest and other public lands and has the highest concentration of National Park Service units in the nation, including such crown jewels of our park system as Grand Canyon, Bryce, Zion, Arches, and Mesa Verde. However, in recent years tourism in this region has literally exploded. Millions of people from around the country and from abroad have made the Colorado Plateau an increasingly more popular tourist destination.

The number of visitors to the 27 National Park Service units on the Colorado Plateau has increased dramatically. For example, between 1980 and 1995 visitor days increased from 9 million to more than 16 million, or about 80 percent. During that same time period visitor days to all National Park Service units increased by only 10 percent. Five million visitors gazed into the gaping jaws of the Grand Canyon in 1996, whereas only half that many visited in 1960; the National Park Service estimates that by 2010 visitation will reach 7 million annually. Restrictions may be placed on visitors within a decade.

Rapid visitor growth has outrun the regional infrastructure's ability to provide accommodations. For example, visitor days in Canyonlands increased by 160 percent between 1981 and 1993; employment in the park during the same period increased by only 43 percent. Similarly, a visitor day increase of 68 percent in Grand Canyon during that time period was accompanied by only a 38 percent increase in employment within the park and an increase of only 23 percent in constant dollars of support.

3. If the world's people decide to strive for this second outcome rather than for the first, the sooner they begin working to attain it, the greater will be their chances of success.

The Limits to Growth model has received considerable attention and criticism throughout the academic world. For example, a group of thirteen scientists set out to critically examine *The Limits to Growth* model and published the results of their analysis in a book entitled *Models of Doom: A Critique of the Limits to Growth*. Their criticism of *The Limits to Growth* model was centered around the following three essential differences (Cole, et al., 1973, 10):

1. *The Limits to Growth* emphasized purely physical limits, whereas the critics felt that a greater emphasis should be placed on the social and political limits to growth.

2. *The Limits to Growth* was criticized because it is said to have underestimated the possibilities of continuous technological progress, which is by its very nature difficult to predict. For example, the critics pointed out that a forecast made in 1870 would have omitted the principal source of energy in 1970, oil. It would have excluded not only all the synthetic materials, fibers, and rubbers, but probably aluminum and sundry other metals as well.

3. The third essential difference is skepticism that world models developed from System Dynamics are useful tools for forecasting and making policy decisions. Golub forcefully argued that the world model approach is inherently dangerous and encourages self-delusion in five ways (Cole, et al., 1973, 12):

   1. By giving the spurious appearance of precise knowledge of quantities and relationships which are unknown and in many cases unknowable.

   2. By encouraging the neglect of factors which are difficult to quantify, such as policy changes or value changes.

   3. By stimulating gross oversimplification, because of the problem of aggregation and the comparative simplicity of our computers and mathematical techniques.

   4. By encouraging the tendency to treat some features of the model as rigid and immutable.

   5. By making it extremely difficult for the nonnumerate or those who do not have access to computers to rebut what are essentially tendentious and rather naive political assumptions.

In a response to *Models of Doom* the authors of *The Limits to Growth* expanded on the disagreements between the two views. A basic difference turns out to be different concepts of people, and they conclude (Cole, et al., 1973, 240) that they see no objective way of resolving these very different views of man and his role in the world. It seems to be possible for either side to look at the same world and find support for its view.

Technological optimists see only rising life expectancies, more comfortable lives, the advance of human knowledge, and improved wheat strains. Malthusians see only rising populations, destruction of the land, extinct species, urban deterioration, and increasing gaps between the rich and the poor.

Computer models provide a way to look at the dynamics of an ecological system. By altering variables and equations it is possible to observe the consequences of user's assumptions or proposed policies (Muir, 1991, 113). In an evaluation of *The Limits to Growth* model, demographer Kingsley Davis (1991, 12) concluded that:

> It is now two decades since *The Limits to Growth* was published. This is too brief a period to test the accuracy of even the short-term theoretical predictions, much less the long-term ones that were the main focus of interest in the project. Nevertheless, one cannot ignore developments in the last two decades that tend to support the study's findings. . . . Thus the grizzly truth may turn out to be that *Limits* was more prophetic than its detractors and even some of its defenders thought possible.

Again the truth probably lies in some intermediate zone between the Technologists and the Malthusians. The relationships between population growth, environmental degradation, and resource depletion are difficult to define and understand, as evidenced by the considerable disagreement between social and physical scientists.

Finally, we sometimes forget that those very same humans who cause so many of our environmental problems are also capable of fixing them, a point made much more frequently by economists than by ecologists. The United States provides an excellent example. In the late 1960s Americans read about rivers on fire—the Kaw in Kansas and the notorious Cuyahoga in Cleveland. Also in the 1960s, eye-searing smog in Los Angeles brought tears even to the eyes of the most vigorous opponent of environmental laws, and in our National Forests timber was clear-cut almost at will. These and a host of similar observations led to the first Earth Day (in 1970) and to creation of the Environmental Protection Agency and a long series of environmental laws, starting with the Clean Air Act (1970) and Clean Water Act (1972). People cared enough to make a difference, and Americans still want an environment that is cleaner and better.

THE GLOBAL 2000 STUDY

Another attempt to look at population, resource, and environmental problems was the Global 2000 Study. This study, prepared at the request of President Carter, used a variety of analytical techniques and computer models in order to focus on trends and changes that would take place ". . . in the world's population, natural resources, and environment through the end of the century" (Barney, 1980, 6).

The conclusions of the Global 2000 Study were similar to those of the Limits to Growth study. If present trends continue, according to the study, then

> . . . the world in 2000 will be more crowded, more polluted, less stable ecologically, and more vulnerable to disruption than the world we live in now. Serious stresses involving population, resources, and environment are clearly visible ahead. Despite greater material output, the world's people will be poorer in many ways than they are today.
>
> For hundreds of millions of the desperately poor, the outlook for food and other necessities of life will be no better. For many it will be worse. Barring revolutionary advances in technology, life for most people on earth will be more precarious in 2000 than it is now—unless the nations of the world act decisively to alter current trends (Barney, 1980, 1).

Unfortunately, this plea for decisive action has not been acted upon and the Reagan and Bush administrations virtually ignored the recommendations of the Global 2000 Study. The Clinton administration did little as well.

Among both scholars and politicians there is a growing interest in "sustainable development"—though many would argue that it is an oxymoron. Former President Clinton established a Presidential Council on Sustainable Development; the United Nations has a Commission on Sustainable Development; and there is even a Business Council for Sustainable Development. So, we might ask, what is it?

The idea is both simple and complex. Simply stated, it involves a world in which economic growth can be achieved without either environmental degradation or impoverishment of a sizable segment of the population. As geographer Thomas Wilbanks (1994, 543) said about sustainable development, "Clearly, the concept revolves around our capacity for meeting the basic needs of the world's population—especially if that population continues to grow—without running into environmental limits." Current trends, many of which we have already discussed in this chapter, are probably unsustainable—the rising price of environmental deterioration may result in curtailing economic progress if we are unable to change our ways.

The concept of sustainable development, important though it may be for the future of humans on the planet, is complicated not only by its ambiguity but also because it runs head long into a diversity of conflicts. Ultimately the idea is political; growth versus conservation and individual freedom versus control are among the potential areas of conflict. Wilbanks (1994, 553) has these final words for geographers:

> In addition to integrating knowledge in order to meet pressing social needs and helping to unify our various traditions as a discipline, sustainable development focuses our attention on a great problem of mutual concern that can help to integrate the various pieces of our individual professional lives—to integrate them in

the interest of a problem that we care enough about to go that extra mile to do extraordinarily well, not only in our scholarship but in every aspect of the ways that we live as experts in something the world needs very badly.

One of the most troubling issues facing acceptance of sustainable development is likely to be environmental protection, already under attack in the United States by a Republican-dominated Congress. A number of environmental concerns were pointed out recently by Heinrich von Lersner (1995), president of Germany's *Umweltbundesamt* (the German equivalent of our Environmental Protection Agency). He suggests that advances in technology, for example, will be absolutely essential to future sustainability; more self-sufficient housing and better transportation systems, for example, will help conserve energy. Agriculture, he suggests, may be the most serious threat; clean water is already getting scarcer. Waste is still another serious problem, though it could be mitigated by better packaging (and often less packaging). He ends up saying that, ". . . the greatest demand in the future will not be for coal, oil or natural gas; it will be for the time we need to adapt our laws, behaviors and technologies to the new requirements." (von Lersner, 1995, 188)

Without finding solutions to the combined threats of future human population growth, environmental degradation, and the potential for divisiveness and violent conflict, the quality of life for most of us can only get worse. As geographer Crispin Tickell (1993, 226) reminds us, "Life itself is so robust that the human experience could soon become no more than a tiny episode. Nature is not fragile. But we are." A somewhat different perspective is suggested by David Quammen, who argues that people will continue to survive, but believes that "What will increase most dramatically as time proceeds . . . won't be generalized misery or futuristic modes of consumption but the gulf between two global classes experiencing those extremes." (Quammen, 1998, 69)

Population and Food Supply

Nowhere do we hear more frequent and emotional echoes of early Malthusian arguments than in discussions of the relationship between population growth and the supply of food. This idea was summarized in his famous Principle of Population: "The power of population is indefinitely greater than the power of the earth to produce subsistence for man." Since Malthus first proposed this principle in 1798, many developments, unforeseen by Malthus, have occurred. For example, world cropland has more than doubled, new agricultural technologies have been developed to more than quadruple yields achieved by traditional farming methods, and international trade networks and communication links have led to a degree of global interdependence inconceivable in the eighteenth century. As in the days of Malthus, however, debate continues. Are there too many people? Can we eliminate hunger? Raw emotion often dominates such discussions, usually obscuring both the known facts and the complexity of people/food relationships.

In an earlier discussion of population growth, it was argued that throughout most of human history the number of people was limited by the supply of food, among other things. With the beginning of the Agricultural Revolution, perhaps 10,000 years ago, humans began to domesticate plants and animals. At the same time they also began a long process of diminishing the variety in their diets (though not necessarily to the extent of today's American teenager!). Of the hundreds of different animals and thousands of different plants that were once consumed, only a selected few were domesticated. With the emergence of cities, variety in the diet was even further diminished, until finally, as Harlan (1976, 89) commented, "The supermarket and quick-food services have drastically restricted the human diet in the U.S., and their influence is beginning to be felt abroad." Golden Arches, Burger Kings, and Pizza Huts are strewn throughout the cityscapes of Europe and parts of Asia, for example. Harlan (1976, 89) also seriously questioned the revolutionary nature of early agricultural changes and went so far as to point out that ". . . agriculture is not an invention or a discovery and is not as revolutionary as we had thought; furthermore, it was adopted slowly and with great reluctance."

One way or another, in response to an increasing ability to wrest a living from their environments, populations began to grow somewhat more rapidly, though growth rates were far below those that we are experiencing in the world today. In turn, population growth created pressures for further increases in food production. Today, even with signs of a decrease in the rate of world population growth, the number of new mouths to feed is rapidly increasing. Each year an additional 79 million or more people are added to a planet on which most physical resources—land, clean water, and breathable air—are obviously finite. The growing number of people, combined with the increasing affluence of many of the world's residents, is generating an enormous increase in the demand for food.

The relationship between more people and increased food needs is intuitively obvious. More subtle, however, is the effect of income increases, which operate through changes in the nature of the diet. Mainly, increasing affluence leads to an increase in meat consumption, as is apparent in Table 10-1.

In turn, then, producing the meat requires more cereals, unless animals are raised entirely on refuse or on the open range (where they can do a considerable amount of environmental damage). Animals, however, are not very efficient converters of grain into calories that people consume. We find that in the United States per capita consumption of cereals is nearly five times what it is in the developing countries, though in the United States most of that cereal is consumed only indirectly, in the form of meat. Furthermore, the linkages between meat consumption and diseases—including heart disease and cancer of the colon—are well established. All of us, as well as the planet, would be better off if humans, especially those in the wealthier countries, ate lower down on the food chain.

New challenges are at hand if world food production is to continue to meet the increasing demands for food. In the remainder of this chapter some of the major prospects for increasing food supplies are discussed, along with some of their associated problems and consequences.

CURRENT TRENDS IN FOOD PRODUCTION

In recent years crops have been grown on perhaps three billion acres of land, approximately 10 percent of the earth's entire land surface. About two-thirds of this cultivated cropland is planted to cereals, which provide just over half of the entire human food-energy intake. Wheat and rice are at the top of the list, as is apparent in Table 10-2;

■■■ TABLE 10-1. World Meat Consumption Selected Countries (Kilograms Carcase Weight per Capita)			
Country	1996/97	1997/98	1998/99*
Australia	105.0	107.0	104.7
Austria	74.5	75.5	75.9
Belgium-Luxembourg	65.7	75.7	81.4
Brazil	67.2	67.2	68.8
Bulgaria	83.5	82.8	85.8
Canada	92.7	96.5	98.1
China; Peoples Republic of	42.0	43.5	44.0
Czech Republic	98.1	98.2	98.5
Denmark	89.9	92.2	94.5
France	85.9	87.6	89.0
Germany	83.8	86.9	88.7
Hong Kong	114.6	125.1	115.2
Hungary	66.7	66.0	70.1
Ireland	65.4	68.8	67.7
Italy	81.5	83.0	82.6
Japan	42.7	42.7	42.6
Korea; Republic of	40.1	39.0	41.6
Mexico	50.7	51.7	51.9
Netherlands	84.8	86.6	89.4
Philippines	21.9	21.8	22.4
Poland	51.7	58.2	59.1
Portugal	48.1	50.4	50.8
Romania	64.2	64.3	61.8
Russian Federation	49.1	42.2	39.0
Singapore	71.5	68.5	58.3
Spain	101.4	106.3	111.4
Sweden	55.8	58.0	58.7
Taiwan	77.5	81.9	82.7
Ukraine	35.6	34.0	32.9
United Kingdom	65.4	69.3	69.0
United States of America	119.1	122.6	126.4

Note: Meat consumption includes pigmeat, beef, veal, lamb, mutton and poultry meat   *preliminary
Source: USDA, ABARE, and the *Australian Pig Industry Handbook.*

cultural preferences and environmental constraints help explain variations in the geographic patterns of these basic crops. Typically, for any country, as income rises cereals make up an increasingly smaller proportion of total calories in the diet, and the share provided by livestock increases.

It is one thing to look at world agricultural output and talk about it in terms of world per capita supplies of cereals and other foods. It is another, and significantly different, thing to talk about geographic differences in the production and consumption of agricultural commodities. As we would expect, and as is apparent in Figure 10-1, there are considerable variations in the average daily calorie supply from one place to the next. The relative abundance of food in the affluent countries is not necessarily a solution to the problems of

■▪■ **TABLE 10-2.** Sources of Humanity's Food Energy

Food	Percentage of Energy Supplied	
Cereals		56
Rice	21	
Wheat	20	
Corn	5	
Other cereals	10	
Roots and tubers		7
Potatoes	5	
Cassava	2	
Fruits, nuts, and vegetables		10
Sugar		7
Fats and oils		9
Livestock products and fish		11
Total		100

Source: Lester R. Brown with Erik P. Eckholm. (1974) *By Bread Alone.* The Overseas Development Council. Reprinted by permission of Praeger Publishers, A. Division of Holt, Rinehart and Winston.

food shortages or famines in the developing world. Nor is the surfeit of food in those countries always healthy; obesity is a growing problem in the United States, for example, as are anorexia and bulimia (which affect females almost exclusively).

That the geographic distribution of food supplies is in turn related to the distribution of wealth is hardly a new idea. The venerable Chinese poet Tu Fu (*A.D.* 712-770), writing twelve centuries ago (though still appropriate today in many lands), remarked that:

> Behind those vermilion gates meat and wine go to waste
>
> While out on the road lie the bones of men frozen to death. . . .

Global food production between 1950 and 1990, usually measured by cereal grain production, increased approximately 2.6 times, exceeding the 2.1 times increase in the population during the same period (2.5 billion to 5.3 billion). More food is produced per person today, on the average, than forty years ago (Figure 10-2).

These average figures, however, are misleading. Since 1984, grain production per capita has actually fallen, and production increases have failed to keep up with population growth in many areas. In the late 1980s global grain production received several setbacks (Ehrlich, 1991, 223) and per capita production had not returned to the 1984-85 level. Also, the apparent progress in increasing food supplies in developing countries can be mainly attributed to significant production increases in one country, China, which accounts for 35 percent of the Third World's food output and 30 percent of its population. If China is removed from this calculation, food production increases in the other developing countries is barely matched by population

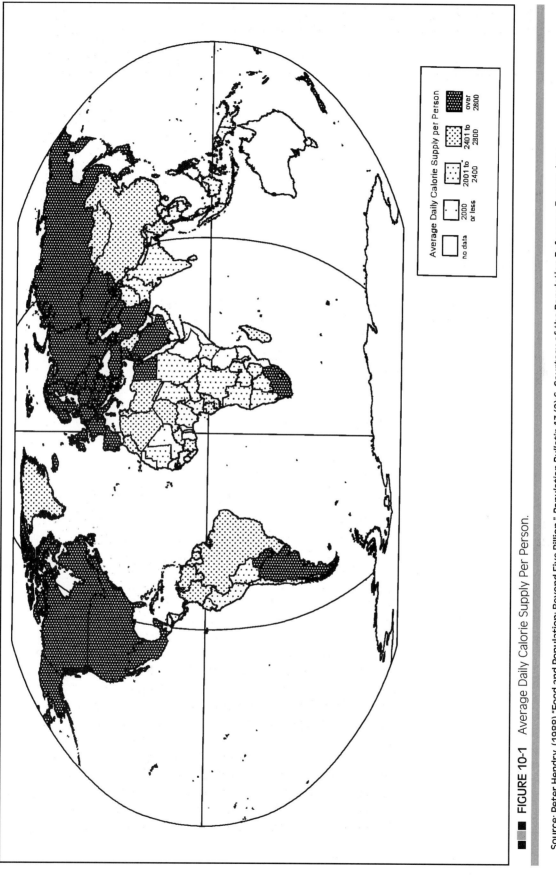

■ **FIGURE 10-1** Average Daily Calorie Supply Per Person.

Source: Peter Hendry, (1988) "Food and Population: Beyond Five Billion," *Population Bulletin* 43 (2):6. Courtesy of the Population Reference Bureau, Inc.

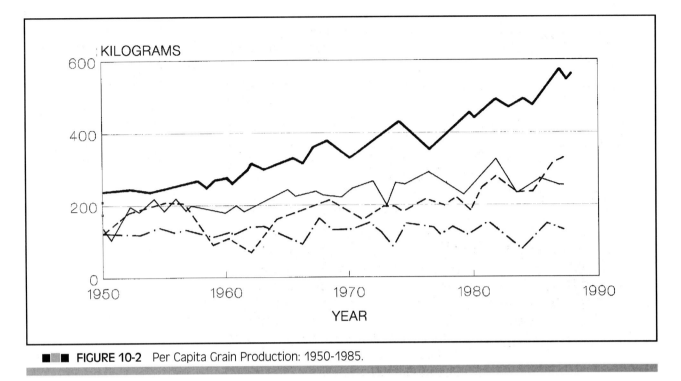

KILOGRAMS

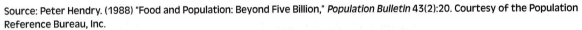

■■■ FIGURE 10-2 Per Capita Grain Production: 1950-1985.

Source: Peter Hendry. (1988) "Food and Population: Beyond Five Billion," *Population Bulletin* 43(2):20. Courtesy of the Population Reference Bureau, Inc.

growth and in some areas it was significantly less than population growth (Hendry, 1988, 7). In the mid-1990s China, with its improving economy and growing population, began to scour the world for grains to import. Providing that Asian giant with additional cereals is likely to affect world grain prices and disturb traditional relationships between countries that have had grain surpluses—for example, Australia, Canada, and the United States—and Third World countries that may be unable to compete with the Chinese in the world grain market (Brown, 1995).

Both a deterioration in diets and an increase in hunger are now realities in sub-Saharan Africa. Both the absolute number and the proportion of hungry people are increasing (Brown and Young, 1990, 77). More than any other world region (Figure 10-3), Africa has seen a significant decline in per capita food production. As Lester Brown (1989, 32) said, "With more hungry people in the world today than when this decade began, there's little to celebrate on the food front as we enter the nineties." As this century begins, the most malnourished countries in the world are Peru, Chad, the Central African Republic, Angola, Ethiopia, Kenya, Afghanistan, and Bangladesh (Tempest, 1997).

Although some writers, Simon (1983) for example, have been led by ideology and "demographic quackery" to believe that there is little relationship between population growth and food availability,

More than any other world region, Africa has been in significant decline in per capita food production.

many demographic scholars believe the real battle to feed the expanding human population may be just beginning. As Hinrichsen and Marshall (1991, 26) observed:

> There are genuine fears that the world may be reaching a watershed in food production. There is not much more good land available for agricultural expansion; water suitable for irrigation is shrinking as demand grows; in many cases current cropland has been pushed to its productive limits; and additional inputs of fertilizers and pesticides are proving counterproductive.

A survey of 93 developing countries, conducted by the Food and Agricultural Organization (FAO), determined that to meet food production requirements at the end of the previous century, another 83 million hectares would have to be added to the existing 770-million-hectare area of farmable land (FAO, 1987). Although such expansion seemed modest, the FAO study cautioned that most of these new **arable lands** contained only marginal soils (Figure 10-4) and some were located in areas of unreliable rainfall. According to Tarrant (1990, 235):

> Pressure to conserve rainforests, difficulties and the expense of developing further agricultural areas, and the loss of agricultural land to other uses, erosion, desertification, and salination, are together likely to mean that an expansion of the area of cultivation will continue to make only a small contribution to extra food production over the next decade.

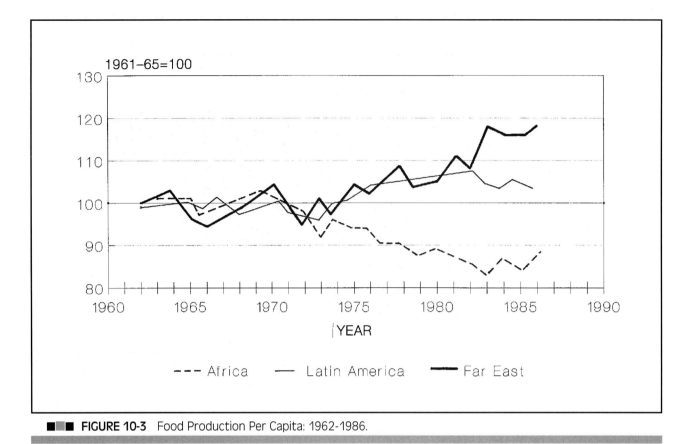

1961–65=100

■■■ **FIGURE 10-3**   Food Production Per Capita: 1962-1986.

Source: John R. Tarrant, "World Food Prospects for the 1990s," *Journal of Geography*, Nov.-Dec. 1990: 235. Reprinted with permission.

Although a few countries, such as Brazil, will be able to add more arable land, most will not; therefore, food for the nearly one billion people that will be added to the world's population during this decade will primarily come from raising land productivity. Most developing countries will have to increase the yield from the same land in order to substantially increase food production (Figure 10-5). In those same countries, of course, pressure to move onto marginal lands will be most intense, natural ecosystems will be most threatened, and pleas for help from the developed countries will be the loudest.

Perhaps the most thorough and balanced look at the world's food needs and ways to meet them in the years ahead was done by geographer Vaclav Smil (2000). He tells us (Smil, 2000, xxvii) that "There are no guarantees that we will succeed: irrational policies and misplaced priorities may lead us astray. . . . But we do have the tools needed to steer a more encouraging course . . . there appear to be no insurmountable biophysical reasons why we could not feed humanity in decades to come while at the same time easing the burden that modern agriculture puts on the biosphere."

Smil arrives at his conclusions only after careful and exhaustive study of the many components of food production, including photosynthesis and crop productivity, land, water, nutrients, agroecosystems and biodiversity, soils, climate change, fertilizer efficiency,

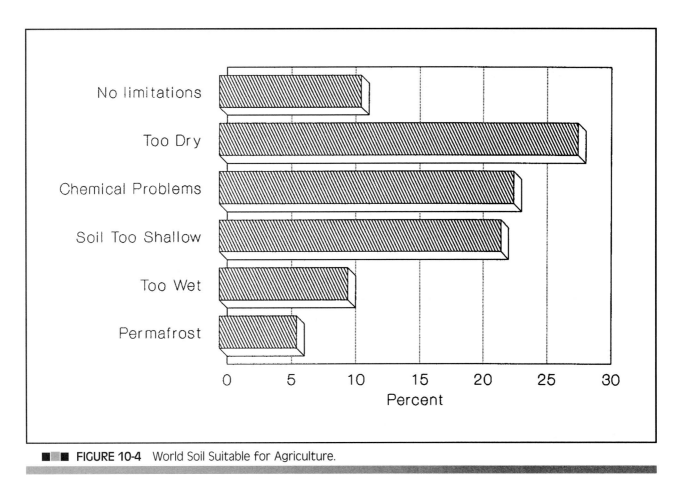

FIGURE 10-4 World Soil Suitable for Agriculture.

Source: United Nations, Food and Agricultural Organization, New York.

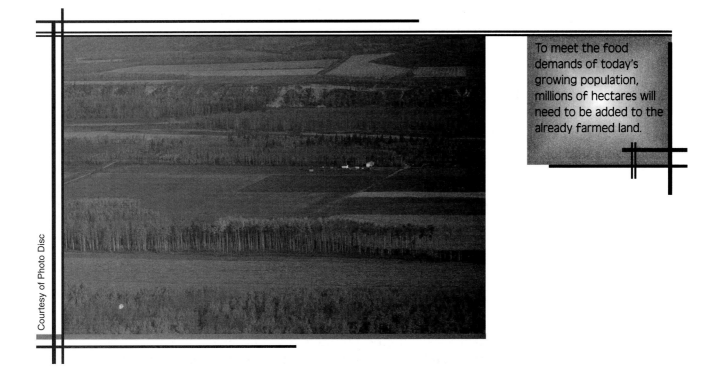

To meet the food demands of today's growing population, millions of hectares will need to be added to the already farmed land.

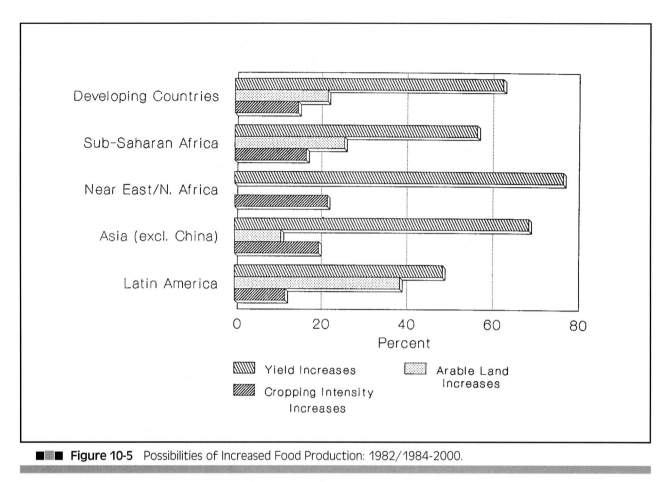

■■■ **Figure 10-5**   Possibilities of Increased Food Production: 1982/1984-2000.

Source: Peter Hendry. (1988) "Food and Population: Beyond Five Billion," *Population Bulletin* 43(2):9. Courtesy of the Population Reference Bureau, Inc.

farming systems, harvest and post-harvest food losses, and human dietary needs. Smil's review of a vast literature suggests that we can, if we wish, make decisions that will aid us in feeding a few billion more people, and feeding them adequate diets.

By better using our current resources, improving the efficiency of agricultural production systems, reducing waste, and encouraging people to make wiser dietary choices, Smil believes that success is possible. It is not, however, guaranteed. Nor is there any guarantee that additional food supplies will reach the world's poorest mouths and stomachs—a reminder that political economy, not food supply limitations, result in a current world in which perhaps 800 million people have either nutritionally inadequate diets or not enough food to eat. Hunger amidst plenty still exists, and it will so long as grinding poverty leaves millions unable to purchase enough food for their needs, even if it is readily available.

An analysis of nutrition in developing countries by the World Bank (Reutlinger, 1985) provided some useful information for determining food security trends. This study was based on Food and Agriculture Organization/World Health Organization (FAO/WHO) calculations that are used to determine the amount of food a person would need in order to function at full capacity in all daily activities. The FAO/WHO looked at people whose food consumption was 80 percent and 90 percent of this established level. The World Bank estimated that 340 million people fell short of the 80 percent consumption level, which results in stunted growth and serious health risks. Women and children are most at risk; malnourished infants are less likely to live than others, and if they do live, are less likely to grow to full size and to develop normal brain capacities. Half of these 340 million people were living on the Indian subcontinent and one-fourth in sub-Saharan Africa. The great majority, nearly four-fifths, lived in poor countries with average incomes less than $400 per capita.

When looking at the 90 percent consumption level, at which growth is not severely stunted but where people do not receive enough calories for a fully productive working life, the World Bank found approximately 730 million people in that category. Of this total, 150 million were in sub-Saharan Africa, 470 million on the Indian subcontinent, and the remaining 110 million in North Africa, the Middle East, and Latin America.

Another important measure of the current world food situation is the food security associated with **carry-over stocks** (Table 10-3)—grain in storage when the new crop begins to come in and readily available if transportation facilities were provided to ship the grain to needed areas. Everyone is in favor of having large carry-over stocks of food but, unfortunately, few are willing to pay for them. Such stocks are very expensive to acquire and maintain. Although the most recent data indicate adequate supplies of carry-over stocks, great fluctuations can occur from year to year and reserves can be rapidly depleted. A discussion of strategies for increasing food security is included in Table 10-4.

Oceanic fisheries and aquaculture play a significant role in feeding humankind also. Between 1940 and 1990 the world fish catch quintupled, though such percentage gains are unlikely to occur again. Oceanic fisheries, which account for around 90 percent of the total catch, may be nearing their maximum sustainable yield (which has already been exceeded for many individual species).

Oceanic fisheries are also far from evenly distributed around the globe. The greatest catches are found in China, Peru, and Chile; Japan, Russia, and the United States are next in line. Around the world controversies about fishing rights have become common as sophisticated trawlers range ever farther in search of their catch. Many fisheries are being depleted, including the salmon runs in California, Oregon, and Washington. Canadian cod fisheries have closed in recent years as well, and a "dead zone" has been identified in the Gulf of Mexico (Beardsley, 1997).

Recent research suggests that the ocean and its inhabitants have long been affected by humans. Commercial fishing and assorted pollutants hurt coastal ecosystems today, but even early humans who settled near the sea had significant impacts on local oceanic populations. With the help of such technology as on-board sonar, today's

Carry-over stocks are grain in storage when the new crop begins to come in and readily available if transportation facilities were provided to ship the grain to needed areas.

Table 10-3. WORLD STOCKS: Estimated Total Carryovers of Cereals 1/

	Crop Years Ending in:						
	1995	1996	1997	1998	1999	2000 estim.	2001 f'cast
	(million tonnes)						
TOTAL CEREALS	**313.2**	**254.5**	**294.5**	**330.5**	**345.1**	**331.0**	**320.9**
held by:							
- main exporters 2/	110.8	75.0	99.5	127.0	154.4	147.4	147.5
- others	202.5	179.5	195.0	203.5	190.7	183.7	173.4
BY GRAINS							
Wheat	**115.4**	**101.7**	**112.7**	**135.2**	**139.8**	**132.9**	**128.8**
held by:							
- main exporters 2/	32.6	28.7	36.6	39.3	51.4	49.7	48.8
- others	82.9	73.0	76.1	95.9	88.3	83.2	80.0
Coarse Grains	**142.8**	**100.4**	**125.5**	**140.0**	**148.5**	**137.9**	**135.7**
held by:							
- main exporters 2/	63.8	31.7	46.1	68.9	84.2	77.9	81.3
- others	79.0	68.6	79.4	71.2	64.3	60.1	54.4
Rice (milled basis)	**55.0**	**52.5**	**56.2**	**55.3**	**56.9**	**60.2**	**56.4**
held by:							
- main exporters 2/	14.5	14.6	16.8	18.8	18.8	19.8	17.3
- others	40.6	37.9	39.5	36.5	38.1	40.4	39.1
BY REGIONS							
Developed Countries	**158.9**	**102.5**	**120.8**	**166.5**	**172.4**	**159.1**	**164.3**
North America	**69.3**	**35.2**	**53.9**	**69.1**	**90.2**	**89.5**	
Canada	9.2	9.8	14.0	10.4	12.4	12.7	
United States	60.2	25.5	39.9	58.7	77.8	76.8	
Others	**89.5**	**67.3**	**66.9**	**97.4**	**82.2**	**69.6**	
Australia	2.6	3.1	4.1	3.7	3.3	3.1	
EC 4/	25.1	22.5	24.2	35.1	41.4	34.6	
Japan	5.5	6.1	6.7	6.8	6.1	5.7	
Russian Fed.	15.9	7.2	6.5	18.0	5.8	4.0	
South Africa	3.2	1.3	1.9	3.3	1.9	1.4	
Developing Countries	**154.4**	**152.0**	**173.7**	**164.0**	**172.7**	**172.0**	**156.6**
Asia	**122.2**	**125.7**	**139.7**	**132.6**	**139.5**	**137.7**	
China 4/	48.2	53.3	63.9	55.9	57.8	52.2	
India 5/	24.1	18.4	10.7	19.0	22.1	25.0	
Indonesia	5.0	6.0	6.4	4.7	5.4	5.5	
Iran, Islamic Rep. of	5.4	4.6	5.5	4.4	4.2	4.5	
Korea, Rep. of	2.4	2.0	2.4	2.5	2.7	2.9	
Pakistan	3.2	3.4	3.7	4.1	4.4	4.1	
Philippines	1.2	1.9	2.0	2.0	2.6	2.8	
Syria	3.0	3.3	3.2	2.2	2.1	1.0	
Turkey	1.9	4.0	5.9	5.9	6.0	3.6	
Africa	**17.9**	**11.4**	**20.1**	**17.3**	**20.1**	**19.7**	
Algeria	2.7	1.5	2.2	1.1	1.9	1.7	
Egypt	1.3	1.6	2.2	2.8	3.0	3.0	
Morocco	2.9	0.6	3.8	2.5	4.4	3.3	
Tunisia	1.5	1.0	2.1	1.9	1.7	1.7	

	Crop Years Ending in:						
	1995	1996	1997	1998	1999	2000 estim.	2001 f'cast
	(million tonnes)						
Central America	4.6	6.3	7.0	6.9	7.0	7.2	
Mexico	2.8	5.0	5.7	5.9	6.1	6.3	
South America	9.5	8.4	6.8	7.0	6.0	7.2	
Argentina	0.7	0.8	1.9	1.7	2.0	2.2	
Brazil	5.8	5.0	2.5	2.8	1.5	2.9	
WORLD STOCKS	(percentage)						
as % of consumption	17.5	13.8	15.7	17.6	18.2	17.4	16.6

Source: FAO
Note: Based on official and unofficial estimates. Totals computed from unrounded data.
1/ Stock data are based on an aggregate of carryovers at the end of national crop years and should not be construed as representing world stock levels at a fixed point in time. 2/ For a list of main exporters of wheat, coarse grains and rice see table A.4. 3/ From 1996, includes 15 member countries. 4/ Including Taiwan Province. 5/ Government stocks only.

■■■ **TABLE 10-4.** Strategies for Increasing Food Security

To stem falling per capita food production and resource degradation and to improve agricultural production, hard-pressed developing countries should undertake the following:

- Establish comprehensive national population programmes;
- Provide integrated planning for future food needs that takes account of population growth, distribution, and rural-urban migration patterns;
- Implement sustainable development strategies that combat soil erosion and impoverishment, deforestation, falling agricultural output, and water mismanagement;
- Establish rural agricultural extension schemes that provide credit, seeds, fertilizers and advice to poorer farmers, whether men or women;
- Provide special agricultural and environmental extension services for and available to women, who do most of the land and water management in poorer areas of the developing world;
- Ensure that women can inherit, buy, and have full legal title to land and that they have access to credit and marketing facilities;
- Emphasize education for women and girls in rural areas. Better educated women are more effective as farmers and environmental managers, and have smaller families;
- Encourage community development strategies, including the setting up of agricultural cooperatives, where essential services such as purchasing, marketing, soil and water conservation, water supply, health care and family planning, sanitation, housing and education can be integrated;
- Establish comprehensive and accessible programmes of maternal and child health care and family planning to reduce the size of families and to improve the health and well-being of the entire community;
- Support research on the integration of traditional and emerging technologies for food production.

Source: Don Hinrichsen and Alex Marshall. (1991) "Population and the Food Crisis," *Populi* 16(2):32.

fishermen have become too efficient, and political issues make the management of oceanic fisheries extremely difficult.

Aquaculture at first glance might seem to be an excellent supplement to our fish supply. However, it is often limited by pollution in populated coastal areas. In addition, if fish are "farmed" they must be fed, and most of that food comes from land sources, including cereals and legumes that could also feed animals or even people. Aside from fish, shrimp aquaculture has grown considerably during the last two decades, though it has run into a number of environmental problems (Boyd and Clay, 1998). McGinn (1998) argued that aquaculture's success rests in large part on resisting the use of chemicals and hormones and minimizing energy inputs.

INCREASING YIELDS ON LAND ALREADY UNDER CULTIVATION

There are four major means of raising crop yields:

1. chemical fertilizers

2. pesticides and herbicides

3. irrigation

4. genetic improvement of plants

These are not new developments, but their use in many places, especially in the developing countries, has come about only in recent decades.

FERTILIZER AND CROP YIELDS

The use of fertilizers for improving yields is hardly new. For centuries farmers have added manure to fields to increase productivity; even human wastes have been used by some farmers. In parts of East Asia it is referred to as "night soil." In 1847, Justus von Liebig, a German agricultural chemist, established the technological foundation for the use of chemical fertilizers when he demonstrated that all the nutrients needed by plants for growth could be added in chemical form. However, the escalating use of chemical fertilizers has mainly occurred in the last 35 years (Figure 10-6), and mainly in the developed countries.

Chemical fertilizers will improve yields rather dramatically under favorable circumstances, though there are decreasing returns as the amount of fertilizer is increased, as is apparent in Figure 10-7. Between 1950 and 1989, world fertilizer use rose from 14 million tons to an estimated 146 million tons (Brown, 1989, 35). Since 1980, there has been a reduction in the rate of growth of fertilizer use. Also, in the 1970s many Third World governments encouraged fertilizer use through subsidies, but these subsidies have recently either been elim-

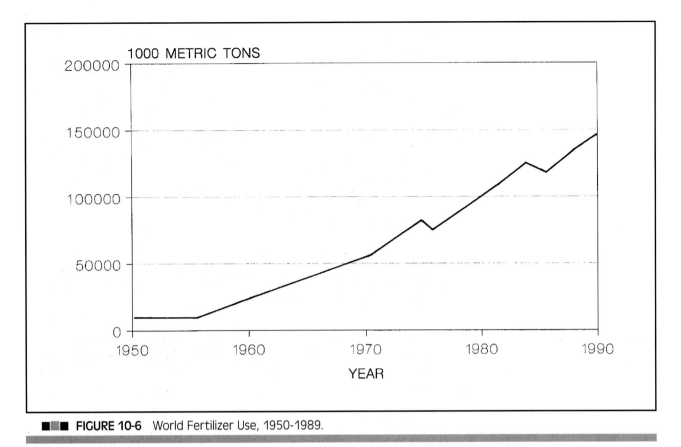

■■■ **FIGURE 10-6** World Fertilizer Use, 1950-1989.

Source: Lester Brown, "Feeding Six Billion," *Worldwatch*, Sep.-Oct. 1989:37. Reprinted with permission.

inated entirely or significantly reduced. Fertilizer prices have risen considerably during the last two decades, making their use in Third World countries ever more difficult.

Obviously, given the nature of the fertilizer response curve shown in Figure 10-7, farmers in the developing countries could benefit considerably from the increased use of chemical fertilizer. Yet, these are the very farmers least able to afford such fertilizers in many cases. A major redistribution of world fertilizer supplies would probably increase world agricultural output, but it is unlikely to occur for both political and economic reasons. Golf courses and lawns in the rich countries will remain lush and green while Indian farmers harvest their meager yields of wheat and rice! Table 10-5 shows data for fertilizer use and world grain production. In 1950, when 14 million tons of fertilizer were used, grain production was 624 million tons, with a response ratio of 46. With the increasing use of fertilizer between 1950 and 1980, the response ratio fell from 46 to 13. During the next six years the ratio was constant, an indication of stability in the grain/fertilizer price relationship. According to Brown (1987, 130), "Growth in world fertilizer use is likely to remain slow in the absence of either a substantial improvement in the grain/fertilizer price ratio or technological advances that boost the fertilizer responsiveness of grain." Developing countries are also starting to experience diminishing

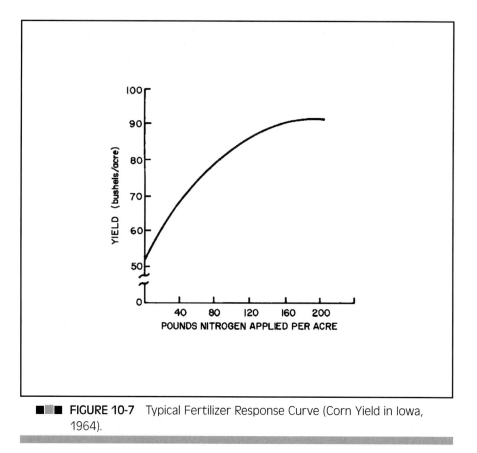

■■■ **FIGURE 10-7** Typical Fertilizer Response Curve (Corn Yield in Iowa, 1964).

Source: U.S. Department of Agriculture.

returns in fertilizer use in some cases. Presently, both India and China have grain-fertilizer response ratios very similar to that of the United States (Brown, 1990, 68).

This reliance on chemical fertilizers will also have an environmental impact in both the near and longer term. Hendry (1988, 25) noted that, "Beginning with Rachel Carson's seminal 'Silent Spring' in 1962, a growing body of literature documented the environmental and health hazards likely to arise from agricultural systems that depended too heavily on chemical inputs." Nonetheless, it is difficult to argue with geographer Vaclav Smil (1998, 81), who recently commented that "Barring some surprising advances in bioengineering, virtually all the protein needed for the growth of another two billion people to be born during the next two generations will come from the same source—the Haber-Bosch synthesis of ammonia."

Herbicides and Pesticides

Some estimates suggest that as much as half of the world's food production is lost, mainly to hungry animals and insects. Chemical control of pests would certainly help save some of this tragic loss, as would more adequate storage facilities. Yet, pesticides raise an environmental dilemma. Many such chemicals remain in the environment;

TABLE 10-5. Ratio of World Grain Production to Fertilizer Use, 1950-1988

Year	Grain Production	Fertilizer Use	Response Ratio
1950	624	14	46
1955	790	18	43
1960	812	27	30
1965	1,002	40	25
1970	1,197	63	19
1975	1,354	82	16
1980	1,509	112	13
1981	1,505	116	13
1982	1,551	115	14
1983	1,474	114	13
1984	1,628	125	13
1985	1,674	130	13
1986	1,661	131	13

Source: Lester R. Brown. (1987) *State of the World 1987.* (New York: W.W. Norton Table 7-3, p. 130.

and some are concentrated as they move up through the food chain. DDT is only one example.

Weeds are another menace to better crop production. Weeds compete with food plants for sunlight and nutrients; hence, their control is another way to improve yields. Again, the major means of control is the use of chemicals, and there is again the possibility of environmental problems. In the developing countries, tradeoffs are necessary. An increase in food supplies is essential and some risks are acceptable.

IRRIGATION

Although many large areas of the world receive sufficient rainfall to meet their agricultural requirements, many of the countries that are the most densely populated are found in the drier climatic regions. Many of these countries are poor and don't have the resources to purchase food from other areas. Therefore, in order for many of these countries to produce enough food to feed their populations, it will be absolutely essential that irrigation facilities be developed. In the several thousand years since irrigation was first developed in the Middle East, it has diffused gradually throughout the world. Much of the increase in irrigated lands, however, has taken place in the past four decades. Between 1950 and 1985 the amount of irrigated land nearly tripled, a major factor in the impressive growth of world food production.

Irrigated agricultural land plays an important role in feeding the world's growing population and in providing an adequate living for rural populations as well. For example, in Pakistan 65 percent of its farmland is under irrigation, producing 80 percent of that country's food supply. In India 30 percent of cultivated land is under irrigation and this area produces 55 percent of the food output. Similarly, 50

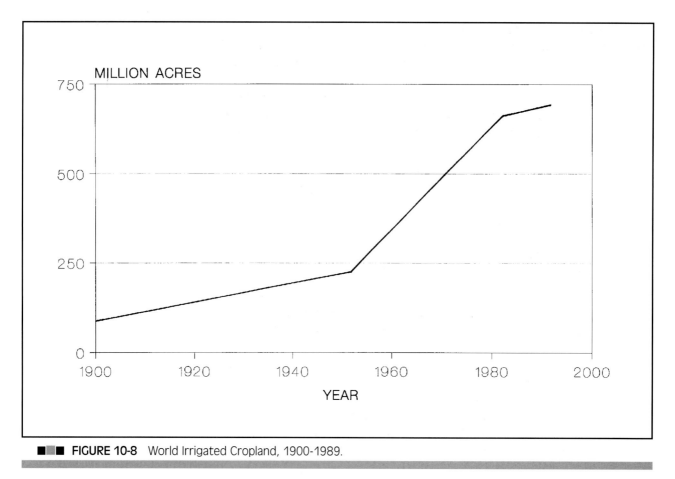

■■■ **FIGURE 10-8**   World Irrigated Cropland, 1900-1989.

Source: Lester Brown, "Feeding Six Billion," *Worldwatch*, Sep.-Oct. 1989:35. Reprinted with permission.

percent of China's croplands are irrigated, producing 70 percent of all the food produced in that nation (Hinrichsen and Marshall, 1991, 30).

The irrigated area grew most rapidly (Figure 10-8) during the fifties and sixties, with an annual rate of almost 4 percent. During the 1970s the rate of expansion slowed and the first half of the 1980s has seen an expansion of irrigated land that is less than 1 percent per year—a continued slow rate of expansion is expected throughout the 1990s. After this initial rapid expansion of irrigated land, the amount of land being taken out of production was rising while there was a decline in the amount of new land being put under irrigation. There has also been a decrease in the irrigated acreage per person on a worldwide basis (Ehrlich, 1991, 225). The reason for taking irrigated land out of production is primarily associated with what the FAO calls the three "silent enemies,"—alkalization, salinization, and waterlogging.

In some regions the amount of irrigated land is actually declining. In Africa, for example, a review of irrigation projects by Biswas (1985, 35) pointed out that ". . . the development of new irrigation areas has barely surpassed the surface of older ones which had to be abandoned."

Only about 12 percent of the earth's cultivated land is currently being irrigated, although approximately one-half of the world's food comes from these areas. Thus, it appears that there is a significant potential for expanding irrigation. However, the expansion of irrigation is limited both by the uneven distribution of water sources and the high capital costs of major irrigation projects such as dams and canals.

Water will, however, be a key factor in raising agricultural output. Many scholars have attributed at least part of the failure of the Green Revolution to inadequate water supplies. According to Falkenmark and Suprapto (1992, 35):

> It will be fundamental to match rapidly enough the reduced per capita water availability and the increasing per capita demands. In order to keep even the present level of water demands, more water-resource structures are needed to make more water accessible for use. The financial aspects of water-resource development projects that are needed to supply the rapidly growing population in low-income countries, require serious attention.

Water from the oceans could conceivably be used to irrigate such salt-tolerant plants as glasswort, which can be consumed by livestock and has seeds that can be pressed for oil (Glenn, Brown, and O'Leary, 1998).

Clean, fresh water is likely to become scarcer in the decades ahead, leading to rising prices and potential conflicts between different users—farmers and urban dwellers for example. Nonetheless, economic reality will go a long ways toward forcing people to adapt, especially if we add another 3 billion people over the next 50 years. As Gleick (2001, 45) tells us:

> Addressing the world's basic water problems requires fundamental changes in how we think about water, and such changes are coming about slowly. Rather than trying endlessly to find enough water to meet hazy projections of future desires, it is time to find a way to meet our present and future needs with the water that is already available, while preserving the ecological cycles that are so integral to human well-being.

NEW STRAINS OF PLANTS AND THE GREEN REVOLUTION

Since the beginning of the Agricultural Revolution, improvements in crops and livestock have made notable contributions to increasing the quantity and improving the quality of the human diet. Revolutions of varying dimensions have occurred many times over the past 10,000 years. Farmers have gradually kept and improved a few major plant species, while myriads of others have long since disappeared from the diets of most of the world's people.

Among such "revolutions" would be found the exchange of crops between the Old and New World, an exchange that began perhaps 500 years ago and continues even today, and one that certainly increased the world's food-producing capacity. (Consider Ireland before the potato, for instance.)

More recently, a number of "revolutions" have resulted from discoveries in plant genetics, discoveries which have led to the

Hybrid corn has long been common in the United States, providing better yields and extending the range of corn production as well.

development of higher-yielding cereals. Hybrid plants may also be developed to better tolerate environmental conditions such as cold or drought, or to be more resistant to problematic diseases, or to be more responsive to fertilizers. Improved strains of corn have been grown in the United States for several decades with good results. The breeding of short-season corn has allowed the northern limit on corn production to be extended as much as 500 miles.

But the **"Green Revolution"** that has attracted the world's attention has been that of the development of so-called "miracle strains" of wheat and rice that were developed in the mid-1960s. These new strains are shorter and stiffer-strawed than traditional varieties. They are also highly responsive to chemical fertilizers, mature earlier than traditional varieties, and have a reduced sensitivity to variations in the length of days. This latter property creates increased opportunities for multiple cropping (raising more than one crop per calendar year—in Pakistan, for example, some farmers get three crops over a 14-month period).

The **"Green Revolution"** that has attracted the world's attention has been that of the development of so-called "miracle strains" of wheat and rice that were developed in the mid-1960s.

The spread of these high-yielding varieties of wheat and rice occurred quite rapidly. Among the major countries to benefit from these new strains were India, Pakistan, the Philippines, Indonesia, and Mexico. Countries sometimes went rapidly from being net grain importers to being net grain exporters.

Throughout the 1970s food production at least kept pace with population growth, thanks mainly to the crops from the new seeds. Some countries, such as India, Pakistan, the Philippines, and Thailand, proved that yields could be substantially increased by peasant farmers when conditions were favorable. By the end of the 1970s new varieties of corn, sorghum, and soybeans were available along with the new strains of rice and wheat, but other crops still await significant improvement.

The Green Revolution was a major advance in international agricultural development, yet the agricultural progress made possible by the Green Revolution was not evenly distributed. Aggregate statistics hide a significant group of farmers who have not benefited from the new technologies. Most of these farmers bypassed by the Green Revolution were subsistence farmers who were raising food for their families on rain-fed, marginal land. According to Dalrymple (1986), only about one-third of the land planted to cereal grains in the Third World uses the high-yielding varieties. The rates of adoption also vary widely by region, with Africa having the lowest adoption rate, at only 1 percent of the grain area, Latin America with 22 percent, and Asia and the Middle East with the highest percentage of adoption, 36 percent of the grain area.

As Brown (1981, 34) noted, "Traditional farmers must also be trained and encouraged to use new techniques: how and when to plant the new seeds; how to control weeds, insects, and disease; and how to conserve and manage water." However, training programs are too often inefficient and poorly financed. Also, many countries still fail to provide sufficient incentives to encourage small farmers to adopt new seeds and more efficient farming techniques. Farmers in Third World countries are often penalized by deliberate government policies that attempt to keep food prices low in their burgeoning cities.

Though it undoubtedly helped in the battle against hunger, the Green Revolution was not a panacea for food problems in the developing countries. Complacency about population growth sometimes resulted from the rapid advances in agricultural production. Populations have continued to grow, often eating up the benefits of increased production in true Malthusian fashion, reducing per capita gains to little or nothing.

Young (1992) believes that although the biotechnology industry could be an important weapon in the future fight to eradicate hunger, the control of the industry is by corporations who have little interest in plant research related to Third World food crops. He noted (1992, 97) that, "Firmer tomatoes and herbicide-tolerant corn, not hardier rice or cassava, are their objectives. These companies are not sinking money into less-profitable plant research that might help close the widening gap between the growth in world food supplies and that in human numbers."

World food production varies from year to year in response to weather, among other factors. Each swing upward or downward in production tends to produce a new wave of optimism or pessimism. The world currently produces enough food to provide everyone with an adequate, or at least nearly adequate diet, yet hundreds of millions, mainly in the developing countries, are undernourished. Even when diets are adequate in terms of calories, they are often inadequate in terms of nutrients, especially protein. Such inadequacies lead to nutrient deficiency diseases such as kwashiorkor, which results from severe protein deficiency.

The specter of Malthus is regularly called upon in discussions of population growth versus food supplies in the developing countries. Although for reasons that are not always in agreement, experts generally agree that slowing the rate of world population growth is, if not an absolute necessity, at least a way of easing other problems. It is tempting at this point to agree with the following statement by Abercrombie and McCormack (1976, 489):

> It is becoming abundantly clear that the last quarter of this century is likely to be one of the most crucial periods in the history of mankind's efforts to feed a growing population. On the one hand, there is general agreement that if the right action is taken, population growth can be reduced, the increase in food production accelerated, and the purchasing power of the poorest people raised sufficiently for the virtual elimination of hunger and malnutrition by the end of the century. This could lay the foundations for a widely shared prosperity in the more distant future. On the other hand, the alternative is not hard to visualize. If sufficient political will cannot be found for the massive action that is necessary, there will not only be widespread starvation but also a gradual drift downward in the nutritional levels of the poorest people. This too might lead to eventual prosperity for a stabilized population, but the reduction in population growth would have resulted from higher infant mortality rather than deliberate fertility limitation.

We have entered a new century; six billion celebrated the beginning of the new millennium. More than 800 million of them were underfed, and the majority were poor and tired.

GENETICALLY MODIFIED FOODS

Though the Green Revolution provided the world with more food as a result of plant hybridization and different agricultural practices, the newer methods of genetically modifying plants represent a new phase in crop alteration. As a result, genetic engineering—the production of transgenic plants—has become a source of considerable controversy, as has the successful cloning of farm animals.

People have modified plants and animals to improve their usefulness to humans for thousands of years. What is new about today's genetically modified organisms (GMOs) is the way in which new genetic traits are introduced into plants or animals. New genes from a variety of sources (including, but not limited to, other plants) can be spliced into existing plants, for example, to make crops that are resistant to droughts, diseases, and pests, or even to improve the nutritional quality of the plant.

Suppose you would like to have a new strain of corn that would be resistant to certain common insect pests? Such a corn has been genetically engineered by isolating a gene from *Bacillus thuringiensis* (a bacterium) that directs cells to produce a protein that is toxic to certain insects. The Bt genes are inserted into cells, along with a "marker" gene, so that Bt's presence in cells can subsequently be identified. Cells that have the Bt gene will then be allowed to grow into corn plants, which will produce the Bt toxin in their cells, in turn killing insects that consume them. It is much neater than applying insecticides to fields to control pests.

However, GMOs have created a firestorm of controversy. Though there are perhaps 110 million acres of genetically altered crops growing today, led by corn, cotton, and soybeans, opponents have grown both more vocal and more numerous in recent years. Commercial planting of genetically altered plants began with tobacco in China in 1992, and two years later the FlavrSavr tomato became the first commercially produced GMO in the United States. By 2000 the United States accounted for about two-thirds of total world production of GMOs.

Critics denounce GMOs as "Frankenfoods" and argue that they are injurious to everything from people and insects to the environment. Among their many concerns are whether harm will come to consumers of genetically altered crops, whether innocent creatures will be harmed by them, whether "superweeds" will arise, and whether such crops might suddenly fail because of new resistant strains of insects and weeds. So far evidence has been scanty, but some environmental links have been identified, including one study that found harm to monarch butterflies that consumed Bt corn. More needs to be done before scientists are convinced, however.

Advocates of GMOs see them as the next essential step in allowing us to feed a burgeoning human population—providing safe (and sometimes more nutritious) foods with less damage to the environment than traditional crops cause. American farmers alone use nearly a billion pounds of pesticides annually, a figure that could be reduced by using Bt crops, especially cotton, corn, and potatoes. A new strain of "golden" rice, rich in beta carotene, has the potential for decreasing vitamin A deficiency in children who consume large quantities of rice as their staple diet. Is it a new "miracle" rice or a "Frankenfood?" Should we keep children in the Third World from having it, especially if an early death may be the alternative for one or two million of them a year?

Unfortunately, sound scientific answers to questions raised by critics of genetically engineered foods are still hard to find because few large-scale tests have been conducted over sufficiently long periods of time. In the meantime indications of the widespread GMO products on the market have become more common. In 2000, for example, Starlink corn (a genetically modified variety not approved for human consumption) was found in the human food supply in the United States. So far, however, no harm has come to humans from consuming it, though it did become clear to everyone that GMOs were more widely distributed than many people had thought. Among concerns for humans are the possible introduction of allergens into crops.

So far, despite numerous protests and discussions, there has not even been a law passed to require labeling products that contain GMOs, though a few manufacturers are now doing the opposite. They are carefully noting that their products do not contain genetically modified foods of any sort. At the same time, genetically modified soybeans, corn, and canola have become so widely distributed that it is unlikely that they would disappear in the short run no matter what we decide. You can be certain, however, that more studies will be done, more protests will be held, and more people will be eating the products of genetic engineering, often without even knowing it.

FOOD AND PEOPLE

Despite the impressive gains in agriculture brought about by better technology, irrigation, genetic engineering, and chemical fertilizers, there still appears, often among knowledgeable and competent scientists, the growing fear of creeping Malthusianism. On the other hand, there are scientists who feel that continued advances will keep the growth in food supplies somewhat ahead of population growth for awhile longer. Most would agree that a decline in the rate of population growth would ease the strain on resources and possibly allow adequate diets for an increasing proportion of the world's people. Some of these differing views are discussed in the following section.

THE PADDOCKS AND TRIAGE

Two decades ago William and Paul Paddock, in a revised edition of an earlier book, made the following statement (1976, 8-9):

> A locomotive is roaring full-throttle down the track. Just around the bend an impenetrable mudslide has oozed across the track. There it lies, inert, static, deadly. Nothing can stop the locomotive in time. Catastrophe is foredoomed. Miles back up the track the locomotive could have been warned and stopped. Years ago the mud-soaked hill could have been shored up to forestall the landslide. Now it is too late.

The locomotive roaring straight at us, of course, is the population explosion. The unmovable landslide across the tracks is the stagnant production of food in the undeveloped nations, the very nations where the population increases are greatest.

The collision is inevitable. The famines are inevitable. After accepting the veracity of the above statement, the Paddocks suggested that the most reasonable solution to the situation is to employ the concept of triage as a means of allocating America's food surplus to the "hungry nations."

The term **triage** has been used mainly in military medicine (remember MASH?). It is a method of assigning treatment priorities to the wounded by classifying them into three categories:

1. those so seriously wounded that they will die regardless of the treatment they receive

2. those who will survive without treatment

3. those who will live providing they receive prompt medical attention

Scarce medical resources, including doctors, may then be best allocated by treating those in the third category, thus allowing doctors to save the maximum number of lives. The Paddocks suggest that such a system should be used to classify the hungry nations of the world into the following three categories:

1. those who cannot be saved because of population growth and the lack of agricultural potential and leadership ability

2. those who have sufficient resources, in agriculture and/or foreign exchange, to adequately deal with their population growth

3. those in which there is an imbalance between population growth and food supply which would be manageable if aid were received

Thus, the Paddocks suggest that America's scarce foreign aid resources can be best allocated if they are directed toward the countries in the third category. As an approach to the solution of the world food crisis, triage seems attractive to a significant number of people. Too often, however, those same people have an inadequate understanding of the relationship between population growth and food supplies. Furthermore, the argument is morally questionable. *The New York Times* referred to this use of triage as "one of the most pessimistic and threadbare intellectual positions to be advanced since the demise of the Third Reich." (Cited in Simon, 1975, 36)

HARDIN AND THE LIFEBOAT ETHIC

Garrett Hardin, an eminent biological scientist, has popularized a view of the rich and the poor countries in an analogy using **lifeboats.** The rich countries are viewed as **lifeboats** filled almost to capacity. The addition of more people threatens to sink them, hence drowning everyone. The poor, of course, are struggling to enter the **lifeboats** and, hence, must be pushed away if the rich countries are to survive. In other words, in Hardin's view, feeding those in the hungry nations threatens those in the **lifeboats,** so it is not the thing to do.

Like triage, the moral nature of this argument is extremely weak in the opinion of many observers. For example, Simon (1975, 36) proclaimed the following challenge:

> The rich countries are viewed as **lifeboats** filled almost to capacity. The addition of more people threatens to sink them, hence drowning everyone. The poor, of course, are struggling to enter the **lifeboats** and, hence, must be pushed away if the rich countries are to survive.

> In Hardin's view, feeding those in the hungry nations threatens those in the **lifeboats,** so it is not the thing to do.

Behind the lifeboat and triage theories lie some frightening, unspoken assumptions about the value of life. Is a hungry, impoverished Asian child less human, with less reason to live and less right to live, than our own children? Playing God in this regard is especially dangerous because we are easily seduced by answers that protect our own advantages, even if they cost lives.

Furthermore, as many have already pointed out, the analogy itself is questionable. Why, for instance, are the rich countries portrayed as lifeboats rather than, as some have suggested, luxury liners?

IS THERE HOPE?

There is no denying that millions in today's world are hungry and malnourished, nor is there any doubt that everywhere it is the poor (and especially women and children) who suffer the most. But is it necessary to take such a dim view of this large share of humanity that we simply turn our heads and disregard its existence? Fortunately, many think not.

No serious view of the future suggests that solving food and population problems would be easy or automatic. However, in his essay on raising agricultural productivity, Wolf (1987, 156) concluded that:

> The world is far from having solved the problems of agricultural productivity. The conventional approach to raising productivity—combining new crop varieties with fertilizers, pesticides, and heavy use of energy—succeeded dramatically in increasing food production in industrial countries and in parts of the Third World. But new approaches are needed to reach farmers who could not afford to follow this path, as well as to correct inequities in the distribution of environmental problems. Complementing the use of conventional resources with innovative biological technologies that maximize agriculture's internal resources can begin to achieve affordable and sustainable gains in agricultural productivity.

Though mass starvation appears unlikely in the near future, the 1990s promise to be a decade in which hunger and malnutrition remain widespread, especially in Third World countries. The specter of Malthusianism is unlikely to disappear in the decades ahead. More people implies an obvious need for more food. Food supplies can be increased, though probably not without threatening the environment. Raising yields on land already under cultivation remains the best alternative for expanding food supplies in the years ahead.

At the same time, as was suggested in the previous chapter, the concept of sustainable development needs to be more carefully spelled out and encouraged. Plucknett and Winkelmann (1995) have discussed various aspects of the technology necessary for sustainable development. Also, geographers Richard Le Heron and Michael Roche (1995) found that in New Zealand the idea of sustainability is being worked through in a way that fits well with current business practices. Finally, in the United States sustainable ideas in agriculture are also receiving considerable attention. In many California vineyards, for example, owls and hawks are replacing pesticides and ground covers are being grown to attract "good" bugs, those that feed on undesirable insect pests.

Though it is scientifically possible to feed a growing population, given today's technology, the principal determinants of success are going to be political, social, and economic variables. Moral, rather than scientific, questions must be answered as well. For example, should the developed countries help feed the less fortunate ones? How can an equitable distribution of food be assured? Does each person in the world have a *right* to food? Such questions cannot be answered by science alone.

Introduction References

Beaujeu-Garnier, Jacqueline (1966) *Geography of Population*. New York: St. Martin's Press.

Bogue, Donald J. (1969) *Principles of Demography*. New York: John Wiley and Sons, Inc.

Brunn, Stanley (1992) "Are We Missing Our 'Forests' and our 'Trees'? It's Time for a Census," *Annals of the Association of American Geographers* 82:1-2.

Camp, Sharon L. (1993) "Population: The Critical Decade," *Foreign Policy* 90(Spring):126-144.

Clarke, J. and Noin, D. eds. (1998) *Population and Environment in Arid Regions*. Paris: UNESCO and Parthenon.

Clarke, John I. (1972) *Population Geography*. Second Edition. Oxford: Pergamon Press.

Clarke, John I. ed. (1984) *Geography and Population: Approaches and Applications*. Oxford: Pergamon Press.

Coleman, David and Salt, John (1992) *The British Population: Patterns, Trends, and Processes*. Oxford: Oxford University Press.

Congdon, P. and Batey, P. eds. (1989) *Advances in Regional Demography*. London: Belhaven Press.

Courgeau, D. (1976) "Quantitative, Demographic, and Geographic Approaches to Internal Migration," *Environment and Planning* A, 8:261-269.

Crook, Nigel and Timaeus, Ian M. (1997) *Principles of Population and Development: With Illustrations from Asia and Africa*. New York, NY: Oxford University Press.

Demko, George J., Rose, Harold M., and Schnell, George A., eds. (1970) *Population Geography: A Reader*. New York: McGraw-Hill Book Company.

Findlay, Allan M. (1991) "Population Geography," *Progress in Human Geography* 15:64-72.

Findlay, Allan M. (1993) "Population Geography: Disorder, Death and Future Directions," *Progress in Human Geography* 17:73-83.

Findlay, Allan M. and Graham, Elspeth (1991) "The Challenge Facing Population Geography," *Progress in Human Geography* 15:149-162.

Gober, Patricia (1992) "Urban Housing Demography," *Progress in Human Geography* 16:171-189.

Gore, Senator Al (1992) *Earth in the Balance: Ecology and the Human Spirit*. Boston: Houghton Mifflin Company.

Hornby, William F. and Jones, Melvyn (1980) *An Introduction to Population Geography*. Cambridge: Cambridge University Press.

Jones, Huw R. (1981) *A Population Geography*. New York: Harper and Row, Publishers.

Murdock, Steve H. and Ellis, David R. (1991) *Applied Demography: An Introduction to Basic Concepts, Methods, and Data*. Boulder, CO: Westview Press.

James, Preston E. (1954) "The Geographic Study of Population," in Preston E. James and Clarence F. Jones, eds., *American Geography: Inventory and Prospect*. Syracuse, N.Y.: Association of American Geographers, pp. 106–122.

Lindahl-Kiessling, Kerstin and Landberg, Hans. eds. (1994) *Population, Economic Development, and the Environment*. New York: Oxford University Press.

McNicoll, Geoffrey (1999) "Population Weights in the International Order," *Population and Development Review* 25(3):411-442.

Menard, Scott W. and Moen, Elizabeth W. (1987). *Perspectives on Population: An Introduction to Concepts and Issues*. New York: Oxford University Press.

Myers, Dowell (1992) *Analysis with Local Census Data: Portraits of Change*. Boston: Academic Press.

Nash, Alan (1994) "Population Geography," *Progress in Human Geography* 18:385-395.

Newman, James L. (1995) *The Peopling of Africa: A Geographic Interpretation*. New Haven: Yale University Press.

Noin, Daniel and Woods, Robert. eds. (1993) *The Changing Population of Europe*. Oxford: Blackwell Publishers.

Overbeek, Johannes (1982) *Population: An Introduction*. New York: Harcourt Brace Jovanovich, Inc.

Pacione, M. ed. (1986) *Population Geography: Progress and Prospect*. London: Croom Helm.

Petersen, William (1975) *Population*. Third Edition. New York: Macmillan Publishing Company, Inc.

Plane, David A. and Rogerson, Peter A. (1994) *The Geographical Analysis of Population, With Applications to Planning and Business*. New York: John Wiley and Sons, Inc.

Schnell, George A. and Monmonier, Mark Stephen (1983) *The Study of Population: Elements, Patterns, Processes*. Columbus, Ohio: Charles E. Merrill Publishing Company.

Stoddart, D. R. (1987) "To Claim the High Ground: Geography for the End of the Century," *Transactions of the Institute of British Geographers* 12:327-336.

Thomlinson, Ralph (1976) *Population Dynamics: Causes and Consequences of World Demographic Change*. Second Edition. New York: Random House, Inc.

Trewartha, Glenn T. (1953) "A Case for Population Geography," *Annals of the Association of American Geographers* 43:71-97.

Trewartha, Glenn T. (1969) *A Geography of Population: World Patterns*. New York: John Wiley and Sons, Inc.

Turner, B.L., II, Hyden, Goran, and Kates, Robert W. eds. (1993) *Population Growth and Agricultural Change in Africa*. Gainesville, FL: University Press of Florida.

Weeks, John R. (1999) *Population: An Introduction to Concepts and Issues*. Seventh Edition. Belmont, California: Wadsworth Publishing Company.

Weller, Robert H. and Bouvier, Leon F. (1981) *Population: Demography and Policy*. New York: St. Martin's Press.

White, Stephen E., et al. (1989) "Population Geography," in Gaile, Gary L. and Willmott, Cort J. eds. *Geography in America*. Columbus, Ohio: Merrill Publishing Co., 258–289.

Woods, Robert (1979) *Population Analysis in Geography*. London and New York: Longman.

Woods, Robert (1982) *Theoretical Population Geography*. London and New York: Longman.

Woods, Robert and Rees, P. eds. (1986) *Population Structure and Models: Developments in Spatial Demography*. London: Allen and Unwin.

Zelinsky, Wilbur (1966) *A Prologue to Population Geography*. Englewood Cliffs, N.J.: Prentice-Hall, Inc.

Zwingle, Erla (1998) "Women and Population," *National Geographic* (4):36-55.

Chapter One References

Ahlburg, Dennis A. and Vaupel, James W. (1990) "Alternative Projections of the U.S. Population," *Demography* 27:639-652.

Barinaga, Marcia (1992) " 'African Eve' Backers Beat a Retreat," *Science* 255:686-687.

Berelson, Bernard and Freedman, Ronald (1974) "The Human Population," in Scientific American, eds. *The Human Population*. San Francisco: W. H. Freeman and Co., pp. 3–11.

Bouvier, Leon F. (1976) "On Population Growth," *Intercom* 4:8-9.

Bouvier, Leon F. (1992) *Peaceful Invasions: Immigration and Changing America*. Lanham, MD: University Press of America, Inc.

Bouvier, Leon F. and DeVita, Carol J. (1991) "The Baby Boom—Entering Midlife" *Population Bulletin* 46(3):1-33.

Callahan, Daniel (1971) *Ethics and Population Limitation*. New York: Population Council.

Campbell, Paul R. (1994) *Population Projections for States, by Age, Sex, Race, and Hispanic Origin: 1993–2020*. Current Population Reports P25–1111. Washington, D.C.: Bureau of the Census.

Cann, R. L., Stoneking, M., and Wilson, A.C. (1987) "Mitochondrial DNA and Human Evolution," *Nature* 325:31-36.

Cipolla, C. M. (1974) *The Economic History of World Population*. Baltimore: Penguin Books.

Darden, Joe T. (1975) "Population Control or a Redistribution of Wealth: A Dilemma of Class and Race," *Antipode* 7:50-52.

Day, Jennifer Cheeseman (1993) *Population Projections of the United States, by Age, Sex, Race, and Hispanic Origin: 1993–2050*. Current Population Reports P25–1104. Washington, D.C.: Bureau of the Census.

Deevey, Edward S. (1960) "The Human Population," *Scientific American* 203:3-9.

Demeny, Paul (1974) "The Populations of the Underdeveloped Countries," in Scientific American, eds. *The Human Population*. San Francisco: W. H. Freeman and Co., pp. 105–115.

Denevan, William M. (1996) "Carl Sauer and Native American Population Size," *Geographical Review* 86(3):385-397.

Diamond, Jared (1997) *Guns, Germs, and Steel: The Fates of Human Societies*. New York: W. W. Norton.

Easterlin, Richard, Wachter, Michael, and Wachter, Susan (1979) "The Coming Upswing in Fertility," *American Demographics* 1:12-15.

England, Robert (1987) "The Senior Citizen Secret," *Insight*, (March 2):8-11.

Francese, Paula A. and Renaghan, Leo M. (1991) "Finding the Customer," *American Demographics* 13(1):48-51.

Gibson, Campbell (1977) "Population Projections for the United States," *Intercom* 5:7-9.

Goldstein, Joshua R. and Schlag, Wilhelm (1999) "Longer Life and Population Growth," *Population and Development Review* 25(4):741-747.

Hardin, Garrett (1999) *The Ostrich Factor: Our Population Myopia*. New York:Oxford University Press.

Harrison, Paul and Pearce, Fred eds. (2000) *AAAS Atlas of Population and Environment*. Berkeley: University of California Press.

Hollmann, F. W., Mulder, T. J., and Kallan, J. E. (2000) *Methodology and Assumptions for the Population Projections of the United States: 1999–2100*. Population Division Working Paper No. 38. Washington, D.C.: U.S. Census Bureau.

Haub, Carl (1987) "Understanding Population Projections," *Population Bulletin* 42(4):1-41.

Haub, Carl (1992) "New UN Projections Show Uncertainty of Future World," *Population Today* 20(2):6-7.

Hooyman, Nancy R. and Kiyak, H. Asuman (1988) *Social Gerentology: A Multidisciplinary Perspective*, Boston, Ma.: Allyn and Bacon.

Hyatt, James (1979) *Changing Demographics and How They Affect the Future of Business*, FOB Series No. 9, Washington, D. C.: Center for Strategic and International Studies.

Irwin, Richard (1977) *Guide for Local Area Population Projections*. United States Bureau of the Census Technical Paper No. 39. Washington, D. C.: United States Government Printing Office.

Johanson, D. C. and Edey, M. A. (1981) *Lucy: The Beginnings of Humankind*. New York: Simon and Schuster.

Laws, Glenda (1991) "Aging Population: A Planning Dilemma," *Earth and Mineral Sciences* 60(2):32-36.

Lazer, William (1985) "Inside the Mature Market," *American Demographics* 7(3):3-25.

Leakey, Meave (1995) "The Dawn of Humans: The Farthest Horizon," *National Geographic* 188(3):38-51.

Leakey, Meave and Walker, Alan (1997) "Early Hominid Fossils from Africa," *Scientific American* 276(6)74-79.

Lee, Ronald (2000) "Long–term Population Projections and the US Social Security System," *Population and Development Review* 26(1):137-143.

Lord, Lewis (1997) "How Many People Were Here Before Columbus?" *U.S. News and World Report* 123(7):68-70.

Mitchell, J. D. (1998) "Before the Next Doubling," *World Watch* 11(1):20-27.

Nagel, John S. (1978) "Mexico's Population Policy Turnaround," *Population Bulletin* 33(5):1-39.

Nortmann, Dorothy, assisted by Hofstatter, Ellen (1975) "Population and Family Planning Programs: A Factbook," *Reports on Population/Family Planning*. No. 2. Seventh Edition. New York: The Population Council.

Ornstein, Robert and Ehrlich, Paul (1989) *New World, New Mind*. New York: Doubleday.

Peters, Gary L. (1980) "Population Projections: An Exercise for Population Geography," *Journal of Geography* 79:269-270.

Soldo, Beth J. and Agree, Emily M. (1988) "America's Elderly," *Population Bulletin* 43(3):1-53.

Spencer, Gregory (1980) "Measuring the Accuracy of U.S. Population Projections," *Intercom* 8(5):8-9.

Stringer, Christopher B. (1990) "The Emergence of Modern Humans," *Scientific American* 263(6):98-104.

Tattersall, Ian (1997) "Out of Africa Again . . . and Again?" *Scientific American* 276(4):60-67.

Tattersall, Ian (2000) "Once We Were Not Alone," *Scientific American* 282(1):56-62.

Tarver, James D. (1995) *The Demography of Africa*. Westport, CT: Praeger Publishers.

Thorne, Alan G. and Wolpoff, Milford H. (1992) "The Multiregional Evolution of Humans," *Scientific American* 266(4):76-83.

Tickell, Crispin (1993) "The Human Species: A Suicidal Success?" *The Geographical Review* 159(2):219-226.

United Nations (1974) Department of Economic and Social Affairs. Population Studies No. 56. *Concise Report on the World Population Situation in 1970–1975 and Its Long-Range Implications.* New York: The United Nations.

United Nations Population Division (2001) *World Population Projections: The 2000 Revision.* New York: Population Division, Department of Economic and Social Affairs, United Nations.

United Nations (1991) *Long-Range World Population Projections: Two Centuries of Population Growth, 1950–2150.* New York: United Nations.

United States Bureau of the Census (1989) *Projections of the Population of the United States, by Age, Sex, and Race: 1988 to 2080.* Current Population Reports, Series P-25, No. 1018. Washington, D.C.: United States Government Printing Office.

United States Bureau of the Census (1990) *Projections of the Population of States by Age, Sex, and Race: 1989 to 2010.* Current Population Reports, Series P-25, No. 1053. Washington, D.C.: United States Government Printing Office.

Waldrop, Judith (1991) "The Baby Boom Turns 45," *American Demographics* 13(1):22-27.

Wallerstein, Immanuel (1999) *The End of the World as We Know It: Social Science for the Twenty-First Century.* Minneapolis: University of Minnesota Press.

Wenke, Robert J. (1990) *Patterns in Prehistory.* Third Edition. Oxford: Oxford University Press.

Wilson, Allan C. and Cann, Rebecca L. (1992) "The Recent African Genesis of Humans," *Scientific American* 266(4):68-73.

Wong, Kate (2000) "Who Were the Neandertals?" *Scientific American* 282(4):98-107.

Woods, Robert (1979) *Population Analysis in Geography.* London: Longman.

Young, Louise B., ed. (1968) *Population in Perspective.* New York: Oxford University Press.

Zinsser, Hans (1967) *Rats, Lice, and History.* New York: Bantam Books, Inc.

Chapter Two References

Anderson, Margo J. (1988) *The American Census: A Social History*. New Haven, CT: Yale University Press.

Anderson, Margo J. and Fienberg, Stephen E. (1999) *Who Counts? The Politics Of Census-Taking in Contemporary America*. New York: Russell Sage.

Blacker, J. G. C. (1969) "Some Unsolved Problems of Census and Demographic Works in Africa," in *International Population Conference*. Vol. 1. London: United Nations.

Bogue, Donald J. (1969) *Principles of Demography*. New York: John Wiley and Sons, Inc.

Caldwell, J. and Igun, A. A. (1971) "An Experiment with Census-Type Age Enumeration in Nigeria," *Population Studies* 25:287-302.

Carr-Saunders, A. M. (1936) *World Population*. Oxford: The Clarendon Press.

Cole, K. C. (1997) "Counting on Inaccuracy in the U.S. Census," *Los Angeles Times* (Oct. 16): B2.

Crispell, Diane (1990) *The Insider's Guide to Demographic Know-How*. Second Edition. Ithaca, NY: American Demographics Press.

Fiore, Faye (1997) "Census Bureau Polishes Plan for Hard-to-Count Americans," *Los Angeles Times* (Dec. 2):A5.

Fitzpatrick, John C. ed. (1939) *The Writings of George Washington*. Washington, D.C.: United States Government Printing Office.

Francese, Peter K. (1979) "The 1980 Census: The Counting of America," *Population Bulletin* 34(4):1-39.

Hauser, Philip M. (1981) "The Census of 1980," *Scientific American* 245(5):53-61.

Hock, Saw Swee (1967) "Errors in Chinese Age Statistics," *Demography* 4:859-875.

Howenstine, Erick (1993) "Measuring Demographic Change: The Split Tract Problem," *The Professional Geographer* 45(4):425-430.

Hughes, James W. and Seneca, Joseph J. eds. (1999) *America's Demographic Tapestry: Baseline for the New Millennium*. New Brunswick, NJ: Rutgers University Press.

Kahn, E. J., Jr. (1974) *The American People*. Baltimore: Penguin Books, Inc.

Kaplan, Charles P. and Van Valey, Thomas L. (1980) *Census 80: Continuing the Factfinder Tradition*. Washington, D.C.: United States Government Printing Office.

Malsawma, Zuali H. (1998) "A Researcher's Guide to Population Information Web Sites," *Population Today* 26(2):4-5.

Myers, Dowell (1992) *Analysis with Local Census Data: Portraits of Change*. Boston: Academic Press.

Nichols, Judith E. (1990) *By the Numbers: Using Demographics and Psychographics for Business Growth in the '90s*. Chicago: Bonus Books, Inc.

Petersen, William (1975) *Population*. Third Edition. New York: MacMillan Publishing Co., Inc.

Prewitt, Kenneth (2000) "The US Decennial Census: Political Questions, Scientific Answers," *Population and Development Review* 26(1):1-16.

Pritzker, Leon and Rothwell, N. D. (1967) "Procedural Difficulties in Taking Past Censuses in Predominantly Negro, Puerto Rican, and Mexican Areas," paper presented for a Conference on Social Statistics and the City.

Robey, Bryant (1989) "Two Hundred Years and Counting: The 1990 Census," *Population Bulletin* 44(1):1-44.

Ryder, N. (1964) "Notes on the Concept of a Population," *American Journal of Sociology* 69:447-463.

Seltzer, William (1973) *Demographic Data Collection: A Summary of Experience*. New York: The Population Council.

Shryock, H. S., Jr. and Siegel, J. (1971) *The Methods and Materials of Demography*. Washington, D.C.: United States Government Printing Office.

Skerry, Peter (2000) *Counting on the Census? Race, Group Identity, and the Evasion of Politics*. Washington, D.C.: Brookings Institution Press.

Thomlinson, Ralph (1976) *Population Dynamics: Causes and Consequences of World Demographic Change*. Second Edition. New York: Random House.

United Nations (1954) Department of Economic Affairs, Statistical Office. *Handbook of Population Methods*. Studies in Methods. Series F, No. 5. New York: United Nations.

United Nations (1967) *Manual IV: Methods of Estimating Basic Demographic Measures from Incomplete Data.* Department of Economic Affairs. Population Studies No. 42. New York: United Nations.

United Nations (1969) *Principles and Recommendations for the 1970 Population Censuses.* Statistical Papers. Series M, No. 44. New York: United Nations.

United States Bureau of the Census (1971) *U.S. Census of Population: 1970.* "Number of Inhabitants: United States Summary." PC(1)-A1. Washington, D.C.: United States Government Printing Office.

United States Bureau of the Census (1990a) *Census '90 Basics.* CPH-I-8. Washington, D.C.: United States Bureau of the Census.

United States Bureau of the Census (1990b) *TIGER: The Coast-to-Coast Digital Map Data Base.* Washington, D.C.: United States Bureau of the Census.

United States Bureau of the Census (1991) *Census Catalog and Guide: 1991.* Washington, D.C.: United States Government Printing Office.

Whetten, Nathan L. (1961) *Guatemala: The Land and the People.* Caribbean Series 4. New Haven: Yale University Press.

Willcox, Walter F. (1940) *Studies in American Demography.* Ithaca, NY: Cornell University Press.

Wolter, Kirk M. (1991) "Accounting for America's Uncounted and Miscounted," *Science* 253:12-15.

Zelinsky, Wilbur (1966) *A Prologue to Population Geography.* Englewood Cliffs, NJ: Prentice-Hall, Inc.

Chapter Three References

Adepoju, Aderanti and Oppong, Christine. eds. (1994) *Gender, Work, and Population in Sub-Saharan Africa*. London: James Currey.

Alba, Richard D., et al (1995) "Neighborhood Change under Conditions of Mass Immigration: The New York City Region, 1970–1990," *International Migration Review* 29(3):625-656.

Allen, James Paul and Turner, Eugene James (1997) *The Ethnic Quilt: Population Diversity in Southern California*. Northridge: The Center for Geographical Studies, California State University, Northridge.

Allen, James Paul and Turner, Eugene James (1988) *We the People: An Atlas of America's Ethnic Diversity*. New York: Macmillan Publishing Company.

Beaujeu-Garnier, J. (1966) *Geography of Population*. New York: St. Martin's Press.

Bondi, Liz (1992) "Gender and Dichotomy," *Progress in Human Geography* 16:98-104.

Bondi, Liz (1993) "Gender and Geography," *Progress in Human Geography* 17:241-246.

Bourne, Larry S. and Ley, David F. (1993) *The Changing Social Geography of Canadian Cities*. Montreal: McGill-Queen's University Press.

Bouvier, Leon, Atlee, Elinore, and McVeigh, Frank (1975) "The Elderly in America," *Population Bulletin* 30(3):1-36.

Bouvier, Leon and de Vita, Carol J. (1991) "The Baby Boom—Entering Midlife," *Population Bulletin* 46(3):1-34.

Bouvier, Leon F. (1992) *Peaceful Invasions: Immigration and Changing America*. Lanham, MD: University Press of America, Inc.

Clarke, John J. (1972) *Population Geography*. Second Edition. Oxford: Pergamon Press.

Coale, Ansley J. (1964) "How a Population Ages or Grows Younger," in Ronald Freedman, ed., *Population: The Vital Revolution*. Garden City, New York: Anchor Books, Doubleday and Co., Inc., pp. 47–58.

Cowgill, Donald O. (1978) "Residential Segregation by Age in American Metropolitan Areas," *Journal of Gerontology* 33:446-453.

Crimmins, Eileen M. (2001) "Americans Living Longer, Not Necessarily Happier, Lives," *Population Today* 29(2):5 and 8.

Doyle, Rodger (1997) "By the Numbers: Female Illiteracy Worldwide," *Scientific American* 276(5):20.

Doyle, Rodger (1998) "By the Numbers: Women in Politics Throughout the World," *Scientific American* 278(1)35.

Eberstadt, Nicholas (2000) *Prosperous Paupers and Other Population Problems*. New Brunswick, NJ: Transaction Publishers.

Fuchs, Lawrence H. (1990) *The American Kaleidoscope: Race, Ethnicity, and the Civic Culture*. Hanover, NH: The University Press of New England.

Glazer, Nathan (1983) *Ethnic Dilemma*. New York: Hawthorne Press.

Gonzalez, Juan (2000) *Harvest of Empire: A History of Latinos in the Americas*. New York: Viking.

Gonzales, Juan L., Jr. (1990) *Racial and Ethnic Groups in America*. Dubuque, IA: Kendall/Hunt Publishing Company.

Gonzales, Juan L., Jr. (1990) *The Lives of Ethnic Americans*. Dubuque, IA: Kendall/Hunt Publishing Company.

Goodman, Allen C. (1987) "Using Lorenz Curves to Characterize Urban Elderly Populations," *Urban Studies* 24:77-80.

Guttentag, Marcia and Secord, Paul F. (1983) *Too Many Women? The Sex Ratio Question*. Beverly Hills, CA: Sage Publications.

Ingoldsby, Bron B. and Smith, Suzanna. eds. (1995) *Families in Multicultural Perspective*. New York: The Guilford Press.

Jaffe, A. J. (1992) *The First Immigrants from Asia: A Population History of the North American Indians*. New York: Plenum Books.

Jones, John Paul III, Nast, Heidi J., and Roberts, Susan M. (1997) *Thresholds in Feminist Geography: Difference, Methodology, Representation*. Lanham, MD: Rowman and Littlefield.

Lamphere, Louise. ed. (1992) *Structuring Diversity: Ethnographic Perspectives on the New Immigration*. Chicago: The University of Chicago Press.

Lee, Sharon M. (1998) "Asian Americans: Diverse and Growing," *Population Bulletin* 53(2):1-40.

McKee, Jesse O. (1985) *Ethnicity in Contemporary America: A Geographical Appraisal.* Dubuque, IA: Kendall/Hunt Publishing Company.

Monk, Janice (1994) "International Perspectives on Feminist Geography," *The Professional Geographer* 46:277-288.

Monk, Janice and Hanson, Susan (1982) "On Not Excluding Half of the Human in Human Geography," *The Professional Geographer* 34:11-23.

Morrill, Richard L. (1990) "Regional Demographic Structure of the United States," *The Professional Geographer* 42:38-53.

Murdock, Steve H. (1995) *An America Challenged: Population Change and the Future of the United States.* Boulder, CO: Westview Press.

Peters, Julie and Wolper, Andrea. eds. (1995) *Women's Rights, Human Rights: International Feminist Perspectives.* New York: Routledge.

Peterson, Peter G. (1999) *Gray Dawn: How the Coming Age Wave Will Transform America—and the World.* New York: Times Books.

Population Reference Bureau (1977) *Intercom* 5(3):6.

Roberts, Sam (1993) *Who We Are: A Portrait of America Based on the Latest U.S. Census.* New York: Times Books.

Rodriguez, Gregory (2001) "The Future Americans," *Los Angeles Times* (March 18):M1 and M6.

Rosenblatt, Roger (1996) "Come Together," *Modern Maturity* 39(1):32-34.

Russell, Cheryl (1998) "Boomers May Go Bust," *American Demographics* (August):14–17.

Smith, T. Lynn and Zopf, Paul E., Jr. (1976) *Demography: Principles and Methods.* Port Washington, NY: Alfred Publishing Co., Inc.

Soldo, Beth J. (1980) "America's Elderly in the 1980s," *Population Bulletin* 35(4):1-47.

Taylor, J. Edward and Martin, Philip L. (2000) "Central Valley Evolving into Patchwork of Poverty and Prosperity," *California Agriculture* 54(1):26-32.

Treas, Judith (1995) "Older Americans in the 1990s and Beyond," *Population Bulletin* 50(2):1-46.

Trewartha, Glenn J. (1969) *A Geography of Population: World Patterns.* New York: John Wiley and Sons, Inc.

Weeks, John R. (1992) *Population: An Introduction to Concepts and Issues.* Fifth Edition. Belmont, CA: Wadsworth Publishing Company.

White, Michael (1986) "Segregation and Diversity Measures in Population Distribution," *Population Index* 52(2):198-221.

Yaukey, David (1985) *Demography: The Study of Human Population.* New York: St. Martin's Press.

Chapter Four References

Abler, Ronald, Adams, John S., and Gould, Peter (1971) *Spatial Organization: The Geographer's View of the World*. Englewood Cliffs, NJ: Prentice-Hall, Inc.

Bach, Robert L. (1980) "The New Cuban Immigrants: Their Background and Prospects," *Monthly Labor Review* 103(10):39-46.

Beaver, Steven E. (1975) *Demographic Transition Theory Reinterpreted*. Lexington, MA.: Lexington Books.

Bianchi, Suzanne (1990) "America's Children: Mixed Prospects," *Population Bulletin* 45(1):1-43.

Boserup, Ester (1965) *The Conditions of Agricultural Growth: The Economics of Agrarian Change under Population Pressure*. Chicago: Aldine Publishing Company.

Boserup, Ester (1981) *Population and Technological Change: A Study of Long-Term Trends*. Chicago: The University of Chicago Press.

Brackett, James W. (1968) "The Evolution of Marxist Theories of Population: Marxism Recognizes the Problem," *Demography* 5:157-173.

Caldwell, John C. (1976) "Toward a Restatement of Demographic Transition Theory," *Population and Development Review* 2:321-366.

Caldwell, John C. (1997) "The Global Fertility Transition: The Need for a Unifying Theory," *Population and Development Review* 23(4):803-812.

Coale, Ansley J. and Watkins, Susan Cotts. Editors. (1986) *The Decline of Fertility in Europe*. Princeton, NJ: Princeton University Press.

Coleman, David and Schofield, Roger (1986) *The State of Population Theory: Forward from Malthus*. Oxford: Basil Blackwell.

Collver, O. Andrew (1965) *Birth Rates in Latin America: New Estimates of Historical Trends and Fluctuations*. Berkeley, CA: Institute of International Studies, University of California.

Danielson, Ross (1979) *Cuban Medicine*. New Brunswick, NJ: Transaction.

Davis, Kingsley. (1963) "The Theory of Change and Response in Modern Demographic History," *Population Index* 29:345-366.

Diaz-Briquets, Sergio (1977) "Mortality in Cuba: Trends and Determinants, 1880–1971," Unpublished Ph.D. dissertation. Philadelphia: University of Pennsylvania.

Diaz-Briquets, Sergio and Perez, Lisandro (1981) "Cuba: The Demography of Revolution," *Population Bulletin* 36(1):1-43.

Downie, Leonard, Jr. (1980) "Scandinavian Inflation Imperils Welfare States," *The Washington Post*, May 10, 1980.

Ehrlich, Paul (1968) *The Population Bomb*. New York: Ballantine Books.

Ehrlich, Paul R. and Ehrlich, Anne H. (1990) *The Population Explosion*. New York: Simon and Schuster.

Eversley, David R. (1959) *Social Theories of Fertility and the Malthusian Debate*. London: Oxford University Press.

Farnos Morejón, Alfonso (1980) "Results of Population Projections," *Revista Cubana de Administracion de Salud* 6:134.

Feeney, G. F., Wang, Fl., Zhou, M. K., and Xiao, B. Y. (1989) "Recent Fertility Dynamics in China: Results from the 1987 One Percent Survey," *Population and Development Review* 15:297-322.

Feng, Wang (1989) "China's One-Child Policy: Who Complies and Why?" (Paper delivered at the Annual Meeting of the Association for Asian Studies, Washington, D.C., March 17–19.)

Freedman, Ronald (1979) "Theories of Fertility Decline: A Reappraisal," in Philip M. Hauser, ed., *World Population and Development: Challenges and Prospects*. Syracuse: Syracuse University Press, pp. 63–79.

Gendell, Murray (1980) "Sweden Faces Zero Population Growth," *Population Bulletin* 35(2):1-44.

Goldstein, S. and Goldstein, A. (1991) *Permanent and Temporary Migration Differentials in China*. Papers, East-West Population Institute 117, Honolulu: East-West Center.

Gonzalez, Gerardo, Correa, German, Errazuriz, Margarita M., and Tapia, Raul (1978) *Development Strategy and Demographic Transition: The Case of Cuba*. Santiago, Chile: Centro Latinoamericano de Demografia.

Greenhalgh, Susan (1990) "Socialism and Fertility in China," *Annals of the American Academy of Political and Social Science* 510:73-86.

Harrison, Paul (1980) "Lessons for the Third World," *People* 7:2-20.

Hernandez, Jose (1974) *People, Power, and Policy: A New View on Population*. Palo Alto, CA: National Press Books.

Heuveline, Patrick (1999) "The Global and Regional Impact of Mortality and Fertility Transitions," *Population and Development Review* 25(4):681-702.

Hofsten, Erland and Lundstrom, Hans (1976) *Swedish Population History: Main Trends from 1750 to 1970*. Stockholm, Sweden: National Central Bureau of Statistics.

Hollerbach, Paula E. (1980) "Recent Trends in Fertility, Abortion, and Contraception in Cuba," *International Family Planning Perspectives* 6:97-106.

Holmberg, Ingvar (1977) "The Population Debate in Sweden in Recent Years," *Current Sweden* 169, Stockholm: The Swedish Institute.

Jacobson, Jodi L. (1991) "China's Baby Budget," in Lester R. Brown ed. *The World Watch Reader on Global Environmental Issues*. New York: W. W. Norton and Company.

Jaffee, Frederick S. and Oakley, Deborah (1978) "Observations on Birth Planning in China," *Family Planning Perspectives* 10:101-108.

Johansson, Sten and Nygren, Ola (1991) "The Missing Girls of China: A New Demographic Account," *Population and Development Review* 17:35-53.

Kennedy, Bingham, Jr. (2001) "Dissecting China's 2000 Census," http://prb.org/Regions/asia_near_east/Dissecting Chinas2000Census.html.

Kindleberger, Charles P. and Herrick, Bruce (1977) *Economic Development*. Third Edition. New York: McGraw-Hill Book Company.

Kuddo, A. (1997) "Determinants of Demographic Change in Transition Estonia," *Nationalities Papers* 25(4):625-642.

Malthus, T. R. (1798) *An Essay on the Principle of Population, as it Affects the Future Improvement of Society*. London: J. Johnson.

Malthus, T. R. (1803) *An Essay on the Principle of Population; or, A View of Its Past and Present Effect on Human Happiness*. London: J. Johnson.

McNicoll, Geoffrey (1995) "On Population Growth and Revisionism: Further Questions," *Population and Development Review* 21(2):307-340.

Merrick, Thomas W., with PRB Staff. (1986) "World Population in Transition," *Population Bulletin*, 41(2):1-51.

National Central Bureau of Statistics (1978) *Population Projection for Sweden, 1978–2025*. Stockholm, Sweden: Information for Prognosfragor.

Noin, Daniel and Woods, Robert. eds. (1993) *The Changing Population of Europe*. Oxford: Blackwell.

Pannell, Clifton W. and Torguson, Jeffrey S. (1991) "Interpreting Spatial Patterns from the 1990 China Census," *The Geographical Review* 81(3):304-317.

Perez, Lisandro (1977) "The Demographic Dimensions of the Educational Problem in Socialist Cuba," *Cuban Studies* 7:33-57.

Petersen, William (1999) *Malthus: Founder of Modern Demography*. New Brunswick, NJ: Transaction Publishers.

Population Reference Bureau (1981) *1981 World Population Data Sheet*. Washington, D.C.: Population Reference Bureau.

Reinhardt, Hazel H. (1979) "The Ups and Downs of Education," *American Demographics* 1(6):9-11.

Robey, Bryant (1993) "The Fertility Decline in Developing Countries," *Scientific American* 269(6):60-67.

Salas, Luis (1979) *Social Control and Deviance in Cuba*. New York: Praeger.

Smil, Vaclav (2000) *Feeding the World: A Challenge for the Twenty-First Century*. Cambridge, MA: MIT Press.

Strangeland, Charles E. (1904) *Pre-Malthusian Doctrines of Population*. New York: Columbia University Press.

State Statistical Bureau (1988) *Tabulations of China One Percent Population Survey, National Volume*. Beijing: China Statistical Press.

State Statistical Bureau (1990) "Geographical Distribution, Density and Natural Growth Rate of China's Population," *Beijing Review* 33(51):25-27.

Tabbarah, Riad B. (1976) "Population Education as a Component of Development Policy," *Studies in Family Planning* 7:197-201.

Teitelbaum, Michael S. (1975) "Relevance of Demographic Transition Theory for Developing Countries," *Science* 188:420-425.

Teitelbaum, Michael S. and Winter, Jay M. Editors. (1989) *Population and Resources in Western Intellectual Traditions*. Cambridge: Cambridge University Press.

Thomlinson, Ralph (1976) *Population Dynamics: Causes and Consequences of World Demographic Change*. Second Edition. New York: Random House.

Tien, H. Yuan (1983) "China: Demographic Billionaire," *Population Bulletin* 38(2):1-43.

Tien, H. Yuan (1988) "A Talk with China's Wang Wei," *Population Today* 16(1):6-8.

Tien, H. Yuan (1990) "China's Population Planning After Tiananmen," *Population Today* 18(9):6-8.

Tomaselli, Sylvana (1989) "Moral Philosophy and Population Questions in Eighteenth Century Europe," in Teitelbaum, Michael S. and Winter, Jay M. Editors. *Population and Resources in Western Intellectual Traditions*. Cambridge: Cambridge University Press, pp. 7–29.

Tomasson, Richard T. (1970) *Sweden: Prototype of Modern Society*, New York: Random House.

van de Walle, Etienne and Knodel, John (1980) "Europe's Fertility Transition: New Evidence and Lessons for Today's Developing World," *Population Bulletin* 34(6):1-43.

van de Walle, Etienne and Muhsam, Helmut V. (1995) "Fatal Secrets and the French Fertility Transition," *Population and Development Review* 21(2):261-279.

Watkins, Susan Cotts (1990) "From Local to National Communities: The Transformation of Demographic Regimes in Western Europe, 1870-1960," *Population and Development Review* 16:241-272.

Weeks, John R. (1992) *Population: An Introduction to Concepts and Issues*. Fifth Edition. Belmont, CA Wadsworth Publishing Company.

Wilson, Chris (2001) "On the Scale of Global Demographic Convergence 1950-2000," *Population and Development Review* 27(1):155-171.

Woods, Robert (1982) *Theoretical Population Geography*. London: Longman.

Wrigley, E. A. (1989) "The Limits to Growth: Malthus and the Classical Economists," in Teitelbaum, Michael S. and Winter, Jay M. Editors. *Population and Resources in Western Intellectual Traditions*. Cambridge: Cambridge University Press, pp. 30–48.

Zeng, Y. (1989) "Population Policy in China: New Challenge and Strategies," in J. M. Eekelaar and D. Pearl (eds.) *An Aging World*. Oxford: Oxford University Press, pp. 61–73.

Zeng, Yi, et al. (1991) "A Demographic Decomposition of the Recent Increase in Crude Birth Rates in China," *Population and Development Review* 17(3):435-459.

Chapter Five References

Anderson, W. French (1995) "Gene Therapy," *Scientific American* 273(3):124-128.

Barclay, George W. (1958) *Techniques of Population Analysis*. New York: John Wiley and Sons, Inc.

Birdsall, Stephen (1991) "Medical Geography," in H. J. deBlij and Peter O. Muller, *Geography: Regions and Concepts*. 6th Edition. New York: John Wiley and Sons, Inc., pp. 392–393.

Boyle, P., Muir, C. S., and Grundman, P. Eds. (1989) *Mapping and Cancer: Recent Results in Cancer Research 114*. Berlin: Springer Verlag.

Brown, Lester R. (1976) *World Population Trends: Signs of Hope, Signs of Stress*. Worldwatch Paper No. 8. Washington, D.C.: Worldwatch Institute.

Brown, J. Larry and Pollitt, Ernesto (1996) "Malnutrition, Poverty and Intellectual Development," *Scientific American* 274(2):38-43.

Buchbinder, Susan (1998) "Avoiding Infection after HIV Exposure," *Scientific American* 279(1):104-105.

Caldwell, John C. and Caldwell, Pat (1996) "The African AIDS Epidemic," *Scientific American* 274(3):62-68.

Caldwell, John C. (2000) "Rethinking the African AIDS Epidemic," *Population and Development Review* 26(1):117-135.

Caldwell, John C. (1997) "The Impact of the African AIDS Epidemic," *Health Transition Review* 7(2):169-188.

Callahan, Daniel (1998) *False Hopes: Why America's Quest for Perfect Health Is a Recipe for Failure*. New York: Simon and Schuster.

Carlson, Beverly A. and Wardlaw, Tessa M. (1990) *A Global, Regional, and Country Assessment of Child Malnutrition*. New York: UNICEF.

Cimons, Marlene (1998) "AIDS Falls Off List of Top 10 Killers in U.S.," *Los Angeles Times* (Oct. 8):A1 and A18.

Clayton, E. C. and Kaldor, J. (1989) "The Role of Advanced Statistical Methods in Cancer Mapping," in Boyle, et al., *Mapping and Cancer*. Berlin: Springer Verlag.

Cliff, A. D. and Haggett, P. (1988) *Atlas of Disease Distributions*. Oxford: Basil Blackwell.

Coates, Thomas J. and Collins, Chris (1998) "Preventing HIV Infection," *Scientific American* 279(1):96-97.

Condran, Gretchen A. and Cheney, Rose A. (1982) "Mortality Trends in Philadelphia: Age- and Cause-Specific Death Rates 1870–1930," *Demography* 18:94-124.

Cotter, John V. and Patrick, Larry L. (1981) "Disease and Ethnicity in an Urban Environment," *Annals of the Association of American Geographers* 71:40-49.

Crimmins, Eileen M. (1981) "The Changing Pattern of American Mortality Decline, 1940–77, and Its Implications for the Future," *Population and Development Review* 7:229-254.

Dutt, Ashok K. et al. (1987) "Geographical Patterns of AIDS in the United States," *The Geographical Review* Vol. 77, No. 4:456–471.

Ezzel, Carol (2000) "Care for a Dying Continent," *Scientific American* 282(5):96-105.

Ezzel, Carol (2000) "AIDS Drugs for Africa," *Scientific American* 283(5):98-103.

Garrett, Laurie (2000) *Betrayal of Trust: The Collapse of Global Public Health*. New York: Hyperion.

Garrett, Laurie (1994) *The Coming Plague: Newly Emerging Diseases in a World out of Balance*. NY: Farrar, Strauss and Giroux.

Gordimer, Nadine (2000) " 'Living with AIDS' Killing World Hopes," *Sacramento Bee* (December 4):B5.

Gould, P., Kabel, J., Gorr, W., and Golub, A. (1991) "AIDS: Predicting the Next Map," *Interfaces* 21:80-92.

Gould, P. (1991) "Editorial Report," *Science* 234:1022.

Grant, James P. (1990) *The State of the World's Children 1990*. New York and Oxford: Oxford University Press for UNICEF.

Gwatkin, Davison R. and Brandel, Sarah K. (1982) "Life Expectancy and Population Growth in the Third World," *Scientific American* 246, No. 5:57-65.

Halstead, Scott B., Walsh, Julia A., and Warren, Kenneth S. Eds. (1985) *Proceedings of a Conference held at the Bellagio Conference Center, Bellagio, Italy, 29 April–3 May 1985*. New York: The Rockefeller Foundation.

Jones, Kelvyn and Moon, Graham (1991) "Medical Geography," *Progress in Human Geography* 15:437-443.

Kearns, Robin A. (1993) "Place and Health: Toward a Reformed Medical Geography," *The Professional Geographer* 45(2):139-147.

Langridge, William H. R. (2000) "Edible Vaccines," *ScientificAmerican* 283(3):66-71.

Le Guenno, Bernard (1995) "Emerging Viruses," *Scientific American* 273(4):56-62.

Madigan, Francis C. (1957) "Are Sex Mortality Differentials Biologically Caused?" *Milbank Memorial Fund Quarterly* 35:202-223.

Mann, Jonathan M. and Tarantola, Daniel J. M. (1998) "HIV 1998: The Global Picture," *Scientific American* 279(1):82-83.

Marx, Jean L. (1987) "Probing the AIDS Virus and Its Relatives," *Science*, 230, (19 June 1987):1523.

McKeown, Thomas and Brown, R. G. (1969) "Medical Evidence Related to English Population Changes in the Eighteenth Century," in Michael Drake, Ed., *Population in Industrialization*. London: Methuen and Co., Ltd.

Mohan, J. (1988) "Restructuring, Privitisation and the Geography of Health Care Provision in England 1983–1987," *Transactions of the Institute of British Geographers* 13:449-465.

Mosley, W. Henry and Cowley, Peter (1991) "The Challenge of World Health," *Population Bulletin* 46:1-39.

National Center for Health Statistics (1990) "Advance Report of Final Natality Statistics, 1988," *Monthly Vital Statistics Report* 39(4):Supplement.

Olshansky, S. Jay and Alt, Brian (1986) "The Fourth Stage of the Epidemiologic Transition: The Stage of Delayed Degenerative Diseases," *The Milbank Quarterly* 64(3):355-391.

Paul, Bimal Kanti (1994) "AIDS in Asia," *The Geographical Review* 84(4):367-379.

Phillips, David R. (1990) *Health and Health Care in the Third World*. New York: John Wiley and Sons, Inc.

Preston, Richard (1994) *The Hot Zone*. New York: Random House.

Pyle, Gerald F. and Patterson, K. David (1987) "The Geography of Influenza," *Focus* 37(3):16-23.

Rockett, Ian R. H. (1999) "Population and Health: An Introduction to Epidemiology," *Population Bulletin* 54(4):1-44.

Rogers, Richard G. and Hackenberg, Robert (1987) "Extending Epidemiologic Transition Theory: A New Stage," *Social Biology* 34(3–4):234-243.

Rutstein, Shea O. (1991) "Levels, Trends, and Differentials in Infant and Child Mortality in the Less Developed Countries." Paper presented at the seminar on Child Survival Interventions: Effectiveness and Efficiency at The Johns Hopkins University School of Hygiene and Public Health, Baltimore, MD.

Ruzicka, Ldao T. and Lopez, Alan D. (1990) "The Use of Cause-of-Death Statistics for Health Situation Assessment: National and International Experiences," *World Health Statistical Quarterly* 43:249-257.

Senior, M. and Williamson, S. (1990) "An Investigation into the Influence of Geographical Factors on Attendance for Cervical Cytology Screening," *Transactions of the Institute of British Geographers* 15:421-434.

United Nations Development Programme (1990) *Human Development Report 1990*. New York: Oxford University Press.

Way, Peter O. and Stanecki, Karen A. (1994) *The Impact of HIV/AIDS on World Population*. Washington, D.C.: U.S. Bureau of the Census.

Wilson, James L. (1993) "Mapping the Geographical Diffusion of a Finnish Smallpox Epidemic from Historical Population Records," *The Professional Geographer* 45(3):276-285.

World Bank (1991) *World Development Report 1991*. New York: Oxford University Press.

Chapter Six References

Alexander, Nancy J. (1995) "Future Contraceptives," *Scientific American* 273(3):136-141.

Bianchi, Suzanne M. and Casper, Lynne M. (2000) "American Families," *Population Bulletin* 55(4):1-43.

Bledsoe, Caroline (1990) "Transformations in Sub-Saharan African Marriage and Fertility," *Annals of the American Academy of Political and Social Science* 510:115-125.

Bongaarts, John and Feeney, Griffith (1998) "On the Quantum and Tempo of Fertility," *Population and Development Review* 24(2):271-291.

Bumpass, Larry L. (1990) "What's Happening to the Family? Interactions Between Demographic and Institutional Change," *Demography* 27:483-498.

Cherlin, Andrew (1990) "Recent Changes in American Fertility, Marriage, and Divorce," *Annals of the American Academy of Political and Social Science* 510:145-154.

Cho, Lee-Jay (1973) *The Demographic Situation in the Republic of Korea.* Papers of the East-West Population Institute. Number 29. Honolulu: East-West Center.

Davis, Kingsley and Blake, Judith (1956) "Social Structure and Fertility: An Analytic Framework," *Economic Development and Cultural Change,* 4:211-235.

Davis, Kingsley, Bernstam, Mikhail S., and Ricardo-Campbell, Rita. eds. (1987) *Below-Replacement Fertility in Industrial Societies: Causes, Consequences, Policies.* Cambridge: Cambridge University Press.

Doring, Gerhard K. (1969) "The Incidence of Anovular Cycles in Women," *Journal of Reproduction and Fertility,* Supplement 6:77-81.

Espenshade, Thomas J. (1977) "The Value and Cost of Children," *Population Bulletin* 32, No. 1:1-47.

Ford, Nicholas and Bowie, Cameron (1989) "Urban-Rural Variations in the Level of Heterosexual Activity of Young People," *Area* 21:237-248.

Foster, Caroline (2000) "The Limits to Low Fertility: A Biosocial Approach," *Population and Development Review* 26(2):209-234.

Gober, Patricia (1994) "Why Abortion Rates Vary: A Geographical Examination of the Supply of and Demand for Abortion Services in the United States in 1988," *Annals of the Association of American Geographers* 84(2):230-250.

Goodstadt, Leo F. (1982) "China's One-Child Family: Policy and Public Response," *Population and Development Review* 8:37-58.

Hajnal, J. (1965) "European Marriage Patterns in Perspective," in David V. Glass and D. E. C. Eversley, eds. *Population in History.* Chicago: Aldine Publishing Company, pp. 101-143.

Hawthorn, Geoffrey (1970) *The Sociology of Fertility.* London: Collier-Macmillian Limited.

Hoffman, Lois W. and Hoffman, Martin L. (1973) "The Value of Children to Parents," in James W. Fawcett, ed. *Psychological Perspectives on Population.* New York: Basic Books, pp. 19-76.

Isaacs, Stephen L. and Holt, Renee J. (1987) "Redefining Procreation: Facing the Issues," *Population Bulletin* 42, No. 3:1-37.

James, W. H. (1966) "The Effect of Altitude on Fertility in Andean Countries," *Population Studies,* 20:97-101.

Jones, Elise F., et al. (1989) *Pregnancy, Contraception, and Family Planning Services in Industrialized Countries.* New Haven, CN: Yale University Press

Jones, Elise F. and Forrest, Jacqueline Darroch (1992) "Underreporting of Abortion in Surveys of U.S. Women: 1976 to 1988," *Demography* 29:113-126.

Kahn, Joan R. and Anderson, Kay E. (1992) "Intergenerational Patterns of Teenage Fertility," *Demography* 29:39-57.

Krishnan, Gopal (1989) "Fertility and Mortality Trends in Indian States," *Geography* 74:53-56.

Kulczycki, Andrzej (1995) "Abortion Policy in Postcommunist Europe: The Conflict in Poland," *Population and Development Review* 21(3):471-505.

Larsen, U. (1997) "Fertility in Tanzania: Do Contraception and Sub-fertility Matter? *Population Studies* 51(2):213-220.

Leete, Richard ed. (1999) *Dynamics of Values in Fertility Change.* Oxford: Oxford University Press.

Leibenstein, Harvey (1963) *Economic Backwardness and Economic Growth*. New York: John Wiley and Sons, Inc.

Lightbourne, Robert, Jr., Singh, Susheela, and Green, Cynthia (1982) "The World Fertility Survey: Charting Global Childbearing," *Population Bulletin* 37(1):1-54.

Lutz, Wolfgang (1989) *Distributional Aspects of Human Fertility: A Global Comparative Study*. New York: Academic Press.

Mamdani, Mahmood (1972) *The Myth of Population Control: Family, Caste, and Class in an Indian Village*. New York: Monthly Review Press.

Matsumoto, Y. Scott, Park, Chai Bin, and Bell, Balla Z. (1971) *Fertility Differentials of Japanese Women in Japan, Hawaii, and California*. Working Papers of the East-West Population Institute. Honolulu: East-West Center.

Mauldin, W. Parker (1976) "Fertility Trends: 1950–1975," *Studies in Family Planning* 7:242-248.

McDonald, Peter (2000) "Gender Equity in Theories of Fertility Transition," *Population and Development Review* 26(3):427-439.

McKenzie, Richard B. and Tullock, Gordon (1989) *The Best of the New World of Economics . . . and then some*. Fifth Edition. Homewood, IL: Richard D. Irwin.

Mott, Frank L. and Mott, Susan H. (1980) "Kenya's Record Population Growth: A Dilemma of Development," *Population Bulletin* 35(3):1-43.

National Research Council (1989) *Contraception and Reproduction: Health Consequences for Women and Children in the Developing World*. Working Group on the Health Consequences of Contraceptive Use and Controlled Fertility. Washington, D.C.: National Academy Press.

Nortman, Dorothy (1977) "Changing Contraceptive Patterns: A Global Perspective," *Population Bulletin* 32(3):1-37.

Orenstein, Peggy (2001) "Japan's 'Parasite Singles,'" *The Sacramento Bee* (July 15):L1 and L6.

Orleans, Leo A. (1976) "China's Population Figures: Can the Contradictions Be Resolved? *Studies in Family Planning* 7:52-57.

Palloni, Alberto (1990) "Fertility and Mortality Decline in Latin America," *Annals of the American Academy of Political and Social Science* 510:126-144.

Population Reference Bureau (1988) "A Billion More People per Decade?" *Population Today* 16(5):3.

Rendall, Michael S. and Bachieva, Raisa A. (1998) "An Old-Age Security Motive for Fertility in the United States?" *Population and Development Review* 24(2):293-307.

Robertson, A. F. (1991) *Beyond the Family: The Social Organization of Reproduction*. Berkeley and Los Angeles: University of California Press.

Robey, Bryant, Rutstein, Shea O., and Morris, Leo (1993) "The Fertility Decline in Developing Countries," *Scientific American* 269(6):60-67.

Sachs, Asron (1994) "Men, Sex, and Parenthood in an Overpopulating World," *World Watch* 7(2):12-19.

Saxton, G. A., Jr. and Serwada, D. M. (1969) "Human Birth Interval in East Africa," *Journal of Reproduction and Fertility*, Supplement 6:83-88.

Schultz, T. Paul. ed. (1995) *Investments in Women's Human Capital*. Chicago: University of Chicago Press.

Segal, Sheldon J. and Nordberg, Olivia Schieffelin (1977) "Fertility Regulation Technology: Status and Prospects," *Population Bulletin* 31(6):1-25.

Teitelbaum, Michael S. and Winter, Jay M. (1985) *The Fear of Population Decline*. Orlando, FL: Academic Press.

Thornton, Arland and Lin, Hui-Sheng (1994) *Social Change and the Family in Taiwan*. Chicago: University of Chicago Press.

Tietze, Christopher (1977) "Legal Abortions in the United States: Rates and Ratios by Race and Age, 1972–1974," *Family Planning Perspectives* 9:12-15.

Tietze, Christopher and Murstein, Marjorie Cooper (1975) "Induced Abortion: 1975 Factbook," *Reports on Population/Family Planning*, No. 14, Second Edition: 75 pages.

Ulmann, Andre, Teutsch, Georges, and Philibert, Daniel (1990) "RU 486," *Scientific American* 262(6):42-48

Chapter Seven References

Ashford, Lori S. (2001) "New Population Policies: Advancing Women's Health and Rights," *Population Bulletin* 56(1):1-43.

Berelson, Bernard (1972) "The President State of Family Planning Programs," in Harrison Brown and Edward Hutchings, eds. *Are Our Descendants Doomed? Technological Change and Population Growth.* New York: The Viking Press, pp. 201–236.

Berelson, Bernard, ed. (1974) *Population Policy in Developed Countries.* New York: McGraw-Hill Book Company.

Bongaarts, John (1986) "The Transition in Reproductive Behavior in the Third World," in Jane Menken, ed. (1986) *World Population and U.S. Policy.* New York: W. W. Norton, pp.105–132.

Bongaarts, John (1991) "The KAP-gap and the Unmet Need for Contraception," *Population and Development Review* 17:293-313.

Bongaarts, John (1997) "Trends in Unwanted Childbearing in the Developing World," *Studies in Family Planning* 28(4):267-277.

Butler, Steven (1998) "Japan's Baby Bust," *U.S. News and World Report* 125(13):42-44.

Caldwell, John C. (1994) "Fertility in Sub-Saharan Africa: Status and Prospects," *Population and Development Review* 20(1):179-187.

Corrêa, Sonia and Reichmann, Rebecca (1994) *Population and Reproductive Rights: Feminist Perspectives from the South.* London: Zed Books.

David, Henry P. (1982) "Eastern Europe: Pronatalist Policies and Private Behavior," *Population Bulletin* 36(6):1-48.

Donaldson, Peter J. and Tsui, Amy Ong (1990) "The International Family Planning Movement," *Population Bulletin* 45:.

Dow, Thomas E., Jr., et al. (1994) "Wealth Flow and Fertility Decline in Kenya, 1981–92," *Population and Development Review* 20(2):343-364.

Ehrlich, Paul R. and Ehrlich, Anne H. (1972) *Population, Resources, Environment: Issues in Human Ecology.* Second Edition. San Francisco: W. H. Freeman and Company.

Freedman, Ronald (1990) "Family Planning Programs in the Third World," *The Annals of the American Academy of Political and Social Science* 510:33-43.

Gillespie, Duff G. and Seltzer, Judith R. (1990) "The Population Assistance Program of the U.S. Agency for International Development," in Helen M. Wallace and Kanti Giri (eds.) *World Population and U.S. Policy.* New York: W. W. Norton and Company, pp. 75–206.

Jain, Anrudh ed. (1998) *Do Population Policies Matter?* New York: The Population Council.

King, Elizabeth M. and Hill, M. Anne. eds. (1993) *Women's Education in Developing Countries: Barriers, Benefits, and Policies.* Baltimore: Johns Hopkins University Press.

Knodel, John, Chamrathrithirong, A., and Debavalya, Nibhon (1987) *Thailand's Reproductive Revolution: Rapid Fertility Decline in a Third World Setting.* Madison, WI: University of Wisconsin Press.

Lapham, Robert J. and Mauldin, W. Parker (1987) "The Effects of Family Planning on Fertility: Research Findings," in Robert J. Lapham and George B. Simmons (eds.) *Organizing Effective Family Planning Programs.* Washington D.C.: National Academy Press, pp. 647–680.

Lee, Luke T. (1990) "Law, Human Rights, and Population Policy," in Godfrey Roberts (ed), *Population Policy: Contemporary Issues.* New York: Praeger.

Leete, Richard and Alam, Iqbal eds. (1993) *The Revolution in Asian Fertility: Dimensions, Causes, and Implications.* Oxford: Clarendon Press.

Li, Jiali (1995) "China's One-Child Policy: How and How Well has it Worked? A Case Study of Hebei Province, 1979–88," *Population and Development Review* 21(3):563-585.

Mackellar, F. L. (1997) "Population and Fairness," *Population and Development Review* 23(2):359-376.

Nortman, Dorothy (1973) "Population and Family Planning: A Factbook," *Reports on Population/Family Planning.* New York: The Population Council.

Nortman, Dorothy; and Hofstatter, Ellen (1976) "Population and Family Planning Programs," *Reports on Population/Family Planning.* No. 2. Eighth Edition. New York: The Population Council.

Nortman, Dorothy (1985) *Population and Family Planning Programs: A Compendium of Data Through 1983.* New York: Population Council.

Potts, Malcolm (2000) "The Unmet Need for Family Planning," *Scientific American* 282(1):88-93.

Presser, Harriet and Sen, Gita (2000) *Women's Empowerment and Demographic Processes: Moving Beyond Cairo.* New York: Oxford University Press.

Pritchett, Lant H. (1994) "Desired Fertility and the Impact of Population Policies," *Population and Development Review* 20(1):1-56.

Roberts, Godfrey Ed. (1990) *Population Policy: Contemporary Issues.* New York: Praeger.

Ross, John A. and Frankenberg, Elizabeth (1993) *Findings from Two Decades of Family Planning Research.* New York: The Population Council.

Ross, John A., et al. (1993) *Family Planning and Population: A Compendium of International Statistics.* New York: The Population Council.

Schroeder, Richard C. (1976) "Policies on Population Around the World," *Population Bulletin* 29(6):1-36.

Simmons, George (1986) "Family Planning Programs," in Jane Menken (ed) *World Population and U.S. Policy.* New York: W. W. Norton and Company, pp. 175-206.

Singh, Harbans (1990) "India's High Fertility Despite Family Planning: An Appraisal," in Godfrey Roberts (ed), *Population Policy: Contemporary Issues.* New York: Praeger, pp.

Thomlinson, Ralph (1976) *Population Dynamics: Causes and Consequences of World Demographic Change.* Second Edition. New York: Random House.

United Nations (1969) "World Population Situation," Note by the Secretary General. Geneva, Switzerland: United Nations Population Commission.

United Nations (1989) *Trends in Population Policy.* Population Studies No. 114, New York: United Nations.

United Nations (1990) *1990 Revision of World Population Prospects: Computerized Data Base and Summary Tables.* New York: United Nations.

van da Kaa, Dirk J. (1987) "Europe's Second Demographic Transition," *Population Bulletin,* 42(1):1-57.

van de Walle, Etienne and Knodel, John (1980) "Europe's Fertility Transition: New Evidence and Lessons for Today's Developing Countries," *Population Bulletin,* 34(6):1-43.

Witte, James C. and Wagner, Gert G. (1995) "Declining Fertility in East Germany After Unification: A Demographic Response to Socioeconomic Change," *Population and Development Review* 21(2):387-397.

Chapter Eight References

Alba, Richard D., et al. (1995) "Neighborhood Change under Conditions of Mass Immigration: The New York City Region, 1970–1990," *International Migration Review* 24(3):625-656.

Appleyard, Reginald ed. (1999) *Migration and Development*. Geneva: International Organization for Migration.

Bala, Raj (1986) *Trends in Urbanisation in India, 1901–1981*. Jaipur, India: Rawat Publication.

Beale, Calvin L. and Johnson, Kenneth M. (1995) "The Rural Rebound Revisited," *American Demographics* 17(7):46-54.

Bean, Frank D., Edmonston, Barry, and Passel, Jeffrey S. eds. (1990) *Undocumented Migration to the United States: IRCA and the Experience of the 1980s*. Santa Monica, CA: Rand Corporation.

Beaujeu-Garnier, J. (1966) *Geography of Population*. New York: St. Martin's Press.

Beier, George J. (1976) "Can Third World Cities Cope?" *Population Bulletin* 31(4):1-36.

Berry, Brian J. L. (1961) "City Size Distributions and Economic Development," *Economic Development and Cultural Change* 9:573-588.

Berry, Brian J. L. (1973) *The Human Consequences of Urbanization*. New York: St. Martin's Press.

Berry, Brian J. L. (1980) "Urbanization and Counter-urbanization in the United States," *Annals of the American Academy of Political and Social Sciences* 451:13-20.

Berry, Brian J. L. (1990) "Urban Systems by the Third Millennium: A Second Look," *Journal of Geography* 81(3):98-101.

Berry, Brian J. L. (1991) "How 'Sticky' is Urbanward Migration? Evidence for the United States, 1850–1980," *Urban Geography* 12(3):283-290.

Berry, Brian J. L. (1993) "Transnational Urbanward Migration, 1830–1980," *Annals of the Association of American Geographers* 83(3):389-405.

Berry, Brian J. L. and Dahmann, Donald C. (1977) "Population Redistribution in the United States in the 1970s," *Population and Development Review* 3:443-471.

Biggar, Jeanne (1979) "The Sunning of America: Migration to the Sunbelt," *Population Bulletin* 34(1):1-42.

Black, Richard (1991) "Refugees and Displaced Persons: Geographical Perspectives and Research Directions," *Progress in Human Geography* 15:281-298.

Blecher, Marc (1988) "Rural Contract Labour in Urban Chinese Industry," in Josef Gugler (ed.) *The Urbanization of the Third World*. Oxford: Oxford University Press.

Bogue, Donald J. (1969) *Principles of Demography*. New York: John Wiley and Sons, Inc.

Bolan, Marc (1997) "The Mobility Experience and Neighborhood Attachment," *Demography* 34(2):225-237.

Borjas, George J. (1999) *Heaven's Door: Immigration Policy and the American Economy*. Princeton, NJ: Princeton University Press.

Borjas, George J. (1994) "The Economics of Immigration," *Journal of Economic Literature* 32:1667-1717.

Bouvier, Leon F. (1992) *Peaceful Invasions: Immigration and Changing America*. Lanham, MD: University Press of America.

Bouvier, Leon F., Shryock, Henry S., and Henderson, Harry W. (1977) "International Migration: Yesterday, Today, and Tomorrow," *Population Bulletin* 32(4):1-42.

Breese, Gerald (1966) *Urbanization in Newly Developing Countries*. Englewood Cliffs, NJ: Prentice-Hall, Inc.

Brimelow, Peter (1995) *Alien Nation: Common Sense About America's Immigration Disaster*. New York: Random House.

Brockerhoff, Martin P. (2000) "An Urbanizing World," *Population Bulletin* 55(3):1-44.

Castles, Stephen and Miller, Mark J. (1993) *The Age of Migration: International Population Movements in the Modern World*. New York: The Guilford Press.

Clark, W. A. V. (1986) *Human Migration*. Beverly Hills: Sage Publications.

Connelly, Matthew and Kennedy, Paul (1994) "Must it be the Rest Against the West?" *The Atlantic Monthly* 274(6):61-91.

Conner, Roger L. (1989) "Answering the Demo-Doomsayers: Five Myths About America's Demographic Future," *The Brookings Review* 7(4):35-39.

Dale, Gareth and Cole, Mike eds. (1999) *The European Union and Migrant Labour*. New York: Berg.

Davin, Delia (1999) *Internal Migration in Contemporary China*. New York: St. Martin's Press.

Davis, Kingsley (1965) "The Urbanization of the Human Population," *Scientific American* 213(3):40-53.

Davis, Kingsley and Bernstam, Mikhail S. eds. (1991) *Resources, Environment, and Population*. New York: Oxford University Press.

DeJong, Gordon F. and Sell, Ralph R. (1977) "Population Redistribution, Migration, and Residential Preferences," *Annals of the American Academy of Political and Social Science* 429:130-144.

Desbarats, Jacqueline (1985) "Indochinese Resettlement in the United States," *Annals of the Association of American Geographers* 75(4):522-538.

Everitt, John (1991) "Refugees on Mainland Middle America," *The Canadian Geographer* 35(2):194-195.

Exline, Christopher H., Peters, Gary L., and Larkin, Robert P. (1982) *The City: Patterns and Processes in the Urban Ecosystem*. Boulder, CO: Westview Press.

Fairchild, Henry P. (1925) *Immigration*. New York: The Macmillan Company.

Farrag, M. (1997) "Managing International Migration in Developing Countries," *International Migration* 35(3):315-336.

Findlay, Allan (1989) "Skilled International Migration: A Research Agenda," *Area* 21(1):3-11.

Flannery, Tim (2001) *The Eternal Frontier: An Ecological History of North America and Its Peoples*. New York: Atlantic Monthly Press.

Fodor, Eben (1999) *Better not Bigger: How to Take Control of Urban Growth And Improve Your Community*. Stony Creek, CN: New Society Publishers.

Folger, John K. and Nam, Charles B. (1967) *Education of the American Population*. A 1960 Census Monograph. Washington, D.C.: United States Government Printing Office.

Fosler, R. Scott, et al. (1990) *Demographic Change and the American Future*. Pittsburgh: University of Pittsburgh Press.

Frey, William H. (1990) "Metropolitan America: Beyond the Transition," *Population Bulletin* 45(2):1-51.

Fuguitt, Glenn V. and Zuiches, James J. (1975) "Residential Preferences and Population Distribution," *Demography* 12:491-504.

Garreau, Joel (1991) *Edge City: Life on the New Frontier*. New York: Doubleday.

Gibson, Campbell (1975) "The Contribution of Immigration to United States Population Growth: 1790–1970," *International Migration Review* 9:157-177.

Giddens, Anthony (1982) *Profiles and Critiques in Social Theory*. Berkeley: University of California Press.

Giddens, Anthony (1984) *The Constitution of Society: Outline of the Theory of Structuration*. Cambridge: Polity Press.

Glasser, Jeff (2001) "A Broken Heartland," *U.S. News and World Report* (May 7):16-22.

Golden, Hilda H. (1981) *Urbanization and Cities: Historical and Comparative Perspectives on Our Urbanizing World*. Lexington, MA.: D. C. Heath.

Goldscheider, Calvin (1971) *Population, Modernization, and Social Structure*. Boston: Little, Brown and Company.

Gordenker, Leon (1987) *Refugees in International Politics*. New York: Columbia University Press.

Goss, Jon and Lindquist, Bruce (1995) "Conceptualizing International Labor Migration: A Structuration Perspective," *International Migration Review* 22(2):317-351.

Grauman, John V. (1976) "Orders of Magnitude of the World's Urban Population in History," *Population Bulletin of the United Nations* 8:16-33.

Gray, Paul (1994) "Looking for Work? Try the World," *Time* (Sept. 19):44-46.

Greenwood, Michael J. (1975) "Research on Internal Migration in the United States: A Survey," *Journal of Economic Literature* 12:397-433.

Gugler, Josef Ed (1988) *The Urbanization of the Third World*. Oxford: Oxford University Press.

Haggett, Peter (1979) *Geography: A Modern Synthesis*. Third Edition. New York: Harper and Row, Publishers.

Halfacree, Keith H. and Boyle, Paul J. (1993) "The Challenge Facing Migration Research: The Case for a Biographical Approach," *Progress in Human Geography* 17(3):333-348.

Hall, Peter (1980) "New Trends in European Urbanization," *Annals of the American Academy of Political and Social Science* 451:45-51.

Hamilton, Nora and Chinchilla, Norma Stoltz (1991) "Central American Migration: A Framework for Analysis," *Latin American Research Review* 26:75-110.

Hall, Peter (1980) "New Trends in European Urbanization," *Annals of the American Academy of Political and Social Science* 451:45-51.

Hart, John Fraser (1991) "The Perimetropolitan Bow Wave," *The Geographical Review* 81(1):35-51.

Hochschild, Arlie Russell (2000) "The Nanny Chain," *The American Prospect* 11(4):32-36.

Howell, Clark F. (1965) *Early Man*. New York: Time-Life Books.

Isbister, John (1996) *The Immigration Debate: Remaking America*. West Hartford, CN: Kumarian Press.

James, Daniel (1991) *Illegal Immigration: An Unfolding Crisis*. Lanham, MD: University Press of America.

Jefferson, Mark (1939) "The Law of the Primate City," *The Geographical Review* 29:226-232.

Keddie, Philip D. and Joseph, Alun E. (1991) "The Turnaround of the Turnaround? Rural Population Change in Canada, 1976 to 1986," *The Canadian Geographer* 35:367-379.

Kenzer, Martin S. (1991) "The African Refugee Crisis in Context," *The Canadian Geographer* 35(2):197-200.

King, Michael C. (1974) "The Malaise of Migrant Workers in Western Europe—with Special Reference to Switzerland," *Migration Today* No. 18:82-93.

Kliot, N. (1987) "The Era of Homeless Man," *Geography* 72:109-121.

Koser, K. and Salt, J. (1997) "The Geography of Highly Skilled International Migration," *International Journal of Population Geography* 3(4):285-303.

Kosinski, Leszek A. and Prothero, R. Mansell. eds. (1974) *People on the Move: Studies on Internal Migration*. London: Methuen and Co., Ltd.

Kunstler, James Howard (1993) *The Geography of Nowhere: The Rise and Decline of America's Man-Made Landscape*. New York: Simon and Schuster.

Lee, Everett S. (1966) "A Theory of Migration," *Demography* 3:47-57.

Leinbach, Thomas R., et al. (1992) "Employment Behavior and the Family in Indonesian Transmigration," *Annals of the Association of American Geographers* 82(1):23-47.

Lewis, Peirce (1985) "Beyond Description," *Annals of the Association of American Geographers* 75:465-477.

Lowry, Ira S. (1991) "World Urbanization in Perspective," in Kingsley Davis and Mikhail S. Bernstam eds. *Resources, Environment, and Population*. New York: Oxford University Press, pp. 148–179.

Luciuk, Lubomyr Y. (1991) "A Landscape of Despair: Comments on the Geography of the Contemporary Afghan Refugee Situation," *The Canadian Geographer* 35(2):195-197.

Martin, Philip and Midgley, Elizabeth (1999) "Immigration to the United States," *Population Bulletin* 54(2):1-44.

Massey, Douglas S. (1999) "International Migration at the Dawn of the Twenty-First Century: The Role of the State," *Population and Development Review* 25(2):303-322.

Massey, Douglas S. (1995) "The New Immigration and Ethnicity in the United States," *Population and Development Review* 21(3):631-651.

McHugh, Kevin E. (1989) "Hispanic Migration and Population Redistribution in the United States," *Professional Geographer* 41(4):429-439.

Meenan, J. F. (1958) "Eire," in Brinley Thomas, ed. *Economics of International Migration*. London: Macmillan.

Moe, Richard and Wilkie, Carter (1997) *Changing Places: Rebuilding Community in the Age of Sprawl*. New York: Henry Holt and Company.

Moore, Eric G. (1972) *Residential Mobility in the City*. Resource Paper No. 13. Washington, D.C.: Association of American Geographers.

Morrill, Richard L. (1979) "Stages in Patterns of Population Concentration and Dispersion," *Professional Geographer*, 31:55-65.

Myers, N. (1997) "Environmental Refugees," *Population and Environment* 19(2)167-182.

Nelson, P. (1959) "Migration, Real Income, and Information," *Journal of Regional Science* 1:43-74.

Newland, Kathleen (1994) "Refugees: The Rising Flood," *World Watch* (May/June):10–20.

Norris, Kathleen (1993) *Dakota: A Spiritual Geography*. New York: Houghton Mifflin Company.

Northam, Ray M. (1979) *Urban Geography*. Second Edition. New York: John Wiley and Sons, Inc.

Oberhauser, Ann M. (1991) "The International Mobility of Labor: North African Migrant Workers in France," *Professional Geographer* 43(4):431-445.

Oberlander, Hans (1992) "Mord an der Seele," *Stern* 14(26.Marz):20-32.

Page, Benjamin I. and Shapiro, Robert Y. (1992) *The Rational Public: Fifty Years of Trends in Americans' Policy Preferences*. Chicago: The University of Chicago Press.

Parfit, Michael (1998) "Human Migration," *National Geographic* (4):6-35.

Parnwell, Mike. (1993) *Population Movements and the Third World*. London: Routledge.

Paul, Bimal K. (1986) "Urban Concentration in Asian Countries: A Temporal Study," *Area* 18:299-306.

Petersen, William (1958) "A General Typology of Migration," *American Sociological Review* 23:256-265.

Petersen, William (1975) *Population*. Third Edition. London: The Macmillan Company.

Phillips, Phillip D. (1978) "New Patterns of American Metropolitan Population Growth in the 1970s," *Journal of Geography* 77:204-213.

Phizacklea, A. ed. (1983) *One Way Ticket? Migration and Female Labor*. London: Routledge and Kegan Paul.

Plane, David A. and Rogerson, Peter A. (1991) "Tracking the Baby Boom, the Baby Bust, and the Echo Generations: How Age Composition Regulates US Migration," *The Professional Geographer* 43(4):416-430.

Population Reference Bureau (1976) *World Population Growth and Response 1965–1975*. Washington, D.C.: Population Reference Bureau, Inc.

Potter, Robert B. (1985) *Urbanisation and Planning in the 3rd World*, New York: St. Martin's Press.

Ravenstein, Edward G. (1889) "The Laws of Migration," *Journal of the Royal Statistical Society* 52:241-305.

Richter, Kerry (1985) "Nonmetropolitan Growth in the late 1970s: The End of the Turnaround?" *Demography* 22:245-264.

Richter, Stephan-Götz (2000) "The Immigration Safety Valve," *Foreign Affairs* 79(2):13-16.

Ritchey, P. Neal (1976) "Explanations of Migration," *Annual Review of Sociology* 2:363-404.

Robinson, Isaac (1986) "Blacks Move Back to the South," *American Demographics* 8(6):40-43.

Rogge, John R. (1991) "Refugees in Southeast Asia: An Uncertain Future?" *The Canadian Geographer* 35(2):190-193.

Roseman, Curtis C. (1971) "Migration as a Spatial and Temporal Process," *Annals of the Association of American Geographers* 61:589-598.

Roseman, Curtis C. (1977) *Changing Migration Patterns Within the United States*. Resource Papers for College Geography No. 77–2. Washington, D.C.: Association of American Geographers.

Rossi, Peter (1955) *Why Families Move*. Glencoe, IL: Free Press.

Rudzitis, Gundars (1991) "Migration, Sense of Place, and Nonmetropolitan Vitality," *Urban Geography* 12(1):80-88.

Sachs, Aaron (1994) "The Last Commodity: Child Prostitution in the Developing World," *World Watch* (July/August):24-30.

Salt, John (1985) "Europe's Foreign Labour Migrants in Transition," *Geography* 70:151-158.

Sassen, Saskia (1988) *The Mobility of Labor: A Study in International Investment and Labor Flow*. Cambridge: Cambridge University Press.

Schacter, Jason (2001) "Geographical Mobility: March 1999 to March 2000," *Current Population Reports* P20–538. Washington, D.C.: U. S. Census Bureau.

Scheer, Robert (1998) "The Dark Side of the New World Order," *Los Angeles Times* (Jan. 13):B7.

Schwind, Paul J. (1971) "Spatial Preferences of Migrants for Regions: The Example of Maine," *Proceedings of the Association of American Geographers* 3:150-156.

Shaw, R. Paul (1975) *Migration Theory and Fact.* Philadelphia: Regional Science Research Institute.

Simkins, Paul D. (1970) "Migration as a Response to Population Pressure: The Case of the Philippines," in Wilbur Zelinsky, Leszek A. Kosinski, and Mansell R. Prothero, eds. *Geography and a Crowding World.* New York: Oxford University Press, pp. 259–267.

Simkins, Paul D. (1978) "Characteristics of Population in the United States and Canada," in Glenn T. Trewartha, ed. *The More Developed Realm: A Geography of Its Population.* Oxford: Pergamon Press, pp. 189–220.

Simon, Julian L. (1999) *The Economic Consequences of Immigration.* 2nd ed. Ann Arbor: University of Michigan Press.

Sjaastad, L. A. (1962) "The Costs and Returns of Human Migration," *The Journal of Political Economy.* 70:80-93.

Sjoberg, Gideon (1960) *The Preindustrial City.* New York: The Free Press.

Stockwell, Edward G. (1968) *Population and People.* Chicago: Quadrangle Books.

Svart, L. M. (1976) "Environmental Preference Migration: A Review, *Geographical Review* 66:314-330.

Tenhula, John (1991) *Voices from Southeast Asia: The Refugee Experience in the United States.* New York: Holmes and Meier.

Thomlinson, Ralph (1976) *Population Dynamics: Causes and Consequences of World Demographic Change.* Second Edition. New York: Random House.

Trewartha, Glenn T. (1969) *A Geography of Population.* New York: John Wiley and Sons, Inc.

Tyler, Charles (1991) "The World's Manacled Millions," *Geographical Magazine* 58(1):30-35.

Tyner, James A. (1996) "Constructions of Filipina Migrant Entertainers," *Gender, Place and Culture* 3(1):77-93.

Tyner, James A. (1998) "Asian Labor Recruitment and the World Wide Web," *The Professional Geographer* 50(3):331-344.

United Nations Population Division (1975) *Trends and Prospects in Urban and Rural Population, 1950–2000, as Assessed in 1973–74.* ESA/P/WP.54 New York: United Nations.

United Nations (1991) *World Urbanization Prospects.* New York: United Nations.

United States Agency for International Development (1973) Office of Population. *Population Program Assistance: Annual Report for Fiscal Year 1973.* Washington, D.C.: United States Agency for International Development.

United States Commission on Population Growth and the American Future (1972) *Population Distribution and Policy.* Sara Mills Mazie, ed. Volume V of Commission Research Reports. Washington, D.C.: United States Government Printing Office.

United States Department of Labor (1977) Employment and Training Administration. *Why Families Move: A Model of the Geographic Mobility of Married Couples.* R and D Monograph 48. Washington, D.C.: United States Government Printing Office.

Vernez, Georges and Ronfeldt, David (1991) "The Current Situation in Mexican Immigration," *Science* 251 (4998):1189-1193.

Vialet, J. C. (1997) "Immigration: Reasons for Growth," *Migration World* 25(5):14-17.

Walker, Robert and Hannan, Michael (1989) "Dynamic Settlement Processes: The Case of US Immigration," *Professional Geographer* 41(2):172-183.

Warren, Robert and Passel, Jeffrey S. (1987) "A Count of the Uncountable: Estimates of Undocumented Aliens Counted in the 1980 Census," *Demography* 24:375-394.

White, Paul; and Woods, Robert eds. (1980) *The Geographical Impact of Migration.* London and New York: Longman.

Wolpert, Julian (1965) "Behavioral Aspects of the Decision to Migrate," *Papers and Proceedings of the Regional Science Association* 15:159-169.

Wolpert, Julian (1966) "Migration as an Adjustment to Environmental Stress," *Journal of Social Issues* 22:92-102.

Yap, Loren (1975) *Internal Migration in Less Developed Countries: A Survey of the Literature*. World Bank Staff Working Paper No. 215. New York: The World Books.

Yaukey, David (1985) *Demography: The Study of Human Population*. New York: St. Martin's Press.

Zelinsky, Wilbur (1971) "The Hypothesis of the Mobility Transition," *The Geographical Review* 61:219-249.

Zipf, George K. (1949) *Human Behavior and the Principle of Least Effort*. New York: Addison-Wesley Press.

Zolberg, Aristide R., Suhrke, Astri, and Aguayo, Sergio (1989) *Escape from Violence: Conflict and the Refugee Crisis in the Developing World*. Oxford: Oxford University Press.

Chapter Nine References

Abernethy, Virginia (1992) "Editorial: Lessons from the Past," *Population and Environment* 13(4):235-236.

Ackerman, Edward A. (1967) "Population, Natural Resources, and Technology," *Annals of the American Academy of Political and Social Science* 369:84-97.

Athanasiou, Tom (1996) *Divided Planet: The Ecology of Rich and Poor.* Boston: Little, Brown.

Barney, Gerald O. Ed. (1980) *The Global 2000 Report to the President of the U.S.* Vol. 1, New York: Pergamon Press.

Barzun, Jacques (2000) *From Dawn to Decadence: 1500 to the Present.* New York: HarperCollins Publishers.

Blaustein, Andrew R. and Wake, David B. (1995) "The Puzzle of Declining Amphibian Populations," *Scientific American* 272(10):52-57.

Boulding, Kenneth E. (1966) *Human Values on the Spaceship Earth.* New York: National Council of Churches.

Brown, Lester R., McGrath, Patricia L., and Stokes, Bruce (1976) *Twenty-Two Dimensions of the Population Problem.* Worldwatch Paper No. 5, New York: Worldwatch Institute.

Bryson, Reid A. and Ross, John E. (1972) "The Climate of the City," in Thomas R. Detwyler and Melvin G. Marcus, eds. *Urbanization and Environment,* Belmont, CA: Duxbury Press, pp.1–68.

Calhoun, John B. (1962) "Population Density and Social Pathology," *Scientific American* 206:139-146.

Cole, H. S. D., Freeman, Christopher, Jahoda, Marie, and Pavitt, K. L. R. Eds. (1973) *Models of Doom: A Critique of the Limits to Growth.* New York: Universe Books.

Collins, Charles O. and Scott, Steven L. (1993) "Air Pollution in the Valley of Mexico," *The Geographical Review* 83(2):119-133.

Commoner, Barry (1990) *Making Peace With The Planet.* New York: Pantheon Books.

Commoner, Barry, Corr, Michael, and Stamler, Paul J. (1971) "The Causes of Pollution," *Environment* 13:2-19.

Cramer, James C. (1998) "Population Growth and Air Quality in California," *Demography* 35(1):45-56.

Cutter, Susan L. and Renwick, William H. (1999) *Exploitation, Conservation, Preservation: A Geographic Perspective on Natural Resource Use.* New York: John Wiley.

Dasmann, Raymond F., Milton, John P., and Freeman, Peter H. (1973) *Ecological Principles for Economic Development.* New York: John Wiley and Sons, Inc.

Davidson, Eric A. (2000) *You Can't Eat GNP: Economics as if Ecology Mattered.* Cambridge, MA: Perseus Publishing.

Davis, Kingsley (1991) "Population and Resources: Fact and Interpretation," in Kingsley Davis and Mikhail S. Bernstam (ed), *Resources, Environment, and Population.* New York: Oxford University Press, pp. 1–25.

de Sherbinin, Alex and Kalish, Susan (1994) "Population-Environment Links: Crucial, but Unwieldy," *Population Today* 22(1):1-2.

Durning, Alan (1991) "Cradles of Life," in Lester R. Brown (ed), *The Worldwatch Reader on Global Environmental Issues.* New York: W. W. Norton, pp. 167–188.

Eckholm, Erik P. (1976) *Losing Ground: Environmental Stress and World Food Prospects.* New York: W. W. Norton and Co.

Eden, Sally (1998) "Environmental Issues: Knowledge, Uncertainty, and the Environment," *Progress in Human Geography* 22(3):425-432.

Ehrlich, Paul R. and Holdren, John P. (1971) "Impact of Population Growth," *Science* 171:1212-1217.

Ehrlich, Paul R., and Ehrlich, Anne H. (1972) *Population, Resources, Environment.* Second Edition. San Francisco: W. H. Freeman and Co.

Ehrlich, Paul R. and Ehrlich, Anne H. (1990) *The Population Explosion.* New York: Simon and Schuster.

Eldredge, Niles (1998) "Life in the Balance," *Natural History* 107(5):42-53.

Espenshade, Thomas J. (1991) "Book Review of *The Population Explosion,*" in *Population and Development Review* 17(2):331-339.

Fagan, Brian (2000) *The Little Ice Age: How Climate Made History 1300–1850.* New York: Basic Books.

Flavin, Christopher (1991) "The Heat is On," in Lester R. Brown (ed) *The Worldwatch Reader on Global Environmental Issues*. New York: W. W. Norton, pp. 75–96.

Freedman, Jonathan L. (1975) *Crowding and Behavior*. San Francisco: W. H. Freeman and Co.

French, Hilary F. (1991) "You Are What You Breathe," in Lester R. Brown (ed) *The Worldwatch Reader on Global Environmental Issues*. New York: W. W. Norton, pp. 97–111.

Geping, Qu and Jinchang, Li (1994) *Population and the Environment in China*. Boulder: Lynne Rienner Publishers.

Gleick, Peter H. (2001) "Making Every Drop Count," *Scientific American* 284(2):40-45.

Goudie, Andrew (1986) *The Human Impact on the National Environment*, Second Edition. Cambridge, MA.: The MIT Press.

Hall, Edward T. (1969) *The Hidden Dimension*. Garden City, NJ: Anchor Books.

Hanink, Dean M. (1995) "The Economic Geography in Environmental Issues: a Spatial-Analytic Approach," *Progress in Human Geography* 19(3):372-387.

Hardoy, Jorge E. and Satterthwaite, David (1984) "Third World Cities and the Environment of Poverty," *Geoforum* 15(3):42-56.

Hern, Warren M. (1990) "Why Are There So Many of Us? Description and Diagnosis of a Planetary Ecopathological Process," *Population and Development* 12(1):9-40.

Hertsgaard, Mark (1997) "Our Real China Problem," *The Atlantic Monthly* 280(5):96-114.

Hill, M. K. (1997) *Understanding Environmental Pollution*. Cambridge: Cambridge University Press.

Hines, Lawrence G. (1973) *Environmental Issues; Population, Pollution and Economics*. New York: W. W. Norton and Co.

Hinrichsen, Don (1991) "The Need to Balance Population with Resources," *Populi* 18(3):27-38.

Homer-Dixon, Thomas F., Boutwell, Jeffrey H., and Rathjens, George W. (1993) "Environmental Change and Violent Conflict," *Scientific American* 268(2):38-45.

Kaplan, Robert D. (2000) *The Coming Anarchy: Shattering the Dreams of the Post Cold War*. New York: Random House.

Kaplan, Robert D. (1994) "The Coming Anarchy," *The Atlantic Monthly* 273(2):44-76.

Kennedy, Paul (1993) *Preparing for the Twenty-First Century*. NY: Random House.

Keyfitz, Nathan (1972) "Population Theory and Doctrine: A Historical Survey," in William Petersen (ed.), *Readings in Population*, New York: Macmillan.

Keyfitz, Nathan (1994) "Demographic Discord," *The Sciences* 34(5):21-27.

Lents, James M. and Kelly, William J. (1993) "Clearing the Air in Los Angeles," *Scientific American* 269(4):32-39.

Li, Jing-Neng (1991) "Population Effects on Deforestation and Soil Erosion in China," in Kingsley Davis and Mikhail S. Bernstam (eds), *Resources, Environment, and Population*. New York: Oxford University Press, pp. 254-258.

Loeb, Penny (1998) "Very Troubled Waters," *U.S. News and World Report* 125(12):39-43.

McKibben, Bill (1995) "An Explosion of Green," *The Atlantic Monthly* 275(4):61-83.

McKibben, Bill (1998) "A Special Moment in History," *The Atlantic Monthly* 281(5):55-78.

McNeill, J. R. (2000) *Something New under the Sun: An Environmental History of the Twentieth-Century World*. New York: W. W. Norton.

Meadows, Donella H., Meadows, Dennis L., Randers, Jorgen, and Behrens, William W., III (1972) *The Limits to Growth*. New York: Universe Books.

Miller, John (1997) *Egotopia: Narcissism and the New American Landscape*. Tuscaloosa: University of Alabama Press.

Moore, Thomas Gale (1998) *Climate of Fear: Why We Shouldn't Worry About Global Warming*. Washington, D.C.: Cato Institute.

Muir, Donald E. (1991) "Using Computers to Explore Ecological Issues: A Simple Limits-to Growth Educational Program," *Population and Environment* 13(2):113-117.

Mull, Illar (1989) "Use Them or Lose Them: A Recipe for Sustainable Use of Tropical Forests," *The UNESCO Courier*, January 1989, pp. 29-33.

National Academy of Sciences (1975) *Understanding Climatic Change*. Washington, D.C.: National Academy of Sciences.

National Academy of Sciences (1991) *Policy Implications of Greenhouse Warming*. Washington, D.C.: National Academy Press.

National Research Council (1985) *Epidemiology and Air Pollution*, Washington, D.C.: National Academy Press.

Park, Chris (1993) "Environmental Issues," *Progress in Physical Geography* 17(4):473-483.

Phillips, Kathryn (1994) *Tracing the Vanishing Tree Frogs: An Ecological Mystery*. New York: St. Martin's Press.

Piel, Gerard (1992) *Only One World: Our Own to Make and to Keep*. New York: W. H. Freeman and Co.

Quammen, David (1998) "Planet of Weeds: Tallying the Losses of Earth's Animals and Plants," *Harper's Magazine* 297(1781):57-69.

Repetto, Robert (1987) "Population, Resources, Environment: An Uncertain Future," *Population Bulletin* 42(2):1-44

Richards, Paul W. (1973) "The Tropical Rain Forest," *Scientific American* 232(6):58-67.

Ridker, Ronald G. (1980) *Resource and Environmental Consequences of Population and Economic Growth*. Reprint 172, Washington, D.C.: Resources for the Future, Inc.

Sadowski, Yahya M. (1998) *The Myth of Global Chaos*. Washington, D.C.: Brookings Institution Press.

Sauri-Punjol, David (1993) "Putting the Environment Back into Human Geography: A Teaching Experience," *Journal of Geography in Higher Education* 17(1):3-9.

Schelling, Thomas C. (1997) "The Cost of Combating Global Warming: Facing the Tradeoffs," *Foreign Affairs* 76(6):8-14.

Schneider, Stephen H. (1990) "Cooling It," *World Monitor*, July 1990, pp. 30–38.

Schneider, David (1998) "Good News for the Greenhouse," *Scientific American* 278(6):14.

Shabecoff, Philip (2000) *Earth Rising: American Environmentalism in the 21st Century*. Washington, D.C.: Island Press.

Shutkin, William A. (2000) *The Land That Could Be: Environmentalism and Democracy in the Twenty-First Century*. Cambridge, MA: MIT Press.

Smil, Vaclav (2000) "Rocky Mountain Visions: A Review Essay," *Population and Development Review* 26(1):163-176.

Tickell, Crispin (1993) "The Human Species: A Suicidal Success," *The Geographical Journal* 159(2):219-226.

United States Office of Technology Assessment (1984) *Acid Rain and Transported Air Pollutants: Implications for Public Policy*, Washington, D.C.: U.S. Government Printing Office.

von Lersner, Heinrich (1995) "Commentary: Outline for an Ecological Economy," *Scientific American* 273(3):188.

Warrick, Richard and Farmer, Graham (1990) "The Greenhouse Effect, Climatic Change and Rising Sea Level: Implications for Development," *Transactions of the Institute of British Geographers* NS15:5-20.

Wilbanks, Thomas J. (1994) " 'Sustainable Development' in Geographic Perspective," *Annals of the Association of American Geographers* 84(4):541-556.

Wilson, Edward O. (1993) *The Diversity of Life*. New York: W. W. Norton.

World Resources Institute and International Institute for Environment and Development (1987) *World Resources 1987*. New York: Basic Books.

Wright, Lawrence (1996) "Silent Sperm," *The New Yorker* (January 15):41-54.

Chapter Ten References

Abercrombie, Keith and McCormack, Arthur (1976) "Population Growth and Food Supplies in Different Time Perspectives," *Population and Development Review* 2:479-498.

Barton, Jonathan R. and Staniford, Don (1998) "Net Deficits and the Case for Aquacultural Geography," *Area* 30(2):145-155.

Beardsley, Tim (1997) "In Focus: Death in the Deep," *Scientific American* 277(5):17-18.

Biswas, Asit K. (1985) "Evaluating Irrigation's Impact: Guidelines for Project Monitoring," *Ceres*, July/August 1985.

Boyd, Claude E. and Clay, Jason W. (1998) "Shrimp Aquaculture and the Environment," *Scientific American* 278(6):58-65.

Brown, Lester R. (1976) *World Population Trends: Signs of Hope, Signs of Stress.* Worldwatch Paper No. 8. Washington, D.C.: Worldwatch Institute.

Brown, Lester R. (1981) "World Food Resources and Population: The Narrowing Margin," *Population Bulletin* 36(3):1-43.

Brown, Lester (1987) *State of the World—1987*, New York: W. W. Norton and Company.

Brown, Lester R. (1989) "Feeding Six Billion," *Worldwatch*, September/October: 32–40.

Brown, Lester R. (1995) *Who Will Feed China?* New York: W. W. Norton and Company.

Brown, Lester R. and Young, John E. (1990) "Feeding the World in the Nineties," in Lester R. Brown, *State of the World 1990.* New York: W. W. Norton and Company, pp. 59–78.

Dalrymple, Dana G. (1986) *Development and Spread of High-yielding Rice Varieties in Developing Countries,* Washington, D.C.: U.S. Agency for International Development.

Durning, Alan and Brough, Holly (1992) "Reforming the Livestock Industry," in Lester R. Brown, (ed) *State of the World 1992.* NY: W.W. Norton and Company, pp. 66–83.

Ehrlich, Anne H. (1991) "People and Food," *Population and Environment* 12(3):221-231.

Falkenmark, Malin and Suprapto, Riga Adiwoso (1992) "Population-Landscape Interactions in Development: A Water Perspective to Environmental Sustainability," *Ambio* 21(1):31-36.

FAO (1987) *Agriculture: Toward 2000.* Rome, Italy: FAO.

Glenn, Edward P., Brown, J. Jed, and O'Leary, James W. "Irrigating Crops with Seawater," *Scientific American* 279(2):76-81.

Grigg, David (1980) *Population Growth and Agrarian Change: An Historical Perspective.* Cambridge: Cambridge University Press.

Grigg, David (1995) "The Geography of Food Consumption," *Progress in Human Geography* 19(3):338-354.

Harlan, Jack R. (1976) "The Plants and Animals that Nourish Man," *Scientific American* 235(3):88-97.

Hendry, Peter (1988) "Food and Population: Beyond Five Billion," *Population Bulletin* 43(2):1-40.

Hinrichsen, Don and Marshall, Alex (1991) "Population and the Food Crisis," *Populi* 16(2):24-34.

Hopper, Gordon R. (1999) "Changing Food Production and Quality of Diet in India, 1947–98," *Population and Development Review* 25(3):443-477.

Knight, C. Gregory and Wilcox, R. Paul (1976) *Triumph or Triage: The World Food Problem in Geographical Perspective.* Resource Paper No. 75–3. Washington, D.C.: Association of American Geographers.

Le Heron, Richard and Roche, Michael (1995) "A 'Fresh' Place in Food's Space," *Area* 27(1):23-33.

McGinn, A. P. (1998) "Blue Revolution: The Promises and Pitfalls of Fish Farming," *World Watch* 11(2):10-19.

Paddock, William and Paddock, Paul (1976) *Time of Famines: America and the World Food Crisis.* Boston: Little, Brown and Company.

Plucknett, Donald L. and Winkelmann, Donald L. (1995) "Technology for Sustainable Agriculture," *Scientific American* 273(3):182-186.

Postel, Sandra (2001) "Growing More Food with Less Water," *Scientific American* 284(2):46-51.

Reid, T.R. (1998) "Feeding the Planet," *National Geographic* (4):56-74.

Reutlinger, Shlomo (1985) "Food Security and Poverty in LDC's," *Finance and Development*, December 1985.

Ronald, Pamela C. (1997) "Making Rice Disease-Resistant," *Scientific American* 277(5):100-105.

Sanderson, Fred H. (1975) "The Great Food Fumble," *Science* 188:503-509.

Scrimshaw, Nevin S. and Taylor, Lance (1980) "Food," *Scientific American* 243(3):78-88.

Simon, Arthur (1975) *Bread for the World*. New York: Paulist Press.

Simon, Julian R. (1983) "Life on Earth is Getting Better, Not Worse," *The Futurist*, August: 7–12.

Smil, Vaclav (1997) "Global Population and the Nitrogen Cycle," *Scientific American* 277(1):76-81.

Smil, Vaclav (2000) *Feeding the World: A Challenge for the Twenty-First Century.* Cambridge, MA: MIT Press.

Smith, David W. (1998) "Urban Food Systems and the Poor in Developing Countries," *Transactions of the Institute of British Geographers* 23(2):207-220.

Tarrant, John R. (1990) "World Food Prospects for the 1990's," *Journal of Geography* 89(6):234-238.

Tempest, Rone (1997) "The Coexistence of Feast, Famine," *Los Angeles Times* (Jan. 22) A1 and A16–A17.

United States Department of Agriculture (1974) Economic Research Service. *The World Food Situation and Prospects to 1985.* Foreign Agricultural Economic Report No. 98. Washington, D.C.: United States Government Printing Office.

Waterlow, J. C., et al. eds. (1998) *Feeding a World Population of More Than Eight Billion People: A Challenge to Science.* Oxford: Oxford University Press.

Wolf, Edward C. (1987) "Raising Agricultural Productivity," in Lester Brown, ed., *State of the World—1987*, New York: W. W. Norton.

Woods, Richard G. ed. (1981) *Future Dimensions of World Food and Population.* A Winrock International Study. Boulder, CO: Westview Press.

Wortman, Sterling (1976) "Food and Agriculture," *Scientific American* 235(3):30-39.

Young, John E. (1992) "Bred for the Hungry," in Robert M. Jackson, (ed) *Global Issues 1992/1993*, Guilford, CT: Dushkin Publishing Company, pp. 97-104.

GLOSSARY

Abortion rate The estimated number of abortions in a given year per 1000 women aged 15–44.

Age-sex structure The composition of a population as determined by the number or portion of males and females in each age category. This information is an essential prerequisite for the description and analysis of many other types of demographic data. See also *population pyramid*.

Age-specific rate Rate of population change obtained for a specific age group (for example, age-specific fertility rate, death rate, marriage rate, illiteracy rate, etc.)

Age structure The percentage of people in various age groups in a population. It is best described by a population pyramid.

Antinatalist policy The policy of a society, social group, or government to slow population growth by attempting to limit the number of births.

Baby boom The period following World War II from 1946–1964 characterized by a rapid increase in fertility rates and in the absolute number of births in the U.S., Canada, Australia, and New Zealand.

Baby bust The period immediately after the "baby boom" characterized by a rapid decline in U.S. fertility rates to record low levels.

Basic demographic equation An equation that ties together population change through the major demographic variables: deaths, births, and migration. The equation views the population of a place at some future date as a function of its present population plus births over the time interval, minus deaths over the time interval, plus in-migrants and minus out-migrants over the time interval.

Birth control Practices adopted by couples that permit sexual intercourse with reduced likelihood of conception. The term is frequently used synonymously with such terms as contraception, family planning, and fertility control.

Brain drain The emigration of a significant proportion of a country's highly educated, highly skilled professional population, usually to other countries offering better social and economic opportunities.

Carrying capacity The number of people who can be sustained by a particular land area at a given technological level.

Child-woman ratio The number of children under 5 years old per 1000 women aged 15–49 years in a population.

Cohort A group of people sharing a common temporal demographic experience who are observed through time. For example, the birth cohort of 1950 would be the people born in that year. There are also marriage cohorts, school class cohorts, etc.

Communist-socialist theory The idea, originally expounded by Karl Marx, that economic problems are not associated with population growth but are the result of the maldistribution of resources.

Completed fertility rate The number of children born per woman to a cohort of women by the end of their childbearing years.

Crude birth rate The annual number of births per one thousand people in a given population.

Crude death rate The annual number of deaths per one thousand people in a given population.

Crude divorce rate The number of divorces per 1000 population in a given year.

Demographic transition The transition from a demographic regime characterized by high birth and death rates to one characterized by low birth and death rates.

Demography The scientific study of the human population, focusing on population composition, change, and distribution.

Dependency ratio   The ratio of persons in the dependent ages (under 20 and over 64 years) to those in the economically productive years (20 to 64 years) in a population.

Developed countries   Countries that have industrialized and have relatively high levels of per capita income and per capita productivity.

Developing countries   Countries that have not industrialized and have relatively low rates of per capita income and per capita productivity.

Doubling time   The number of years required for a population of an area to double its present size, given the current rate of population growth. It is possible to closely approximate the doubling time for a population by dividing the annual rate of population growth into the number 70.

Emigration   The process of leaving one country to take up residence in another.

Emigration rate   The number of emigrants departing an area of origin per 1000 inhabitants at that area of origin in a given year.

Environmental population theory   Attempts to explain demographic behavior by a theory centered on cultural processes.

Epidemiologic transition   The transition a society goes through in which death rates decline and the major causes of death shift from communicative diseases to degenerative diseases.

Family planning   Policies and programs designed to help families achieve their desired family size.

Fecundity   The physiological capacity of a woman, man, or couple to produce a child.

Fertility   The actual reproductive performance of an individual, a couple, a group, or a population.

Fertility trend   Any observed or predicted course of fertility over a specified period of time.

Forced migration   A movement of individuals or groups of people in which they participated without choice. The slave trade is a major example.

General fertility rate   The annual number of live births per 1000 women in the childbearing age group (ages 15–44).

Green revolution   A rapid increase in agricultural output resulting from the application of new agricultural practices and the use of "miracle strains" of wheat and rice.

Gross migration   The total sum of all the people who enter and leave an area.

Gross reproduction rate   The average number of daughters a woman would bear if she passed through her entire reproductive life at the prevailing age-specific fertility rates.

Growth rate   The rate at which a population is increasing (or decreasing) in a given year due to natural increase and net migration, expressed as a percentage of the base population.

Growth syndrome   It involves three primary characteristics: continuing growth of the human population in both developed and developing countries, rapid increase in the production and consumption of commodities, and misapplication of old and new technologies, leading to an increasing abuse and pillage of the human habitat.

Immigration   The movement of people into one country from another.

Infant mortality rate   The number of deaths among infants under one year of age in a given year per 1000 live births in that year.

Laissez-faire solution   A solution to population issues based on individual choice. Laissez-faire population exponents feel that those best able to decide on the costs and benefits of children are those who are contemplating having them.

Lee's migration model   A model developed to explain migration as the result of four sets of factors: (1) factors associated with place of origin, (2) factors associated with place of destination, (3) intervening obstacles, and (4) personal factors.

Life expectancy   The average number of additional years a person would live if current mortality trends were to continue.

Life table   A table that gives life expectancy and the probability of dying at each age for a given population, according to the age-specific death rates prevailing at that time.

Median age   The age that divides a population into two numerically equal groups with half the people younger and half older than this age.

Migration   The movement of people from place to place, usually across some political boundary, for the purpose of changing their permanent place of residence.

Migration differentials Characteristics that make migrants different from the general population. Migrants are not random samples of the population from which they came but are "selected" by certain characteristics such as age or educational level.

Migration interval The time interval over which migration is studied or observed.

Migration stream An established movement of people between two places. For example, there is a migration stream between New York and Miami.

Morbidity The frequency of disease and illness in a population.

Mortality Deaths as a component of population change.

Multiple-round survey A sample survey that is longitudinal in nature, whereby information about a specific group or problem is collected on a continuing basis.

Natality Births as a component of population change.

Natural increase The balance between births and deaths in a population in a given time period.

Neo-Malthusian A current revival of interest in the ideas of Malthus. Neo-Malthusians believe that population growth will outrun the food supply and that the world cannot continue to support a growing population.

Neonatal mortality rate The number of deaths to infants under 28 days of age in a given year per 1000 live births in that year.

Net migration The difference between immigration and emigration for an area's population in a given time period; generally expressed as an increase or decrease.

Net reproduction rate The average number of daughters that would be born to a woman if she passed through her lifetime from birth at the prevailing age-specific fertility rates, adjusted for current female mortality rates.

Physiological density The ratio of population to arable land.

Population A group of objects or organisms of the same kind.

Population density Generally refers to the number of people per unit of land.

Population distribution The pattern of settlement and dispersal of a population.

Population equilibrium The level at which the population is in balance with the natural resource base of an area. Because of the changing nature of technology and the resource base, it is difficult to translate population equilibrium into a precise number.

Population policy An official government policy specifically designed to affect the size and growth rate of a population, the distribution of a population, or its composition. See *population program*.

Population program The various means and measures that must be utilized in order to achieve the objectives of a population policy. See *population policy*.

Population projection An estimate of the population of an area for some future date, based on a set of assumptions about the future course of births, deaths, and migration.

Population pyramid A graph showing the age and sex structure of a population. It may be constructed for nations, states, cities, or even smaller areas.

Primary population theory A theory that explains demographic behavior patterns by focusing on specific factors directly related to fertility, migration, or mortality.

Primate city A city that dominates the state in size and importance. It is commonly at least twice as large as the next largest city, and it is the educational, political, industrial, and commercial center.

Primitive migration The type of migration associated with groups that migrate because they are unable to cope with natural forces related to their physical environment.

Sample survey A canvass of selected persons or households in a population usually used to infer demographic characteristics or trends for a larger segment or all of the population.

Secondary population theory A theory that explains demographic behavior patterns by analyzing a broad class of social and physical phenomena that have demographic implications.

Sex ratio The number of males per 100 females in a population.

Total fertility rate The average number of children that would be born to a woman during her lifetime if she were to pass through her childbearing years at the prevailing age-specific fertility rates.

Triage The idea that the world's nations can be put into three categories relative to their food resources: (1) those nations that cannot be saved regardless of food aid from other countries, (2) those nations that will survive without food aid from other countries, and (3) those nations that could feed their people with some aid from the developed countries.

Urbanism The characteristics or conditions of the way of life of those who live in cities.

Urbanization The percentage of a population that lives in urban areas and the process by which this number increases over time.

Vital registrations The recording and compilation of vital statistics, at or near their time of occurrence. They usually include such events as deaths, fetal deaths, births, marriages, divorces, and at times disease and illness.

Vital statistics Demographic data on births, deaths, fetal deaths, marriages, and divorces.

Zero population growth (ZPG) An equilibrium population with a growth rate of zero, achieved when births plus immigration equal deaths plus emigration.

Appendix A:
Census 2000 Form

United States Census 2000

U.S. Department of Commerce
Bureau of the Census

This is the official form for all the people at this address. It is quick and easy, and your answers are protected by law. Complete the Census and help your community get what it needs — today and in the future!

Start Here

Please use a black or blue pen.

● **How many people were living or staying in this house, apartment, or mobile home on April 1, 2000?**

| | Number of people

INCLUDE in this number:

- foster children, roomers, or housemates
- people staying here on April 1, 2000 who have no other permanent place to stay
- people living here most of the time while working, even if they have another place to live

DO NOT INCLUDE in this number:

- college students living away while attending college
- people in a correctional facility, nursing home, or mental hospital on April 1, 2000
- Armed Forces personnel living somewhere else
- people who live or stay at another place most of the time

● **Please turn the page and print the names of all the people living or staying here on April 1, 2000.**

If you need help completing this form, *call 1–800–471–9424 between 8:00 a.m. and 9:00 p.m., 7 days a week. The telephone call is free.*

TDD – *Telephone display device for the hearing impaired. Call 1–800–582–8330 between 8:00 a.m. and 9:00 p.m., 7 days a week. The telephone call is free.*

¿NECESITA AYUDA? *Si usted necesita ayuda para completar este cuestionario llame al 1–800–471–8642 entre las 8:00 a.m. y las 9:00 p.m., 7 días a la semana. La llamada telefónica es gratis.*

The Census Bureau estimates that, for the average household, this form will take about 38 minutes to complete, including the time for reviewing the instructions and answers. Comments about the estimate should be directed to the Associate Director for Finance and Administration, Attn: Paperwork Reduction Project 0607-0856, Room 3104, Federal Building 3, Bureau of the Census, Washington, DC 20233.

Respondents are not required to respond to any information collection unless it displays a valid approval number from the Office of Management and Budget.

OMB No. 0607-0856: Approval Expires 12/31/2000

Please be sure you answered question 1 on the front page before continuing.

Please print the names of all the people who you indicated in question 1 were living or staying here on April 1, 2000.

Example — Last Name

| J | O | H | N | S | O | N | | | | | | | | |

First Name — MI

| R | O | B | I | N | | | | | | | | | | | J |

Start with the person, or one of the people living here who owns, is buying, or rents this house, apartment, or mobile home. If there is no such person, start with any adult living or staying here.

Person 1 — Last Name

| | | | | | | | | | | | | | | | |

First Name — MI

| | | | | | | | | | | | | | | | |

Person 2 — Last Name

| | | | | | | | | | | | | | | | |

First Name — MI

| | | | | | | | | | | | | | | | |

Person 3 — Last Name

| | | | | | | | | | | | | | | | |

First Name — MI

| | | | | | | | | | | | | | | | |

Person 4 — Last Name

| | | | | | | | | | | | | | | | |

First Name — MI

| | | | | | | | | | | | | | | | |

Person 5 — Last Name

| | | | | | | | | | | | | | | | |

First Name — MI

| | | | | | | | | | | | | | | | |

Person 6 — Last Name

| | | | | | | | | | | | | | | | |

First Name — MI

| | | | | | | | | | | | | | | | |

Person 7 — Last Name

| | | | | | | | | | | | | | | | |

First Name — MI

| | | | | | | | | | | | | | | | |

Person 8 — Last Name

| | | | | | | | | | | | | | | | |

First Name — MI

| | | | | | | | | | | | | | | | |

Person 9 — Last Name

| | | | | | | | | | | | | | | | |

First Name — MI

| | | | | | | | | | | | | | | | |

Person 10 — Last Name

| | | | | | | | | | | | | | | | |

First Name — MI

| | | | | | | | | | | | | | | | |

Person 11 — Last Name

| | | | | | | | | | | | | | | | |

First Name — MI

| | | | | | | | | | | | | | | | |

Person 12 — Last Name

| | | | | | | | | | | | | | | | |

First Name — MI

| | | | | | | | | | | | | | | | |

Next, answer questions about Person 1.

Form D-2

Your answers are important! Every person in the Census counts.

What is this person's name? *Print the name of Person 1 from page 2.*

Last Name

| | | | | | | | | | | | | | | | |

First Name   MI

| | | | | | | | | | | | | | | | | |

What is this person's telephone number? *We may contact this person if we don't understand an answer.*

Area Code + Number

| | | | – | | | | – | | | |

What is this person's sex? *Mark* ☒ *ONE box.*

☐ Male
☐ Female

What is this person's age and what is this person's date of birth?

Age on April 1, 2000

| | | |

Print numbers in boxes.

Month   Day   Year of birth

| | | | | | | | | |

NOTE: Please answer BOTH Questions 5 and 6.

Is this person Spanish/Hispanic/Latino? *Mark* ☒ *the "No" box if **not** Spanish/Hispanic/Latino.*

☐ **No**, not Spanish/Hispanic/Latino
☐ Yes, Mexican, Mexican Am., Chicano
☐ Yes, Puerto Rican
☐ Yes, Cuban
☐ Yes, other Spanish/Hispanic/Latino — *Print group.* ↘

| | | | | | | | | | | | | | | |
| | | | | | | | | | | | | | | |

What is this person's race? *Mark* ☒ *one or more races to indicate what this person considers himself/herself to be.*

☐ White
☐ Black, African Am., or Negro
☐ American Indian or Alaska Native — *Print name of enrolled or principal tribe.* ↘

| | | | | | | | | | | | | | | | |
| | | | | | | | | | | | | | | | |

☐ Asian Indian   ☐ Native Hawaiian
☐ Chinese   ☐ Guamanian or Chamorro
☐ Filipino
☐ Japanese   ☐ Samoan
☐ Korean   ☐ Other Pacific Islander —
☐ Vietnamese   *Print race.* ↘
☐ Other Asian — *Print race.* ↘

| | | | | | | | | | | | | | | | |
| | | | | | | | | | | | | | | | |

☐ Some other race — *Print race.* ↘

| | | | | | | | | | | | | | | | |
| | | | | | | | | | | | | | | | |

What is this person's marital status?

☐ Now married
☐ Widowed
☐ Divorced
☐ Separated
☐ Never married

a. At any time since February 1, 2000, has this person attended regular school or college? *Include only nursery school or preschool, kindergarten, elementary school, and schooling which leads to a high school diploma or a college degree.*

☐ No, has not attended since February 1 → *Skip to 9*
☐ Yes, public school, public college
☐ Yes, private school, private college

2043

Form D-2

b. What grade or level was this person attending?
Mark ☒ ONE box.

- ☐ Nursery school, preschool
- ☐ Kindergarten
- ☐ Grade 1 to grade 4
- ☐ Grade 5 to grade 8
- ☐ Grade 9 to grade 12
- ☐ College undergraduate years (freshman to senior)
- ☐ Graduate or professional school *(for example: medical, dental, or law school)*

What is the highest degree or level of school this person has COMPLETED? *Mark ☒ ONE box.*
If currently enrolled, mark the previous grade or highest degree received.

- ☐ No schooling completed
- ☐ Nursery school to 4th grade
- ☐ 5th grade or 6th grade
- ☐ 7th grade or 8th grade
- ☐ 9th grade
- ☐ 10th grade
- ☐ 11th grade
- ☐ 12th grade, **NO DIPLOMA**
- ☐ **HIGH SCHOOL GRADUATE** — high school DIPLOMA or the equivalent *(for example: GED)*
- ☐ Some college credit, but less than 1 year
- ☐ 1 or more years of college, no degree
- ☐ Associate degree *(for example: AA, AS)*
- ☐ Bachelor's degree *(for example: BA, AB, BS)*
- ☐ Master's degree *(for example: MA, MS, MEng, MEd, MSW, MBA)*
- ☐ Professional degree *(for example: MD, DDS, DVM, LLB, JD)*
- ☐ Doctorate degree *(for example: PhD, EdD)*

What is this person's ancestry or ethnic origin?

(For example: Italian, Jamaican, African Am., Cambodian, Cape Verdean, Norwegian, Dominican, French Canadian, Haitian, Korean, Lebanese, Polish, Nigerian, Mexican, Taiwanese, Ukrainian, and so on.)

a. Does this person speak a language other than English at home?

- ☐ Yes
- ☐ No → *Skip to 12*

b. What is this language?

(For example: Korean, Italian, Spanish, Vietnamese)

c. How well does this person speak English?

- ☐ Very well
- ☐ Well
- ☐ Not well
- ☐ Not at all

Where was this person born?

- ☐ In the United States — *Print name of state.*

- ☐ Outside the United States — *Print name of foreign country, or Puerto Rico, Guam, etc.*

Is this person a CITIZEN of the United States?

- ☐ Yes, born in the United States → *Skip to 15a*
- ☐ Yes, born in Puerto Rico, Guam, the U.S. Virgin Islands, or Northern Marianas
- ☐ Yes, born abroad of American parent or parents
- ☐ Yes, a U.S. citizen by naturalization
- ☐ No, not a citizen of the United States

When did this person come to live in the United States? *Print numbers in boxes.*

Year

a. Did this person live in this house or apartment 5 years ago (on April 1, 1995)?

- ☐ Person is under 5 years old → *Skip to 33*
- ☐ Yes, this house → *Skip to 16*
- ☐ No, outside the United States — *Print name of foreign country, or Puerto Rico, Guam, etc., below; then skip to 16.*

- ☐ No, different house in the United States

Form D-2

b. Where did this person live 5 years ago?

Name of city, town, or post office

| | | | | | | | | | | | | | | | | | | | |

Did this person live inside the limits of the city or town?

☐ Yes
☐ No, outside the city/town limits

Name of county

| | | | | | | | | | | | | | | | | | | | |

Name of state

| | | | | | | | | | | | | | | | | | | | |

ZIP Code

| | | | | |

Does this person have any of the following long-lasting conditions:

	Yes	No
a. Blindness, deafness, or a severe vision or hearing impairment?	☐	☐
b. A condition that substantially limits one or more basic physical activities such as walking, climbing stairs, reaching, lifting, or carrying?	☐	☐

Because of a physical, mental, or emotional condition lasting 6 months or more, does this person have any difficulty in doing any of the following activities:

	Yes	No
a. Learning, remembering, or concentrating?	☐	☐
b. Dressing, bathing, or getting around inside the home?	☐	☐
c. (Answer if this person is 16 YEARS OLD OR OVER.) Going outside the home alone to shop or visit a doctor's office?	☐	☐
d. (Answer if this person is 16 YEARS OLD OR OVER.) Working at a job or business?	☐	☐

Was this person under 15 years of age on April 1, 2000?

☐ Yes → *Skip to 33*
☐ No

a. Does this person have any of his/her own grandchildren under the age of 18 living in this house or apartment?

☐ Yes
☐ No → *Skip to 20a*

b. Is this grandparent currently responsible for most of the basic needs of any grandchild(ren) under the age of 18 who live(s) in this house or apartment?

☐ Yes
☐ No → *Skip to 20a*

c. How long has this grandparent been responsible for the(se) grandchild(ren)? *If the grandparent is financially responsible for more than one grandchild, answer the question for the grandchild for whom the grandparent has been responsible for the longest period of time.*

☐ Less than 6 months
☐ 6 to 11 months
☐ 1 or 2 years
☐ 3 or 4 years
☐ 5 years or more

a. Has this person ever served on active duty in the U.S. Armed Forces, military Reserves, or National Guard? *Active duty does not include training for the Reserves or National Guard, but DOES include activation, for example, for the Persian Gulf War.*

☐ Yes, now on active duty
☐ Yes, on active duty in past, but not now
☐ No, training for Reserves or National Guard only → *Skip to 21*
☐ No, never served in the military → *Skip to 21*

b. When did this person serve on active duty in the U.S. Armed Forces? *Mark ☒ a box for EACH period in which this person served.*

☐ April 1995 or later
☐ August 1990 to March 1995 (including Persian Gulf War)
☐ September 1980 to July 1990
☐ May 1975 to August 1980
☐ Vietnam era (August 1964—April 1975)
☐ February 1955 to July 1964
☐ Korean conflict (June 1950—January 1955)
☐ World War II (September 1940—July 1947)
☐ Some other time

c. In total, how many years of active-duty military service has this person had?

☐ Less than 2 years
☐ 2 years or more

2045

Form D-2

LAST WEEK, did this person do ANY work for either pay or profit? *Mark ☒ the "Yes" box even if the person worked only 1 hour, or helped without pay in a family business or farm for 15 hours or more, or was on active duty in the Armed Forces.*

☐ Yes
☐ No → *Skip to 25a*

At what location did this person work LAST WEEK? *If this person worked at more than one location, print where he or she worked most last week.*

a. Address (Number and street name)

| | | | | | | | | | | | | | | | | | | | | | |

| | | | | | | | | | | | | | | | | | | | | | |

(If the exact address is not known, give a description of the location such as the building name or the nearest street or intersection.)

b. Name of city, town, or post office

| | | | | | | | | | | | | | | | | | | | | | |

c. Is the work location inside the limits of that city or town?

☐ Yes
☐ No, outside the city/town limits

d. Name of county

| | | | | | | | | | | | | | | | | | | | | | |

e. Name of U.S. state or foreign country

| | | | | | | | | | | | | | | | | | | | | |

f. ZIP Code

| | | | | |

a. How did this person usually get to work LAST WEEK? *If this person usually used more than one method of transportation during the trip, mark ☒ the box of the one used for most of the distance.*

☐ Car, truck, or van
☐ Bus or trolley bus
☐ Streetcar or trolley car
☐ Subway or elevated
☐ Railroad
☐ Ferryboat
☐ Taxicab
☐ Motorcycle
☐ Bicycle
☐ Walked
☐ Worked at home → *Skip to 27*
☐ Other method

If "Car, truck, or van" is marked in 23a, go to 23b. Otherwise, skip to 24a.

b. How many people, including this person, usually rode to work in the car, truck, or van LAST WEEK?

☐ Drove alone
☐ 2 people
☐ 3 people
☐ 4 people
☐ 5 or 6 people
☐ 7 or more people

a. What time did this person usually leave home to go to work LAST WEEK?

| | : | | ☐ a.m. ☐ p.m.

b. How many minutes did it usually take this person to get from home to work LAST WEEK?

Minutes

| | |

Answer questions 25–26 for persons who did not work for pay or profit last week. Others skip to 27.

a. LAST WEEK, was this person on layoff from a job?

☐ Yes → *Skip to 25c*
☐ No

b. LAST WEEK, was this person TEMPORARILY absent from a job or business?

☐ Yes, on vacation, temporary illness, labor dispute, etc. → *Skip to 26*
☐ No → *Skip to 25d*

c. Has this person been informed that he or she will be recalled to work within the next 6 months OR been given a date to return to work?

☐ Yes → *Skip to 25e*
☐ No

d. Has this person been looking for work during the last 4 weeks?

☐ Yes
☐ No → *Skip to 26*

e. LAST WEEK, could this person have started a job if offered one, or returned to work if recalled?

☐ Yes, could have gone to work
☐ No, because of own temporary illness
☐ No, because of all other reasons *(in school, etc.)*

When did this person last work, even for a few days?

☐ 1995 to 2000
☐ 1994 or earlier, or never worked → *Skip to 31*

Form D-2

Industry or Employer — *Describe clearly this person's chief job activity or business last week. If this person had more than one job, describe the one at which this person worked the most hours. If this person had no job or business last week, give the information for his/her last job or business since 1995.*

a. For whom did this person work? *If now on active duty in the Armed Forces, mark* ☒ *this box* → ☐ *and print the branch of the Armed Forces.*

Name of company, business, or other employer

b. What kind of business or industry was this?
Describe the activity at location where employed. (For example: hospital, newspaper publishing, mail order house, auto repair shop, bank)

c. Is this mainly — *Mark* ☒ *ONE box.*

☐ Manufacturing?
☐ Wholesale trade?
☐ Retail trade?
☐ Other *(agriculture, construction, service, government, etc.)?*

Occupation

a. What kind of work was this person doing?
(For example: registered nurse, personnel manager, supervisor of order department, auto mechanic, accountant)

b. What were this person's most important activities or duties? *(For example: patient care, directing hiring policies, supervising order clerks, repairing automobiles, reconciling financial records)*

Was this person — *Mark* ☒ *ONE box.*

☐ Employee of a PRIVATE-FOR-PROFIT company or business or of an individual, for wages, salary, or commissions
☐ Employee of a PRIVATE NOT-FOR-PROFIT, tax-exempt, or charitable organization
☐ Local GOVERNMENT employee *(city, county, etc.)*
☐ State GOVERNMENT employee
☐ Federal GOVERNMENT employee
☐ SELF-EMPLOYED in own NOT INCORPORATED business, professional practice, or farm
☐ SELF-EMPLOYED in own INCORPORATED business, professional practice, or farm
☐ Working WITHOUT PAY in family business or farm

a. LAST YEAR, 1999, did this person work at a job or business at any time?

☐ Yes
☐ No → *Skip to 31*

b. How many weeks did this person work in 1999?
Count paid vacation, paid sick leave, and military service.
Weeks

c. During the weeks WORKED in 1999, how many hours did this person usually work each WEEK?
Usual hours worked each WEEK

INCOME IN 1999 — *Mark* ☒ *the "Yes" box for each income source received during 1999 and enter the total amount received during 1999 to a maximum of $999,999. Otherwise, mark* ☒ *the "No" box.*

If net income was a loss, enter the amount and mark ☒ *the "Loss" box next to the dollar amount.*

For income received jointly, report, if possible, the appropriate share for each person; otherwise, report the whole amount for only one person and mark ☒ *the "No" box for the other person. If exact amount is not known, please give best estimate.*

a. Wages, salary, commissions, bonuses, or tips from all jobs — *Report amount before deductions for taxes, bonds, dues, or other items.*

☐ Yes   Annual amount — *Dollars*

$ | | | | , | | | .00

☐ No

b. Self-employment income from own nonfarm businesses or farm businesses, including proprietorships and partnerships — *Report NET income after business expenses.*

☐ Yes   Annual amount — *Dollars*

$ | | | | , | | | .00   ☐ Loss

☐ No

2047

Form D-2

c. Interest, dividends, net rental income, royalty income, or income from estates and trusts — *Report even small amounts credited to an account.*

☐ Yes   Annual amount — *Dollars*

$ | | | ¦ | | | .00   ☐ Loss

☐ No

d. Social Security or Railroad Retirement

☐ Yes   Annual amount — *Dollars*

$ | | ¦ | | | .00

☐ No

e. Supplemental Security Income (SSI)

☐ Yes   Annual amount — *Dollars*

$ | | ¦ | | | .00

☐ No

f. Any public assistance or welfare payments from the state or local welfare office

☐ Yes   Annual amount — *Dollars*

$ | | ¦ | | | .00

☐ No

g. Retirement, survivor, or disability pensions — *Do NOT include Social Security.*

☐ Yes   Annual amount — *Dollars*

$ | | | ¦ | | | .00

☐ No

h. Any other sources of income received regularly such as Veterans' (VA) payments, unemployment compensation, child support, or alimony — *Do NOT include lump-sum payments such as money from an inheritance or sale of a home.*

☐ Yes   Annual amount — *Dollars*

$ | | | ¦ | | | .00

☐ No

What was this person's total income in 1999? *Add entries in questions 31a—31h; subtract any losses. If net income was a loss, enter the amount and mark ☒ the "Loss" box next to the dollar amount.*

Annual amount — *Dollars*

☐ None   OR $ | | | ¦ | | | .00   ☐ Loss

Now, please answer questions 33—53 about your household.

Is this house, apartment, or mobile home —

☐ Owned by you or someone in this household with a mortgage or loan?
☐ Owned by you or someone in this household free and clear (without a mortgage or loan)?
☐ Rented for cash rent?
☐ Occupied without payment of cash rent?

Which best describes this building? *Include all apartments, flats, etc., even if vacant.*

☐ A mobile home
☐ A one-family house detached from any other house
☐ A one-family house attached to one or more houses
☐ A building with 2 apartments
☐ A building with 3 or 4 apartments
☐ A building with 5 to 9 apartments
☐ A building with 10 to 19 apartments
☐ A building with 20 to 49 apartments
☐ A building with 50 or more apartments
☐ Boat, RV, van, etc.

About when was this building first built?

☐ 1999 or 2000
☐ 1995 to 1998
☐ 1990 to 1994
☐ 1980 to 1989
☐ 1970 to 1979
☐ 1960 to 1969
☐ 1950 to 1959
☐ 1940 to 1949
☐ 1939 or earlier

When did this person move into this house, apartment, or mobile home?

☐ 1999 or 2000
☐ 1995 to 1998
☐ 1990 to 1994
☐ 1980 to 1989
☐ 1970 to 1979
☐ 1969 or earlier

How many rooms do you have in this house, apartment, or mobile home? *Do NOT count bathrooms, porches, balconies, foyers, halls, or half-rooms.*

☐ 1 room   ☐ 6 rooms
☐ 2 rooms   ☐ 7 rooms
☐ 3 rooms   ☐ 8 rooms
☐ 4 rooms   ☐ 9 or more rooms
☐ 5 rooms

Form D-2

How many bedrooms do you have; that is, how many bedrooms would you list if this house, apartment, or mobile home were on the market for sale or rent?

- ☐ No bedroom
- ☐ 1 bedroom
- ☐ 2 bedrooms
- ☐ 3 bedrooms
- ☐ 4 bedrooms
- ☐ 5 or more bedrooms

Do you have COMPLETE plumbing facilities in this house, apartment, or mobile home; that is, 1) hot and cold piped water, 2) a flush toilet, and 3) a bathtub or shower?

- ☐ Yes, have all three facilities
- ☐ No

Do you have COMPLETE kitchen facilities in this house, apartment, or mobile home; that is, 1) a sink with piped water, 2) a range or stove, and 3) a refrigerator?

- ☐ Yes, have all three facilities
- ☐ No

Is there telephone service available in this house, apartment, or mobile home from which you can both make and receive calls?

- ☐ Yes
- ☐ No

Which FUEL is used MOST for heating this house, apartment, or mobile home?

- ☐ Gas: from underground pipes serving the neighborhood
- ☐ Gas: bottled, tank, or LP
- ☐ Electricity
- ☐ Fuel oil, kerosene, etc.
- ☐ Coal or coke
- ☐ Wood
- ☐ Solar energy
- ☐ Other fuel
- ☐ No fuel used

How many automobiles, vans, and trucks of one-ton capacity or less are kept at home for use by members of your household?

- ☐ None
- ☐ 1
- ☐ 2
- ☐ 3
- ☐ 4
- ☐ 5
- ☐ 6 or more

Answer ONLY if this is a ONE-FAMILY HOUSE OR MOBILE HOME — All others skip to 45.

a. Is there a business (such as a store or barber shop) or a medical office on this property?

- ☐ Yes
- ☐ No

b. How many acres is this house or mobile home on?

- ☐ Less than 1 acre → *Skip to 45*
- ☐ 1 to 9.9 acres
- ☐ 10 or more acres

c. In 1999, what were the actual sales of all agricultural products from this property?

- ☐ None
- ☐ $1 to $999
- ☐ $1,000 to $2,499
- ☐ $2,500 to $4,999
- ☐ $5,000 to $9,999
- ☐ $10,000 or more

What are the annual costs of utilities and fuels for this house, apartment, or mobile home? *If you have lived here less than 1 year, estimate the annual cost.*

a. Electricity

Annual cost — *Dollars*

$ | , | | | .00

OR

- ☐ Included in rent or in condominium fee
- ☐ No charge or electricity not used

b. Gas

Annual cost — *Dollars*

$ | , | | | .00

OR

- ☐ Included in rent or in condominium fee
- ☐ No charge or gas not used

c. Water and sewer

Annual cost — *Dollars*

$ | , | | | .00

OR

- ☐ Included in rent or in condominium fee
- ☐ No charge

d. Oil, coal, kerosene, wood, etc.

Annual cost — *Dollars*

$ | , | | | .00

OR

- ☐ Included in rent or in condominium fee
- ☐ No charge or these fuels not used

2049

Form D-2

Answer ONLY if you PAY RENT for this house, apartment, or mobile home — All others skip to 47.

a. What is the monthly rent?

Monthly amount — *Dollars*

$ | | ¦ | | |.00

b. Does the monthly rent include any meals?

☐ Yes
☐ No

Answer questions 47a—53 if you or someone in this household owns or is buying this house, apartment, or mobile home; otherwise, skip to questions for Person 2.

a. Do you have a mortgage, deed of trust, contract to purchase, or similar debt on THIS property?

☐ Yes, mortgage, deed of trust, or similar debt
☐ Yes, contract to purchase
☐ No → *Skip to 48a*

b. How much is your regular monthly mortgage payment on THIS property? *Include payment only on first mortgage or contract to purchase.*

Monthly amount — *Dollars*

$ | | ¦ | | |.00

OR

☐ No regular payment required → *Skip to 48a*

c. Does your regular monthly mortgage payment include payments for real estate taxes on THIS property?

☐ Yes, taxes included in mortgage payment
☐ No, taxes paid separately or taxes not required

d. Does your regular monthly mortgage payment include payments for fire, hazard, or flood insurance on THIS property?

☐ Yes, insurance included in mortgage payment
☐ No, insurance paid separately or no insurance

a. Do you have a second mortgage or a home equity loan on THIS property? *Mark ☒ all boxes that apply.*

☐ Yes, a second mortgage
☐ Yes, a home equity loan
☐ No → *Skip to 49*

b. How much is your regular monthly payment on all second or junior mortgages and all home equity loans on THIS property?

Monthly amount — *Dollars*

$ | | ¦ | | |.00

OR

☐ No regular payment required

What were the real estate taxes on THIS PROPERTY last year?

Yearly amount — *Dollars*

$ | | ¦ | | |.00

OR

☐ None

What was the annual payment for fire, hazard, and flood insurance on THIS property?

Annual amount — *Dollars*

$ | | ¦ | | |.00

OR

☐ None

What is the value of this property; that is, how much do you think this house and lot, apartment, or mobile home and lot would sell for if it were for sale?

☐ Less than $10,000
☐ $10,000 to $14,999
☐ $15,000 to $19,999
☐ $20,000 to $24,999
☐ $25,000 to $29,999
☐ $30,000 to $34,999
☐ $35,000 to $39,999
☐ $40,000 to $49,999
☐ $50,000 to $59,999
☐ $60,000 to $69,999
☐ $70,000 to $79,999
☐ $80,000 to $89,999
☐ $90,000 to $99,999
☐ $100,000 to $124,999
☐ $125,000 to $149,999
☐ $150,000 to $174,999
☐ $175,000 to $199,999
☐ $200,000 to $249,999
☐ $250,000 to $299,999
☐ $300,000 to $399,999
☐ $400,000 to $499,999
☐ $500,000 to $749,999
☐ $750,000 to $999,999
☐ $1,000,000 or more

Answer ONLY if this is a CONDOMINIUM —

What is the monthly condominium fee?

Monthly amount — *Dollars*

$ | | ¦ | | |.00

Answer ONLY if this is a MOBILE HOME —

a. Do you have an installment loan or contract on THIS mobile home?

☐ Yes
☐ No

b. What was the total cost for installment loan payments, personal property taxes, site rent, registration fees, and license fees on THIS mobile home and its site last year? *Exclude real estate taxes.*

Yearly amount — *Dollars*

$ | | ¦ | | |.00

Are there more people living here? If yes, continue with Person 2.

Form D-2

Census information helps your community get financial assistance for roads, hospitals, schools and more.

What is this person's name? *Print the name of Person 2 from page 2.*

Last Name

First Name   MI

How is this person related to Person 1?
Mark X ONE box.

- ☐ Husband/wife
- ☐ Natural-born son/daughter
- ☐ Adopted son/daughter
- ☐ Stepson/stepdaughter
- ☐ Brother/sister
- ☐ Father/mother
- ☐ Grandchild
- ☐ Parent-in-law
- ☐ Son-in-law/daughter-in-law
- ☐ Other relative — *Print exact relationship.*

If NOT RELATED to Person 1:

- ☐ Roomer, boarder
- ☐ Housemate, roommate
- ☐ Unmarried partner
- ☐ Foster child
- ☐ Other nonrelative

What is this person's sex?

- ☐ Male
- ☐ Female

What is this person's age and what is this person's date of birth?

Age on April 1, 2000

Print numbers in boxes.

Month   Day   Year of birth

NOTE: Please answer BOTH Questions 5 and 6.

Is this person Spanish/Hispanic/Latino? *Mark X the* **"No"** *box if* **not** *Spanish/Hispanic/Latino.*

- ☐ **No,** not Spanish/Hispanic/Latino
- ☐ Yes, Mexican, Mexican Am., Chicano
- ☐ Yes, Puerto Rican
- ☐ Yes, Cuban
- ☐ Yes, other Spanish/Hispanic/Latino — *Print group.*

What is this person's race? *Mark X one or more races to indicate what this person considers himself/herself to be.*

- ☐ White
- ☐ Black, African Am., or Negro
- ☐ American Indian or Alaska Native — *Print name of enrolled or principal tribe.*

- ☐ Asian Indian
- ☐ Chinese
- ☐ Filipino
- ☐ Japanese
- ☐ Korean
- ☐ Vietnamese
- ☐ Other Asian — *Print race.*
- ☐ Native Hawaiian
- ☐ Guamanian or Chamorro
- ☐ Samoan
- ☐ Other Pacific Islander — *Print race.*

- ☐ Some other race — *Print race.*

What is this person's marital status?

- ☐ Now married
- ☐ Widowed
- ☐ Divorced
- ☐ Separated
- ☐ Never married

2051

Form D-2

INDEX

C

D

population density of, 58–61
resource classification theory about, 318–319
sex ratio in, 70
urbanization in, 288
Evolution. *see* Primitive humans
Exurbanization, 280

F

Factor mobility model, **249**–250
Family Planning and Population Research Act (1970), 231
Family planning programs, 24, 111, **211,** 235–236. *see also* Contraception; Fertility; Population control
in China, 126–133
demographic transition and, 106–107
in developed countries, 230–235
in developing countries, 222–230
ethics of population control and, 26–28
financial support for, 226
history of, 217–219
implementation of, 220–221
law and, 235
national population policy, **212**–217
projected use of contraception and, 225, 228
value of children and, 194–197
Famine, 13. *see also* Food supply
Farming. *see* Agriculture
Fecundity, **179,** 183–184. *see also* Fertility
Female infanticide, 129
Feminism, 65, 72–73. *see also* Women
Fertility, 106, **179.** *see also* Baby boom; Children; Contraception; Family planning programs
abortion and, 189–192
age and, 205–206
among Hispanics, 90
behavior and, 115–116
biological determinants of, 183–184
in China, 127–128
contraception and, 188–189, 222–230

in Cuba, 120–124
demographic transition and, 106–107, 110–111, 113–116
depicted in population pyramids, 76
differentials, **201**–206
economic determinants of, 192–194
government policies of, 214, 216–217 (*see also* Family planning programs)
in Japan, 198–199
in Kenya, 199–201
marriage and, 115, 185–188
measures of, 179–183
population control and, 26–28
population growth projections and, 19
replacement level, 16, 30–31, 116, 182, 210
social determinants of, 184–185
spatial patterns, 206–210
sperm counts, 313
in Sweden, 135
Fertilizer, 340–343
"Fighting tigers," 60
Filipina migrant entertainers, study of, 254
Fishing, commercial, 337–340
Food and Agricultural Organization (FAO), 311, 333, 336–337, 344
Food supply. *see also* Agriculture
Boserup on, 103–104
fertilizer for, 340–343
genetic engineering, 345–348
irrigation for, 343–345
Malthus on, 100–101
mortality and, 176–178
Neo-Malthusians on, 102–103
production, 328–340
triage, 350–**351**
Forced migration, **244**–245, 264, 265
Forecasting, business, 30–31
"Frankenfoods," 349
Free-individual migration, **242**

G

Gambia, 51
Garrett, Laurie, 169
Gender. *see also* Men; Women
AIDS and, 168–169

migration and, 253–254, 257, 262
preference for male children, 196
selection by abortion, 191
sex ratio, **64**–74, 129, 161–164
General fertility rate, **180**
Genetic engineering, 345–350
Genocide, 25–**26**
Geographical Abstracts, xviii
"Global care chains," 254
Global warming, 305–309
"Golden" rice, 349
Gore, Albert, xviii
Grain production, 337, 343, 346
Great Leap Forward, 126
Great Plains, 283
Greece, 206
"Greenhouse effect," 306
"Green Revolution," 345–348
Gross migration, **239**
Gross national product (GNP), 31, 208
of China, 60–61
crude birth rate and, 194
influence on mortality rates, 149–151
Gross reproduction rate, **183**
Group migration, **242**
Growth. *see* Population growth

H

Hall, Edward, 302
Han nationality (China), 131
Hardin, Garrett, 102–103, 351–352
Hawaii, 68
Heart disease, 84, 151
Hemorrhagic fever viruses, 167–169
Herbicides, 342–343
Highways, 281
Hispanic population, in United States, 56, 88–92. *see also* Ethnicity; Race
Homo erectus, **9,** 262–263
Homo habilis, **9**
Homo sapiens, **9**–11
Hot Zone, The (Preston), 168
Housing demography, **xv**
Hua Guofeng, 127
Human capital model, **250**
Human rights, xviii

I

Illegal aliens, 54, 271
Illegal Immigration Reform and
    Immigrant Responsibility Act
    (IIRIRA) (1996), **274–275**
*Illustrations and Proofs of the
    Principle of Population*
    (Place), 217
Immigrant relative, **241**
Immigration. *see* Migration
Immigration Act (1990), 274
Immigration Reform and Control
    Act (IRCA) (1986), 272–276
Immunization, 158
Impelled migration, **244–245**
Income
    fertility and, 201–202
    for Hispanics, 91
    migration and, 249–250, 253
"In-Depth Fertility Survey"
    (Chinese government), 128
India, 16
    contraceptive use in, 224
    crude birth rate of, 209
    family planning programs in,
        212–213, 227–229
    famine in, 13
    gender selection in, 191
    malnutrition in, 337
    migration, 264
    pollution in, 300
    resource classification theory
        about, 318–319
    urbanization of, 40, 289
Indonesia, 224
Industrial Revolution, 6, **14–16**
    demographic transition and,
        106
    Malthusian theory and,
        101–102
    migration and, 279–280
    spatial distribution and, 63–64
Infant and child mortality rate
    (ICMR), 156–158
Infanticide, 129
Infant mortality, **140–142**
    mapping of, 176
    risk factors for, 156–161
    sex ratio and, 161–163
    in United States, 155–156
    world patterns of, 155–161
In-migration, **239**
Intergovernmental Panel for
    Climate Change (IPCC), 307
International immigration, 241.
    see also Migration

International Institute for
    Environment and
    Development, 300
International Planned Parenthood
    Federation (IPPF), **218**, 220
Internet sites about population
    geography, 41–42
Interval, migration, **240**
Intervening obstacles, **247**
Iraq, 215
Ireland, 13, 260, 261
Irrigation, 343–345
Italy, 206, 231, 260–261

J

Janata Party, 228
Japan
    abortion in, 117, 118
    arithmetic density of, 62
    family planning program
        in, 234
    life expectancy in, 146, 148
    lifestyle of women in, 198–199
Java, 303
Jews, forced migration of, 244
Journals about population,
    xviii–xix. *see also individual
    journal names*
Justinian Plague, 13

K

Kampuchea, 215
Kenya, 199–201
Korea, 223–224
Kyoto Protocol, 308

L

Labor force, 31–32
    children in, 198, 200
    forced migration for,
        244–245, 263
    illegal migrants in, 272–273
    migration and, 253–254, 266
Laissez-faire argument, **25**
Laissez-faire population
    exponent, **25**

Laos, 215
Latin America. *see also individual
    country names*
    age structure in, 78
    contraceptive use in, 222–223,
        227
    crude birth rate of, 209
    demographic transition in, 108
    food supply in, 336
    migration, 264
    urban areas defined by, 40
"Laws of Migration"
    (Ravenstein), 245–246
Lee, Everett, 245–247
Lifeboats theory, **351–352**
Life expectancy, 147, 150. *see also*
    Death rates; Mortality
    average remaining life
        measure, 143
    diet and, 153–155
    epidemiologic transition and,
        146
    mapping of, 176
    in Sweden, 136
Life table mortality rate, **143**
Life tables, **142–145**, 163
Limits to Growth, The (Meadows,
    et al.), **319–323**
Linguistic diversity, 116
Literacy, 73, 149–150
Literature about population,
    xviii–xix. *see also individual
    journal names*
Longitudinal surveys, 50–51

M

Malignant neoplasms, 149
Malnutrition, 159–160, 183, 332,
    337, 352. *see also* Food supply
Malthus, Thomas Robert, 99–102,
    328, 348
Mao Zedong, 127
Maps
    of AIDS epidemic, 171–173
    of diseases, 166
    of mortality rates, 176
Mariel sealift (1980), 124, 125
Marketing, 28–29
Marriage
    age of, 186–187
    in China, 130
    in Cuba, 122
    demographic transition
        and, 106

O

N

P

Pennsylvania, 172–173
People's Republic of China. *see* China
Permanent Census Act (1902), 48
Peru, 40
Pesticides, 342–343
Pharmaceuticals, 171, 174
Philadelphia, study of mortality in, 153
Physiological density, **61**–62. *see also* Population density
Place, Francis, 217
Poland, 190
Pollution, **300**–302
Population Bomb, The (Ehrlich), 103, 220
Population change. *see* Population growth; Theories, of population change
Population control. *see also* Contraception; Family planning programs; Fertility in China, 126–133
ethics of, 26–28
reasons for, 105
Population Council, 216, 218
Population density. *see also* Population distribution; Spatial distribution
in China, 131
crowding and violence, 302–305
as measure of population pressure, 315
measuring, 59, 61–63
Urbanized Area definition, 41
Population distribution, 278–279. *see also* Population density; Spatial distribution
factors of, 63–64
family planning programs and, 232
in United States, 255, 278–283
urbanization and, 286
Population geography, **xii**–xiv
Population growth, **1**. *see also* Environment; Population projection; Theories, of population change
during agricultural revolution, 12–13
business and, 28–32
climate and, 308
in Cuba, 119
during cultural revolution, 8–11
environmental degradation from, **299**

food scarcity and, 104
during Industrial Revolution, 14–16
measures of, 1–3
pollution and, 300–302
spatial distribution and, 57–58
in Sweden, 134–138
in United States, 55–56
urbanization and, 297
violence from, 302–305
Population Index, xviii
Population projection, **17, 18**
small area, **21,** 33
for United States, 20–25, 83
for world, 19–20
Population pyramids, **74**–77, 84. *see also* Age structure
of Cuba, 123–124
of Sweden, 137–138
Population Reference Bureau, xviii, 78, 210
Population-resource region, **318**–319
Positive checks, 101
Postneonatal mortality rate, **142**
Poverty, 287, 352
Preston, Richard, 168
Preventive checks, 101
Primary Metropolitan Statistical Area (PMSA), **40**–41
Primary theories, **97**
Primitive humans, 8–11, 241–245, 262–263, 297, 328
Primitive migration, **241**–245
Principle of Population (Malthus), 100–101, 328, 348
Program of Action, 225–227
Pronatalist policies, 215, 230, 234–235
Prostitution, 167, 169, 244–245
Protease inhibitors, 171
Puberty, 183–184
Push-pull models of migration, 245–**247,** 288

R

Race, 56, 87–88, 93–94, 261–262
Rainforests, 311–312
Rate of natural increase (RNI), **2**–3
Rate of population growth, **3**
Ravenstein, Edward, 245–246
Reagan, Ronald, 220
Refugees, 264, **266**–268, 304–305

Registration system, **43**. *see also* Vital registration
Reproduction. *see* Fertility
Resources, natural, 318–325. *see also* Food supply
population pressure on, 304–305
triage, 350–**351**
Restricted immigration, 269–272
Restricted migration, **242**–243
Return migration, **240**–241
Rockefeller, John D., 3rd, 218
Rockefeller Foundation, 149
Romania, 190
RU 486 (contraceptive), 189
Rural areas. *see also* Urbanization
in China, 128–129
fertility differentials, 201
rural-urban migration patterns, 281–**284**
Russia, 13, 232, 245

S

Sample surveys, **49**–51, 143
Sanger, Margaret, 218
Saudi Arabia, 215
Scandinavia, 45. *see also individual country names*
Schools. *see* Education
Science, 315
Scientific-industrial revolution. *see* Industrial Revolution
Secondary theories, **97**
Segregation, 93–94
Sex ratio, **64**–69. *see also* Gender
in China, 129
international variations in, 69–74
mortality and, 161–164
Sexually transmitted diseases (STDs), 168, 183
Shelter, 101
Sickle-cell anemia, 163
Simon, Julius, 296
Singapore, 223–224
Sjaastad, L.A., 250
Slavery, 244–245, 263
Small area population projections, **21,** 33
Smil, Vaclav, 334–336, 342
Socialism. *see* Communist-socialist theory
Soil, suitable for agriculture, 335
South Korea, 229–230

Spain, 231
Spatial distribution, xv, 57–58. *see also* Population density; Population distribution
of AIDS, 171–173
in China, 131
of elderly, 84–86
global patterns, 58–60
Sri Lanka, 149
Standard land unit, 62–63
Standard Metropolitan Statistical Area (SMSA), **40**–41
Standard of living. *see* Economic development
Starlink corn, 349–350
State of the World, xviii
Statistical Abstract of the United States (U.S. Bureau of the Census), 56
Statistical Atlas of the United States, 40
Sterilization, 222–223, 228–229
Stern, 244
Stopes, Marie, 218
Strategic Demographic Initiative (SDI), 127
Streams, migration, **240,** 254
Structuralist view, of migration, **251**
Structuration approach, of migration, **251**–252
Suburbanization, 280
Sunbelt, 280–281
"Superweeds," 349
Survival table, **143**
Sustainable development, 352–353
Sweden, 134–138

T

Taiwan, 223–224
Tanzania, 78
Thailand, 224, 229–230, 245
Theories, of population change, 98–99. *see also* Population growth
by Boserup, 103–104
classification of, 97–98
demographic transition theory, **105**–116
economic theory of fertility, 192–194
environmental *vs.* naturalistic, 98

environment and, 102–103, 220, 296, **315**–317
epidemiologic transition, 146–149
Limits to Growth, The, **319**–323
by Malthus, 99–102, 328, 348
Marxism, 104–105, 118–119, 126
Neo-Malthusian, **102**–103, 296
resource classification, 318–319
role of, **96**–97
Third World countries. *see* Developing countries
Toolmaking revolution, **8.** *see also* Cultural revolution
Total fertility rate, **181**–182
Total population, 38–41
Transportation, 101, 263, 281
Triage, 350–**351**

U

Uganda, 174–176, 206
United Nations, 266
age defined by, 37
on census enumerations, 44
Climate Change Conference, 308
Demographic Yearbook, 56
Fund for Population Activities (UNFPA), 218–221
Genocide Convention, 26
International Conference on Population and Development, 225–227
Population Commission, 218
population projections by, 19, 21–25, 210
UNICEF, 155
Universal Declaration of Human Rights, xviii
United Socialist Soviet Republic (U.S.S.R.), former, 61
United States. *see also individual agency names*
abortion in, 190–193
age structure in, 81–86
AIDS in, 168–173
baby boom, **29**–30, 79–81
census practices by, 39, 45, 53–56, 64
child care in, 254
crude birth rate of, 209
crude death rate of, 2–3, 164
death rates in, 141, 151
elderly in, **81**–86

emigration from, 259
family planning in, 231
fertility in, 202
financial support for family planning programs, 220
immigration to, 20, 242–243, 268–277
infant mortality in, 155–156
internal migration, 278–283
labor force in, 31–32
life tables for, 144–145, 163
marriage age in, 186
migration projections for, 293–294
National Parks of, 321–322
out-migration, 260
population distribution in, 255
racial and ethnic composition of, 56, 87–88
resource classification theory about, 318–319
sex ratio of, 65–69
total fertility rate in, 182
urbanization in, 40–41, 291–293
vital registration in, 47–48
Universal Almanac, The (Wright), 56
Urbanization, **284**–285
in China, 128–129, 132–133
city size preferences and, 255
crude birth rate and, 195
in developing countries, 285–290
impact of migration on, 290–291
mortality studied in, 153–154
patterns of, 287–289
race and ethnicity patterns, 261–262
spatial distribution of elderly and, 85–86
in United States, 291–293
urban, defined, **39**–41
U.S. Agency for International Development (AID), 218, 259
U.S. Bureau of the Census, 41, 45–48, 53–56. *see also* Census
projections by, 21, 83
publications of, 51
U.S. Department of Labor, 249–250
U.S. National Center for Health Statistics, 48
U.S. Office of Technology Assessment, 301
Utah, 67–68
Utopians, 100–101